Int… Tro… Agr… Ser…

General Editor

W. J. A. Payne
Consultant in Tropical Livestock Production

Tropical Soils and Fertiliser Use

Peter Martin Ahn, Ph.D. (Cantab), Dr. Sc (Ghent)
Consultant in Tropical Soils and Agronomy

Formerly

International Board for Soil Research and Management (IBSRAM), Bangkok (1987–90)

Head, Farming Systems Division, Institut des Savanes, Bouaké, Ivory Coast (1984–87)

Professor of Soil Science, Faculty of Agriculture, University of Nairobi, Kenya (1972–81)

Longman Scientific & Technical

Longman Scientific and Technical

Longman Group UK Limited,
Longman House, Burnt Mill, Harlow,
Essex CM20 2JE, England
and Associated Companies throughout the world

First published 1993
Set in 9.5/11pt Times Roman

Printed in Malaysia
by TCP

ISBN 0 582 77507 8

British Library Cataloguing-in-Publication Data.
A catalogue record for this book is available from the British Library.

Other titles in the Intermediate Tropical Agriculture Series

Already published:

J. C. Abbott and J. P. Makeham, Agricultural Economics and Marketing in the Tropics

M. E. Adams, Agricultural Extension in Developing Countries

P. M. Ahn, Tropical Soils and Fertiliser Use

A. Chamberlin, Milk Production in the Tropics

D. Y. Coy, Accounting and Finance for Managers in Tropical Agriculture

C. Devendra and G. B. McLeroy, Goat and Sheep Production in the Tropics

J. A. Eusebio, Pig Production in the Tropics

D. Gibbon and A. Pain, Crops of the Drier Regions of the Tropics

H. T. B. Hall, Diseases and Parasites of Livestock in the Tropics, Second Edition

H. F. Heady and E. B. Heady, Range and Wildlife Management in the Tropics

E. Heath and S. Olusanya, Anatomy and Physiology of Tropical Livestock

D. H. Hill, Cattle and Buffalo Meat Production in the Tropics

D. S. Hill and J. M. Waller, Pests and Diseases of Tropical Crops Volume 1: Principles and Methods of Control

D. S. Hill and J. M. Waller, Pests and Diseases of Tropical Crops. Volume 2: Field Handbook

R. L. Humphreys, Tropical Pastures and Fodder Crops, Second Edition

C. N. Williams, W. Y. Chew and J. H. Rajaratnam, Tree and Field Crops of the Wetter Regions of the Tropics

Titles in preparation

J. F. Redhead and J. B. Hall, Tropical Forestry
F. Inns, Agricultural Mechanisation

Acknowledgements

The Publishers are grateful to the following for their permission to reproduce photographs:

J. Allan Cash Photo Library for fig. 7.6; John Cleare for fig. 7.11; R. H. V. Corley, Unilever Plantations for figs. 16.6 a and b; Dr. Stephen Dowdle, Canpotex, Singapore (Potash and Phosphate Institute) for figs. 13.2b and 15.7; F.A.O./Peyton Johnson for fig. 19.2a; I.C.R.A.F./A. Njenga, Kenya for fig. 19.2b; I.C.R.I.S.A.T., India for figs. 18.2 and 15.4; International Fertiliser Development Center, USA for fig. 16.7; International Rice Research Institute, Philippines for fig. 7.10; International Soil Reference and Information Centre, Wageningen, The Netherlands for figs. 1.9b and 3.1; K. E. Lee and T. G. Wood. authors of *Termites and Soils*, 1971 for fig. 7.7; Mauritius Sugar Industry Research Institute for figs. 15.6a and b; Panos Pictures for figs. 1.7 (Marcos Santilli), 1.9a (Adrian Evans) and 14.1 (Ron Gilling); Photo Information Côte d'Ivoire for fig. 16.4; Dr. Sam Portch/Potash and Phosphate Institute for figs. 12.1, 13.2a, 14.2, 16.1 and 16.2; Rubber Growers Association (Malaysia) Berhad for fig. 16.5; Shell Photographic for fig. 15.3; Dr. R. E. White, Massey University, New Zealand for fig. 2.5 and Zimbabwe Information Services for fig. 15.5.

Fig. 18.1 is © Longman Group and all other photographs not acknowledged have been supplied by the author.

The cover photograph has been reproduced by kind permission of International Soil Reference and Information Centre, Wageningen, The Netherlands.

Contents

Preface

In writing this introduction to tropical soils and their management the aim has been simply to include as much relevant information as the size and scope of the book allows, and present that information in as clear a way as possible to help its understanding and use by the very wide range of people who take an interest in tropical soils or earn a livelihood by cultivating them.

Selection and compression of material inevitably imply both omissions and a personal choice of emphasis. As a guide I have tried to supply that information which I would have liked to have had at my disposal when, as a newly graduated young man armed only with a degree from the University of Cambridge, I first came to the tropics more than 30 years ago. I was promptly instructed to survey the soils and land use of several thousand square kilometres of tropical forest, much of it virgin, and make appropriate agricultural recommendations.

While the book draws freely on the writings and experience of innumerable soil scientists, agronomists and farmers, the emphasis remains my own. I am conscious of the dangers of relying uncritically on soil science texts written by and for workers in temperate zones, and of transferring experience from one zone to another, and have striven for an orientation in keeping with tropical realities. More space has been devoted than in most books of this type to such increasingly important topics as sulphur in the soil, the role of termites, and the management of tropical soils in relation to various cropping systems. This is the only book to my knowledge that combines a basic soil science text with summaries of fertiliser experience of the main tropical crops. It thus offers the reader a particularly wide range of topics, but does so in the belief that this range meets the needs of the many who look for a handbook that is both fundamental and practical.

Agriculture is arguably the noblest and most satisfying of professions, and on it our civilisation rests. Soil science and agronomy range from the frontiers of scientific research to the practical tasks of preparing the soil and applying fertiliser. The hope is that this book will not only provide useful facts and guidance, but will also encourage the reader to pursue the subject further. To this end, a section on further reading has been included at the end of the text.

P.M. Ahn
1992

Glossary

A horizon The surface horizon of a mineral soil. It normally has maximum organic matter content and biological activity but in some cases is the horizon of maximum eluviation. Typically dark coloured owing to humus, and with a crumb structure.

AC profile Soil profile with A and C horizons but lacking a B horizon. Characteristic of some immature soils such as lithosols developed in resistant parent materials on steep slopes.

acid rocks Rocks (usually igneous or metamorphic) with over 66% silica and containing free quartz.

actinomycetes Term applied to a group of organisms with characteristics intermediate between bacteria and true fungi.

aerobic organisms Organisms growing only in the presence of molecular oxygen.

agroforestry Agricultural system in which trees are combined with annual crops and/or pasture and in which there is an economic and biological reaction between tree and non-tree components.

air dry Dryness of a soil at equilibrium with the moisture content of the surrounding air.

albic horizon Pale-coloured mineral soil horizon in which clay and free iron oxides have been removed to reveal the colour of the primary sand and silt particles.

Alfisol (US Soil Taxonomy) Soil order comprising soils that contain an *argillic horizon* or *natric horizon* and have a base saturation of over 35% in the subsoil to a depth of at least either 1.8 m below the soil surface or 1.25 m below the upper boundary of the argillic horizon unless there is a lithic or paralithic contact at shallower depths. Soils more desaturated than this are *Ultisols*.

ammonification Processes by which ammonium (NH_4^+) is released from nitrogen-containing organic compounds.

anaerobic Growing or occurring in the absence of molecular oxygen.

anion Negatively charged ion, such as nitrate (NO_3^-) or sulphate (SO_4^{-2}).

anion exchange capacity Sum total of exchangeable anions that a soil can absorb, expressed as m.e. per 100 g of soil or other material.

aquic moisture regime Soil moisture regime in which the soil is saturated with water long enough to give reducing conditions.

argillic horizon Subsurface diagnostic horizon characterised by an accumulation of layer silicate clays. Some of the clay has been eluviated out of an upper horizon and illuviated into the argillic horizon. Clay skins are often present. The criteria used for the definition of an argillic horizon in the US Soil Taxonomy include definitions of the minimum increase in clay content relative to an overlying horizon. If the overlying eluvial horizon has less than 15% clay, there must be a minimum increase of at least 3% over a distance of 30 cm. If the eluvial horizon has 15–40% clay, the argillic horizon must have at least 1.2 times this amount. If the eluvial horizon has over 40% clay, the argillic horizon must have at least 8% more. Argillic horizons are a required feature of Alfisols and Ultisols.

aridic moisture regime Soil moisture regime with very short wet seasons associated with dry and desert climates, in which the *control section* of the soil is moist for less than 90 consecutive days and dry for over 180 cumulative days.

Aridisols (US Soil Taxonomy) Soil order comprising soils that have an *aridic moisture regime* and an argillic or natric horizon or, alternatively, have no argillic or natric horizon but have a petrocalcic, calcic, gypsic, petrogypsic or cambic horizon whose upper boundary is within 1 m of the soil surface.

aromatic Cyclic hydrocarbon or its derivative containing one or more benzene rings. Benzene rings contain six carbon atoms linked by single or double bonds to form a symmetrical hexagon.

autotrophic Capable of using carbon dioxide and/or carbonates as a sole or major carbon source.

available moisture Moisture available to plants. Commonly equated with the moisture held between *field capacity* and *wilting point*, i.e. at tensions between one-third and fifteen atmospheres, though gravitational water held at lesser tensions than this, if present, is also available to plants.

base saturation Extent, expressed as a percentage, to which the adsorption complex of a soil (the cation exchange capacity) is saturated with exchangeable cations other than hydrogen.

basic rocks Rocks containing between 45% and 52% silica. Normally applied to igneous and metamorphic rocks. Basic rocks do not contain any free quartz. See also *acid rocks*.

B horizon Soil horizon, normally below the A horizon or topsoil, that has some structure, generally blocky or prismatic, and in which there has been an accumulation of clay, iron or aluminium from the horizon above. In many cases the B horizon is equivalent to the more general term 'subsoil'. See *A horizon* and *C horizon*.

biotite See *mica*.

broadcasting Method of applying chemical fertilisers involving their application fairly uniformly to the soil surface before the crop is planted. The fertiliser may or may not be incorporated into the soil.

bulk density The weight of dry soil in relation to its volume. Normally expressed as grams per cubic centimetre (g/cm^3) (see Table 2.4). The bulk density in relation to the specific gravity of the soil particles reflects the porosity of the soil.

C horizon Also called the weathered substratum. A layer of unconsolidated soil material relatively little affected by soil-forming processes, which underlies the B horizon or solum. It may or may not resemble the material from which the B horizon or solum is thought to have formed.

calcareous soil Soil containing enough calcium carbonate or calcium – magnesium carbonate to effervesce visibly when cold 0.1 M hydrochloric acid is applied.

calcic horizon Mineral soil horizon over 15 cm thick enriched with secondary carbonates, with a calcium carbonate equivalent which is over 15%, and at least 5% above that of the underlying C horizon.

cambic horizon Diagnostic subsurface horizon that has undergone certain changes due to the processes of soil formation but which does not qualify as *oxic, argillic*, natric, etc. A cambic horizon has some soil structure, contains some weatherable minerals, and is characterised by the alteration and/or removal of mineral material as shown by mottling or grey colours, redder hues than an underlying horizon, or the removal of carbonates. Cambic horizons lack cementation or induration.

carbon – nitrogen (C/N) ratio Ratio of weight of organic carbon to the weight of total nitrogen (both mineral and organic) in soil or organic material.

catena A sequence of soils related to relief.

cation A positively charged ion, such as Ca^{2+}, Mg^{2+}, K^{+}.

cation exchange Interchange between a cation in the soil solution and another cation more or less on or attracted to the surface of a generally colloidal material, such as clay or humus, which has a cation exchange capacity due to exposed negative sites.

cation exchange capacity (CEC) Sum total of exchangeable cations that a soil or other material such as clay or organic matter can absorb, expressed as m.e. per 100 g soil or other material.

clay (size fraction) Mineral soil particles less than 2 micrometres (2 micron) – that is, less than 0.002 mm in diameter. The clay fraction normally includes *clay minerals* and because of its large surface area is the most active fraction of the soil.

clay minerals Minerals including *kaolin*, *montmorillonite* and *illite*, generally of clay size, with properties related to their crystalline phyllosilicate structure.

clay skins Coatings of clay on the surfaces of soil peds and mineral grains, and in soil pores. Also called clay films, cutans and argillans. Clay skins may form when clay in suspension in soil moisture moves into a horizon, and are thus evidence of clay illuviation, as in an *argillic horizon*.

coarse fraction That part of the soil that is too coarse to pass through a 2 mm sieve. It includes gravel-sized particles and stones.

columnar structure Type of soil structure in which the peds are prismatic in shape and have rounded caps, particularly associated with salt-affected soils.

consistence, consistency The handling properties of the soil with reference to its adhesion or cohesion and its resistance to deformation or rupture. Soil consistency varies, sometimes markedly, according to whether the soil is dry, moist or wet. It is normally described for moist soils (friability). It may also be described for dry and wet soils (hardness, stickiness, plasticity) if consistency properties change markedly with moisture content, as is the case with soils, such as Vertisols, containing montmorillonite.

control section Part of the soil profile used for determining the *soil moisture regime*, defined as being between upper and lower limits corresponding to the depth reached by 2.5 and 7.5 cm of water applied to the soil surface in 24 and 48 hours respectively. Depths of the control section vary with soil texture but are approximately 10–30 cm for clay soils, 20–60 cm for loamy soils and 30–90 cm for sandy soils.

cover crop Crop used to cover and protect the soil surface, to decrease erosion, and shade the ground. Tropical cover crops include leguminous species such as *Pueraria* and *Stylosanthes*.

crumbs Structural units or peds with crumb-like shape and generally porous, bound mainly by organic matter and typically associated with topsoils.

denitrification Biochemical reduction of nitrite or nitrate to gaseous nitrogen either as molecular dinitrogen (N_2) or as an oxide of nitrogen.

diagnostic horizon Soil horizon meeting certain defined criteria used in soil classification.

diffusion Movement of nutrient ions in the soil solution independent of *mass flow*.

E horizon (FAO) Eluvial horizon, high in silt and sand, usually light in colour.

(EC) Electrical conductivity Electrical conductivity of an extract from saturated soil, usually expressed in millimhos per cm at 25 °C. Used particularly to detect salts in soils and in connection with irrigation.

ectotrophic mycorrhiza Mycorrhizal association in which the fungal hyphae form a mantle on the root surface, with strands of mycelia extending inwards into the root between cortical cells and outwards into the surrounding soil.

E_h value Voltage difference at an inert electrode immersed in a reversible oxidation – reduction system used to measure the oxidation – reduction (redox) potential.

eluvial horizon Soil horizon formed by or having been modified by eluviation.

eluviation Removal of soil material in suspension, or in both suspension and solution, from a soil layer. Removal in solution is usually referred to as *leaching*.

endotrophic Receiving nourishment from within, as when mycorrhizal fungi receive nourishment from the plant root.

Entisols (US Soil Taxonomy) Soil order comprising very young soils with very little profile development and no distinct subsurface soil horizons within 1 m of the surface.

epipedon Soil horizon formed at the surface, usually darkened by organic matter or eluviated. Normally equivalent to the A horizon but may include a part of a B horizon darkened by organic matter.

equivalent diameter Diameter of a spherical particle which falls with the same velocity as an irregularly shaped particle of the same density (term used in the mechanical analysis of soils). Diameter of a pore of round cross-section with the same transmission characteristics as a pore of irregular cross-section.

equivalent weight Weight of a substance that combines with or replaces one gm of hydrogen or eight grams of oxygen; the atomic or molecular weight divided by the valence.

evapotranspiration Loss of water from the soil – plant system by both evaporation and plant transpiration.

evergreen forest Forest formation with species not normally shedding leaves during part of the year, associated with high rainfall areas of equatorial climate lacking marked dry seasons. Often replaced by *semi-deciduous* forest where rainfall is lower or more seasonal.

exchangeable potassium Potassium held by the absorption complex (*cation exchange capacity*) of the soil easily exchanged with other cations in a neutral salt solution. Exchangeable potassium is considered freely available to plants.

exchangeable sodium percentage (ESP) Percentage of the *cation exchange capacity* of the soil occupied by sodium.

facies Subdivision of a vegetation association or geological formation.

farmyard manure Mixture of animal dung, urine and straw or litter used as manure.

feldspar Silicate mineral with a tectosilicate (block) structure which forms an important constituent of many igneous and metamorphic rocks. Feldspars show a continuous range of composition between the potassium and sodium feldspars, and between the sodium and calcium feldspars. Of the group, calcium feldspars are the most easily weathered and potassium feldspars the least easily weathered.

ferromagnesian mineral Mineral, normally an aluminium silicate, containing iron and magnesium in the crystal structure.

fertiliser grade Guaranteed minimum percentage by weight in a fertiliser of the major plant nutrients traditionally expressed as the percentages of N, P_2O_5 and K_2O but increasingly as the percentages of elemental N, P and K.

fertirrigation The application of chemical fertilisers in irrigation water.

field capacity The maximum amount of water that can be held by the soil against gravity. In practice usually equated with the water held at a tension of one-third of an atmosphere expressed as a percentage by weight of the oven-dry soil. A free-draining soil that is saturated usually reaches field capacity after about 1–2 days. Of the water held at field capacity, part is available to plants but that part still present in the soil at *wilting point* is not available.

fixation Soil processes by which soil nutrients such as phosphorus or potassium are converted from an available to a less available form.

foliar or leaf analysis Analysis of the chemical composition of selected plant parts in order to estimate mineral nutrient deficiencies or imbalances as a guide to fertiliser use.

friable soil *consistency term* used to describe moist soils which indicates the ease with which soils are crumbled.

FYM Farmyard manure.

gilgai Soil microrelief due to expansion and contraction related to changes in moisture content. Often features a succession of microbasins and microknolls. Occurs with soils that shrink and swell considerably when drying or wetting, such as Vertisols.

grey soil Soil affected by *gleying*.

gleying Soil changes or processes resulting from lack of oxygen due to waterlogging, as evidenced by formation of mottles and grey, blue or green colours.

gneiss Coarse-grained crystalline *metamorphic rock* with characteristic coarsely banded appearance.

green manuring The ploughing-in of a growing leafy crop as a manure.

greywacke Siliceous, non-laminated, generally dark-coloured *sedimentary rock* formed of angular mineral grains (mostly *quartz* and *feldspars*) cemented in a clay-sized matrix.

gypsic horizon Soil horizon affected by the presence of gypsum. In the US Soil Taxonomy a gypsic horizon is at least 15 cm thick and has at least 5% more gypsum than the C horizon or underlying substratum.

H horizon (FAO) Organic horizon with over 30% organic matter if texture is clay and over 20% if it is lighter than clay. Sometimes saturated with water.

histic epipedon (US Soil Taxonomy) Thin organic soil surface horizon saturated with water at some period of the year.

Histosols (US Soil Taxonomy) Soil order comprising soils that have organic soil materials that extend to certain defined depths, and which generally occupy most of the upper 80 cm of the profile unless there is a *lithic* or *paralithic contact* at shallower depths.

horizon A soil horizon is a layer in the soil more or less parallel to the surface which differs in physical, chemical and biological properties from the layers above and below it. Differences may include those of colour, texture, structure, consistency, content of organic matter and coarse material, reaction, presence of roots and organisms and other properties.

hornblende Dark-coloured silicate mineral of the *inosilicate* group, containing iron and magnesium, found in *igneous* and *metamorphic* rocks. Hornblende weathers relatively easily.

humification Processes involved in the decomposition of organic matter which lead to the formation of humus.

humous Containing humus.

humus More or less stable fraction of the soil organic matter which remains after the major portion of the plant and animal residues added to the soil have decomposed.

hydraulic conductivity A measure of the readiness of water to flow through the soil in response to a water potential gradient.

hydrolysis Mechanism of chemical weathering of rocks and minerals involving the replacement of constituent ions by hydrogen ions.

hydrous mica Synonymous with illite.

hygroscopic Having the ability to attract water from the atmosphere.

hygroscopic water Soil water absorbed from the atmosphere, so firmly held that it is incapable of movement and unavailable to plants. Remains in soil after air drying. About equivalent to water held at pF 4.7 (approximately 50 atmospheres).

igneous rock Crystalline rock formed by the cooling of molten magma.

illite Silicate clay with a 2:1 lattice structure, synonymous with hydrous mica. Isomorphous substitution has resulted in aluminium replacing a proportion of the silicon in the silica tetrahedra, and potassium ions serve as an additional bonding between the layers. Particles are larger than those of montmorrillonite, and cation exchange capacity is lower.

immature soil A soil with relatively little profile development.

Inceptisols (US Soil Taxonomy) Soil order comprising soils with only a moderate degree of profile development. May contain a cambic horizon in which minerals have been altered or removed but other diagnostic subsurface horizons such as *argillic* and *oxic* are absent.

indurated Hardened. In tropical soils induration is often due to iron and aluminium oxides (see *sesquioxides*, *plinthite*, *ironstone*).

infiltration Downward entry of water into the soil at the surface and subsequent movement in the profile.

inosilicates Silicate minerals in which the silica tetrahedra join to form chains or fibres. In the pyroxene family they form single chains, and in the amphiboles (of which the most important is hornblende) they form double chains which run the length of the crystal and are linked to each other by calcium and magnesium cations.

intercropping Cropping system in which two or more crops are grown at the same time (see *mixed*, *row* and *strip intercropping*) or sequentially in the same year (see *sequential cropping*).

ironstone, ironpan Very hard soil layer indurated and cemented by iron oxides and other materials.

isomorphous substitution Substitution of an ion in the crystal structure of a mineral by another of approximately the same size.

kaolin Clay mineral of the 1:1 layer lattice type, i.e. with one silicon tetrahedral layer and one aluminium oxide-hydroxide octahedral layer. Also applied more generally to the 1:1 group of clay minerals. Kaolin is the dominant clay mineral in most highly weathered tropical soils. It has a lower cation exchange capacity than *illite* or *montmorillonite*.

kaolinitic clay Clay consisting mainly of *kaolin*.

laterite Term coined by Buchanan in 1807 to describe hardened iron-rich soil material he saw used as building blocks in southern India; subsequently used more loosely to describe a wide variety of indurated or reddish tropical soils and soil materials. Largely replaced by more recent terms such as *ironstone* and *plinthite*.

Latosols general term, now superseded, coined by the American soil scientist Charles Kellogg, to refer to highly weathered iron-rich soils of the tropics.

lattice Three-dimensional grid of lines connecting points representing the centres of atoms or ions in a crystal.

leaching The washing out or removal of substances from a soil or other material in solution by percolating water. (See *eluviation*.)

lime requirement Weight of lime (or equivalent of other liming materials) needed to raise the pH of the soil to a specific value under field conditions.

lithic contact Boundary between soil and continuous, coherent underlying rock.

mass flow The movement of solutes associated with the net movement of water.

matric potential The force (or suction) with which water is held in the soil.

metamorphic rock Rock formed from a previous *sedimentary, igneous* or metamorphic rock fundamentally changed by heat and pressure.

mica Phyllosilicate mineral consisting of layers, in each of which two sheets of silica tetrahedra are held together by shared cations. In white mica, which is colourless rather than white, the double layers are held together by aluminium and hydroxyl ions, and weakly bonded to each other by potassium ions. In black mica, also known as biotite, iron and magnesium replace the internal aluminium bonds. Both types of mica release potassium on weathering, but biotite also releases iron and magnesium. Biotite weathers relatively easily, and white mica much more slowly.

microclimate Climate of a small area resulting from modifications of the general climatic conditions by local differences in elevation, exposure, etc., including, for example, the microclimate within a cocoa farm.

micrometre Unit of length equivalent to one millionth of a metre (or 0.001 mm), also referred to as a micron (μm). referred to as a micron (μm).

micron See *micrometre*.

micronutrient Nutrient required in very small amounts, synonymous with *trace element*.

milliequivalent (m.e.) The thousandth part of the *equivalent weight*. The *cation* or *anion exchange capacity* of the soil or a colloid is normally expressed in milliequivalents per 100 g.

mineralisation Decomposition of humus into its constituent ions.

mixed fertiliser Chemical fertiliser containing two or more major nutrient elements.

mixed intercropping Intercropping system in which two or more crops are mixed together in a random way.

mollic epipedon Diagnostic surface horizon that is dark coloured and usually at least 25 cm thick, with a base saturation of over 50%. Typically, mollic epipedons have a well-developed crumb structure and organic matter contents of at least 1%.

Mollisols (US Soil Taxonomy) Soil order comprising soils with a mollic epipedon overlying a subsoil which has a base saturation of at least 50% to a depth of at least 1.25 m unless there is a lithic or paralithic contact shallower than this.

montmorillonite Clay mineral with an expanding 2:1 lattice clay structure, i.e. with two layers of silicon tetrahedra enclosing an aluminium octahedral layer. The mineral can expand considerably when water moves to positions between the silica layers of adjoining units. The cation exchange capacity of montmorillonite is relatively high, being much higher than that of *kaolin* and higher than that of *illite*.

morphology Soil morphology refers to the thickness, properties and arrangement of the horizons in a soil profile.

muscovite Another name for white mica (see *mica*), usually occurring in relatively large flakes.

mottle Spot or blotch of different colour or shade from the dominant soil colour. In soil description mottles are described according to their frequency, size, shape, distinctness and colour.
muck Highly decomposed organic material in which original plant parts are no longer recognisable.
mycorrhiza Association, usually symbiotic, of specific fungi with plant roots.
natric horizon Soil horizon affected by the presence of sodium. In the US Soil Taxonomy, a natric horizon is an *argillic horizon* in which the presence of sodium is shown by the prismatic or *columnar structure* (usually in the upper part), and by a *sodium absorption ratio* (SAR) of 13 or more in some subhorizons within 40 cm of the upper boundary of the horizon.
nesosilicates Structural group of silicates in which the silica tetrahedra do not share oxygen atoms with adjacent tetrahedra but are linked by intermediate cations. Nesosilicates include the *olivine* group, ferromagnesian minerals in which the tetrahedra are linked by shared iron and magnesium cations.
nitrification Biological oxidation of ammonium (NH_4^+) to nitrite and nitrate (NO_3^-).
nitrogen cycle Sequence of biochemical changes undergone by nitrogen involving in turn its use by a living organism, liberation upon death and decomposition of that organism, and conversion to its original state of oxidation.
O horizon Organic horizon in mineral soils. In the FAO/Unesco world soil map legend, defined as an organic horizon with over 35% organic matter not usually saturated with water.
ochric epipedon Surface horizon of a mineral soil generally too light in colour and low in organic matter to qualify as a *mollic, umbric* or *histic epipedon*, or which is both hard and massive when dry.
olivine *Nesosilicate* mineral, dark in colour, containing iron and magnesium, which weathers relatively easily.
order The highest category in soil classification.
oven-dry soil Soil dried at 105 °C until it reaches constant weight.
oxic horizon Diagnostic mineral soil horizon at least 30 cm thick with the following properties:
1 A low *cation exchange capacity* (the fine earth fraction retains less than 10 m.e. NH_4^+ per 100 g clay from an unbuffered M NH_4 Cl solution, and the apparent cation exchange capacity of the fine earth fraction as measured with ammonium acetate is less than 16 m.e.).
2 No more than traces of primary weatherable silicate minerals such as *feldspars, micas* and ferromagnesian minerals.
3 Has more than 15% clay and a texture of sandy loam or finer.
4 Has gradual or diffuse boundaries.
5 Has little or no traces of rock structure (less than 5% by volume).
Oxisols (US Soil Taxonomy) Order of soils defined as containing an *oxic horizon* within 2 m of the surface (or *plinthite* as a continuous phase within 30 cm) and without a *natric* or *argillic horizon* above the oxic horizon.
paralithic contact Similar to a *lithic contact* but material below contact is not very hard (hardness less than 3 on the Moh scale) and gravel-sized fragments broken out will partially disperse with 15 hours shaking with water or with water containing sodium hexametaphosphate.
parent material In a broad sense, any material in which or from which a soil profile develops. Sometimes limited to the unconsolidated and more or less weathered mineral or organic matter from which the soil solum is directly developed by soil-forming processes.
particle density The mass per unit volume of the soil particles. Normally expressed as g/cm^3.
particle size analysis Determination of the relative proportions by weight of the different particle size classes (sand, slit, clay) in the fine earth fraction of the soil, usually by a combination of sieving and sedimentation after the soil sample has been dispersed into its ultimate particles. Also called mechanical analysis of the soil.
parts per million (ppm) Weight units per million equivalent weight units, or, in the case of solutions, the weight units of solute per million weight units of solution.
ped A soil structural aggregate or unit, such as a crumb, prism, block or granule, formed by natural processes.
pedon The minimal three-dimensional block of soil large enough to permit the study of the shape and relationship of soil horizons: its area is usually between 1–10 m^2.
percolation The downward movement of soil water through the soil.
permeability Ease with which liquids, gases or plant roots pass through or penetrate the soil or a specific soil horizon.
petrocalcic horizon Continuous, indurated horizon cemented by calcium or magnesium carbonate; impenetrable to roots, or with a spade or auger. Dry fragments do not slake in water.
petrogypsic horizon Massive horizon strongly cemented by calcium sulphate. Impenetrable to roots. Dry fragments do not slake in water.
pF Negative logarithm of the matric potential expressed as g/cm^2. Thus *wilting point* corresponds to a pF of 4.2, the negative logarithm of 15 000 g/cm^2 (15 atmospheres).
pH dependent charge That portion of the *cation exchange* or *anion exchange capacity* of the soil which varies with pH.
pH of the soil Scale of from 1 to 14 related to the acidity or alkalinity of the soil, with 7.0 as neutral, defined as the negative logarithm of the hydrogen ion activity of the soil. Measured by an electrode or colour indicator at a specified soil–water ratio (e.g. 1:2 soil:water).
phenolic Containing phenol (C_6H_5OH).
phyllosilicates Group of silicate rocks in which the silica tetrahedra share all the three basal oxygen atoms with adjoining tetrahedra to form sheets. Phyllosilicates include the *micas* and the *clay minerals*.
placement Positioning of fertiliser by insertion into the soil in a specific location in relation to plant roots.

plastic Soil *consistency* descriptive term indicating a soil that can be moulded or deformed continuously when wet with only moderate pressure.

plinthite Mottled soil material, often reddish, containing iron and aluminium oxides and capable of hardening irreversibly on exposure and repeated wetting and drying. Term coined recently to replace older terms now lacking precise definition, such as *laterite*. (See also *ironstone*.)

podsol or podzol Soil group (now obsolete) consisting of soils characterised by a highly leached greyish-white lower A horizon, the A2, and usually by a dark and sometimes indurated horizon of accumulation below. Podsols are typically soils of cooler latitudes associated with coniferous forests whose acid leaves play a part in the formation of the characteristic A2, but tropical podsols, associated with other types of vegetation and sometimes very deep, occur in areas of generally sandy soils and high rainfall, including parts of West Africa, Suriname and Sarawak.

prism Structural aggregate (*ped*) in which the vertical development is greater than the horizontal. Prisms are bounded by flat vertical faces. Those with rounded caps are referred to as columns.

profile The soil profile is a vertical cross-section of the soil showing the various horizons of which it is composed and extending into the parent material.

quartz Silicate mineral, usually colourless, with the formula SiO_2 with a regular block structure (tectosilicate) and no isomorphous substitution which under normal conditions in the soil is extremely resistant to chemical weathering; a major constituent of acid igneous rocks such as granite.

ratoon crop Second crop produced from same root system after the first has been cut or harvested, e.g. sugarcane or rice.

reaction (of the soil) Degree of acidity or alkalinity of the soil. (See *pH*.)

response Increase in crop growth or yield as a result of the use of a fertiliser, amendment or other cultural practice.

rhizobia Bacteria capable of living symbiotically in legume roots and of fixing atmospheric nitrogen; collective common name for the genus *Rhizobium*.

rhizosphere The soil zone close to plant roots in which roots influence the microbial population both quantitatively and qualitatively.

row intercropping Form of *intercropping* in which one or more of the crops are in regular rows.

runoff (surface runoff) Proportion of rain received which runs off the surface without entering the soil.

salic horizon Mineral soil horizon enriched with secondary salts more soluble in cold water than gypsum. Must be at least 15 cm thick and have at least 2% salt.

saline soil Non-sodic soil with enough soluble salt to lower its productivity. Electrical conductivity of the saturation extract is more than 2 mmhos per cm at 25 °C.

saline–sodic soil Soil (formerly called a saline – alkali soil) with a combination of soluble salts and exchangeable sodium sufficient to impair the growth of most crop plants. The *electrical conductivity* of the saturation extract is above 2 mmhos cm at 25 °C, and its *sodium absorption ratio* above 15. The pH of the saturated soil paste is usually less than 8.5.

salinization Process whereby soluble salts accumulate in a soil.

sand Soil particle size class usually defined as mineral particles with a diameter of 50–2000 (μm) (US scale) or 60–2000 μm (international scale). Sometimes divided into very fine, fine, medium coarse and very coarse sand (US scale) or into fine, medium and coarse sand (international scale).

schist *Metamorphic rock* in which subparallel layers of minerals, including micas, are oriented at right angles to the main pressure source, giving a characteristic fine-banded (schistose) structure.

sediment Material transported by wind or water and then deposited.

sedimentary rock Rock formed from sediment often compacted cemented but not fundamentally changed as in a *metamorphic* rock.

semi-deciduous forest Forest formation in which at least some species shed some of their leaves at some season of the year; usually associated with climates that are less wet or less continuously wet that those associated with *evergreen forest*.

sequential cropping Form of *intercropping* in which two or more crops are grown on the same land within a year, one after the other, but without overlap.

sesquioxides Metallic compounds in which two molecules of the metal combine with three of oxygen, particular Fe_2O_3 and Al_2O_3. Sesquioxides are resistant to leaching and tend to accumulate in highly weathered tropical soils or soil horizons (see *ironstone, laterite, plinthite*).

side-dressing Method of applying chemical fertilisers in which fertiliser is applied along or between rows or around individual plants and trees, so that it is more localised than in the case of top-dressing.

silica–alumina ratio The ratio by weight of silicon dioxide (SiO_2) to aluminium oxide (Al_2O_3) in a soil or clay mineral. Since silica is weathered and removed less slowly than alumina, the silica–alumina ratio is used as an index of weathering, becoming lower as soils lose silica and accumulate alumina as they become progressively more highly weathered.

silica–sesquioxide ratio The ratio by weight of silicon dioxide (SiO_2) to aluminium and ferric oxide (Al_2O_3 and Fe_2O_3) in a soil or a clay fraction. Like the *silica–alumina ratio*, this is used as an index of soil weathering.

silt Soil particle size class intermediate between *clay* and *sand*.

single fertiliser Chemical fertiliser containing only one major nutrient element. (See *mixed fertiliser*.)

slickensides Polished and grooved surfaces, common in Vertisols, produced by soil expansion and contraction involving one mass sliding past another.

smectite Group of 2:1 lattice clay minerals of which the most important is *montmorillonite*.

sodic soil Soil with enough exchangeable sodium to impair the growth of most crop plants. Alternatively, a soil in which the sodium absorption ratio of the saturation extract is 15 or more.

sodium absorption ratio (SAR) Ratio between soluble sodium and soluble divalent cations (calcium and magnesium) used to predict the *exchangeable sodium percentage* of a soil equilibrated with a given solution.

$$\text{SAR} = \frac{\text{sodium, mmoles/litre}}{\text{(calcium plus magnesium) mmoles/litre}}$$

soil association Soils associated in nature to form a characteristic pattern; a soil mapping unit showing soils thus associated. In tropical areas soil associations are often equivalent to soil catenas, mapped as associations when the scale is too small to show the individual soil series.

soil-forming factor Environmental factor that influences soil formation and properties. The five main soil-forming factors are parent rock or parent material, relief, climate, vegetation and soil fauna, and time. Human activity is sometimes thought of as a sixth soil-forming factor.

soil-moisture regime Moisture regime of the soil defined in terms of the length of time that the *control section* of the soil profile is moist, i.e. contains moisture available to plants (held at a tension of less than 15 atmospheres). (See *aridic, aquic, udic and ustic moisture regimes.*)

solum The upper part of the soil profile; the A and B horizons.

split application Application of the same fertiliser more than once to a crop in a year.

spodic horizon The dark subsurface horizon of a podsol: a mineral horizon at least 2.5 cm thick continuously cemented by organic matter in combination with iron and/or aluminium, Texture is usually sandy or coarse-loamy and the sand grains covered with cracked dark coatings of the cementing material. Alternatively there are subhorizons which meet certain criteria defining the ratios between the extractable iron and aluminium and the clay content, or the accumulation of amorphous material.

Spodosols (US Soil Taxonomy) Soil order including soils with a *spodic horizon* within 2 m of the surface.

strip intercropping Form of *intercropping* in which the crops are in strips narrow enough to influence each other.

tectosilicates Structural group of silicate minerals in which the silica tetrahedra share all four oxygen atoms with adjoining tetrahedra to link up to form a three-dimensional block structure. *Quartz* and the *feldspars* form examples of tectosilicates.

texture The relative proportions of *sand*, *silt* and *clay* in the fine earth fraction of the soil.

tilth Term denoting the physical condition of the soil as related to ease of tillage, fitness as a seedbed, and resistance to seedling emergence and root penetration.

top-dressing Method of fertiliser application involving the *broadcasting* of fertilisers onto a growing crop.

toposequence Sequence of soils related to relief.

topsoil General term denoting the upper part of the soil or the layer affected by cultivation or the *A horizon*. More generally, soil material containing humus applied agriculturally as a top-dressing.

trace element Essential element needed by plants in only very small quantities.

udic moisture regime Soil moisture regime in which the *control section* of the soil is dry for less than 90 cumulative days.

Ultisols (US Soil Taxonomy) Soil order comprising soils with an *argillic horizon* in which the subsoil has a base saturation of less than 35%. Otherwise similar soils with a higher base saturation are *Alfisols*.

umbric epipedon Dark-coloured surface horizon rich in organic matter and similar to a mollic epipedon except that base saturation is less than 50%.

ustic moisture regime Soil moisture regime intermediate between the arid and udic regimes. The soil moisture *control section* is dry in most years in some or all parts for 90 or more cumulative days, but is moist in some part for at least 180 cumulative days or continuously moist in some part for at least 90 consecutive days. Ustic regimes are essentially the regimes of seasonal rainfall areas.

Vertisol A term generally applied to dark-coloured, cracking clays, widespread in the tropics.

Vertisols (US Soil Taxonomy) A soil order consisting of soils defined as having a clay content of at least 30% in all horizons down to at least 50 cm and cracks at a depth of 50 cm which in most years are at least 1 cm wide and extend upwards to the surface or the plough layer. In addition Vertisols must have either *gilgai* surface relief or *slickensides* or wedge-shaped structural aggregates with their long axes tilted 10–60° from the horizontal at between 25 cm and 1 m.

water stable aggregate Soil aggregate stable in water as evidence by stability in a wet-sieving analysis.

water table The upper limit of saturated soil or of ground water.

weathering Physical and chemical changes produced in rocks and minerals by atmospheric and biological agents.

wilting point Soil moisture content at which most plants wilt permanently and die. Commonly equated to the amount of water held in the soil at a tension of over 15 atmospheres. This water is assumed unavailable to most plants, and the *available water* is defined as the difference between the water held at *field capacity* and wilting point.

zero tillage Cultural practice in which there is no soil tillage, and seeds are simply inserted in a planting hole, usually after weeds have been killed by a weedicide and left as a mulch.

1 The tropical environment

Extent and relief

Geographically the tropics are those areas of the world between latitudes 23.5° N and 23.5° S; that is, those areas where the sun is directly overhead at some part of the year. Thus defined, the tropics include some 38% of the earth's land surface and 45% of its population. This geographical definition of the tropics includes areas of cold tropical highlands, particularly in East Africa and along the Andes Mountains in South America, as well as the more extensive hot lowlands. An important unifying feature of the tropics is that average temperatures are relatively constant throughout the year and do not vary by more than 5 °C between the averages of the three warmest and the three coolest months. This applies both to the hotter lowlands and to the cooler highlands.

As would be expected from such a large area, both the geology and the relief of the tropics are very varied. In overall terms, about 77% of the tropics can be regarded as being lowland, with altitudes less than 900 m. A further 20% lies between 900 and 1800 m, while only about 3% of the tropics is higher than 1800 m.

Climate

Solar radiation and day length

The tropics receive more solar radiation than the temperate regions, and there is thus more energy available for photosynthesis. This is because the sun is overhead or nearly so, and the rays therefore pass through a lesser thickness of atmosphere and lose less radiation to it. In areas with distinct wet and dry seasons there is considerable variation in solar radiation from season to season, largely related to differences in the amount of cloud cover. The highest radiations are received in arid areas towards the edges of the tropics, such as in the Sudan and the Sahara Desert. On the basis of solar radiation and growing season length, it has been calculated that tropical areas have approximately twice the theoretical potential yield of temperate areas, though in practice yields depend on many other factors apart from radiation.

Day length changes less in the tropics than in temperate regions, and during the summer, days are longer in the temperate zone than in the tropics. At the equator, day length is 12.1 hours at all seasons, and at 25° N and S it ranges from 10.6 to 13.7 hours. Tropical plants are considered short-day plants, but many are none the less very sensitive to small changes in day length. Initiation of flowering is particularly affected, and this fact is sometimes made use of by farmers. In West Africa, for example, sorghum growers use day-length-sensitive varieties of sorghum to ensure that the crop matures in the early part of the dry season, on stored soil moisture, thus avoiding the fungal diseases associated with high humidity at the end of the rains. In Asia, selection and breeding have produced rice varieties that vary between being very sensitive to day length and being day length insensitive. The latter varieties are more easily moved from one latitude to another without their growing period being affected.

Temperature

Temperatures in the tropics at a given location vary relatively little either between seasons or from year to year. A general characteristic of tropical climates is thus that temperatures are relatively constant, and a climatic definition of the tropics would be that the tropics are those areas where the mean monthly temperature variation between the average of the three warmest and the three coolest months is less than than 5 °C. The least seasonal temperature variation occurs near the equator, and the most seasonal variation is found in dry areas near the Tropics of Cancer and Capricorn.

Temperature variations between tropical areas in similar latitudes are largely a function of height above sea level. Although minor variations are found, there is an average decrease in temperature of 0.65 °C for every 100 m increase in elevation.

The life cycle of any particular crop variety is closely related to temperature, and is slowed down when it is grown at higher altitudes than usual. In lowland East Java, for example, rice takes 90–100 days from transplanting to maturity, but at 1600 m the same cultivar takes about 220 days. As a result the cropping system changes with altitude. In the lowlands three rice crops a year can be grown as against only one at the higher altitude. Because crops have a certain minimum temperature below which they cease to grow, they also have a maximum elevation. For coffee it is about 1500 m. In the highlands of Colombia, maize is grown up to 4000 m but the growing season under those conditions can be a year or more. The pyrethrum plant, of which Kenya is the leading producer, provides an example of a crop whose distribution is limited by a minimum elevation, also related to temperature. Flower initiation is triggered by a cold spell, and flowering does not occur at elevations of less than about 1800 m.

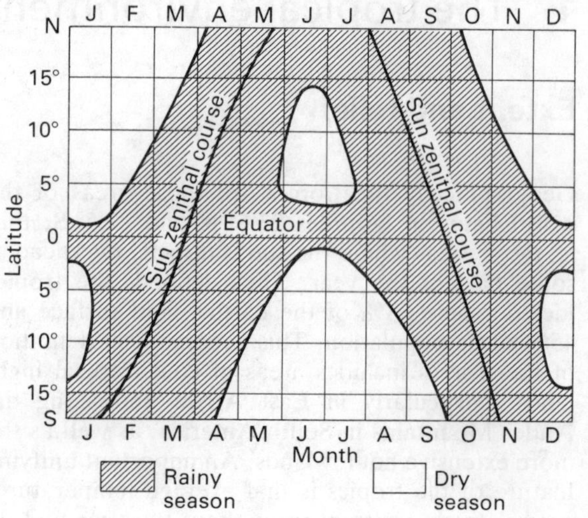

Fig. 1.1 Simplified diagram of wet and dry seasons in the tropics and their relationship to the apparent movement of the sun. In lower latitudes there are two wet seasons separated by two dry seasons, generally of unequal length. Further away from the equator the two wet seasons merge to give a single wet season.

Rainfall

Rainfall is the most important climatic parameter for tropical agriculture. Because of the relative uniformity during the year of the temperature, it is the rainfall distribution that distinguishes the seasons.

Rain in the tropics is associated with the seasonal movement of the Inter-Tropical Convergence Zone (ITCZ), which is a moist body of air within which various phenomena that cause rainfall are likely to occur. It moves with the apparent movement of the sun, bringing rain with it. This movement and the rainy seasons associated with it are shown diagrammatically in Fig. 1.1, while some representative rainfall distribution patterns in relation to potential evapotranspiration are given in Fig. 1.2.

At the equator (Fig. 1.1), rain falls during all or most months. These rainy climates occupy about one-quarter of the tropics. Major areas include the upper Amazon and the Congo basins, most of Indonesia and Malaysia, and parts of the Philippines. The climax vegetation of these areas is typically evergreen tropical forest.

Further away from the equator, both north and south of it, the rainfall becomes more seasonal. Up to about 14° N and S there are typically two wet seasons separated by a longer dry season in the winter months and a short dry season in the summer. With increasing distance from the equator the short summer break in the rains disappears and the two wet seasons merge into one single wet season separated by one dry season. This single wet season gets shorter and less wet as the tropic is approached and the seasonal climate merges into a dry climate. This regular change of rainfall with latitude is well demonstrated by the situation in West Africa (Fig. 1.3).

Fig. 1.2 Some representative rainfall graphs of aridic, udic and ustic climates in the tropics. The solid line indicates rainfall, and the broken line potential evapotranspiration. The figures show total annual rainfall (PPT) and potential evapotranspiration (PET) in mm. (Redrawn from Sanchez, 1976.)

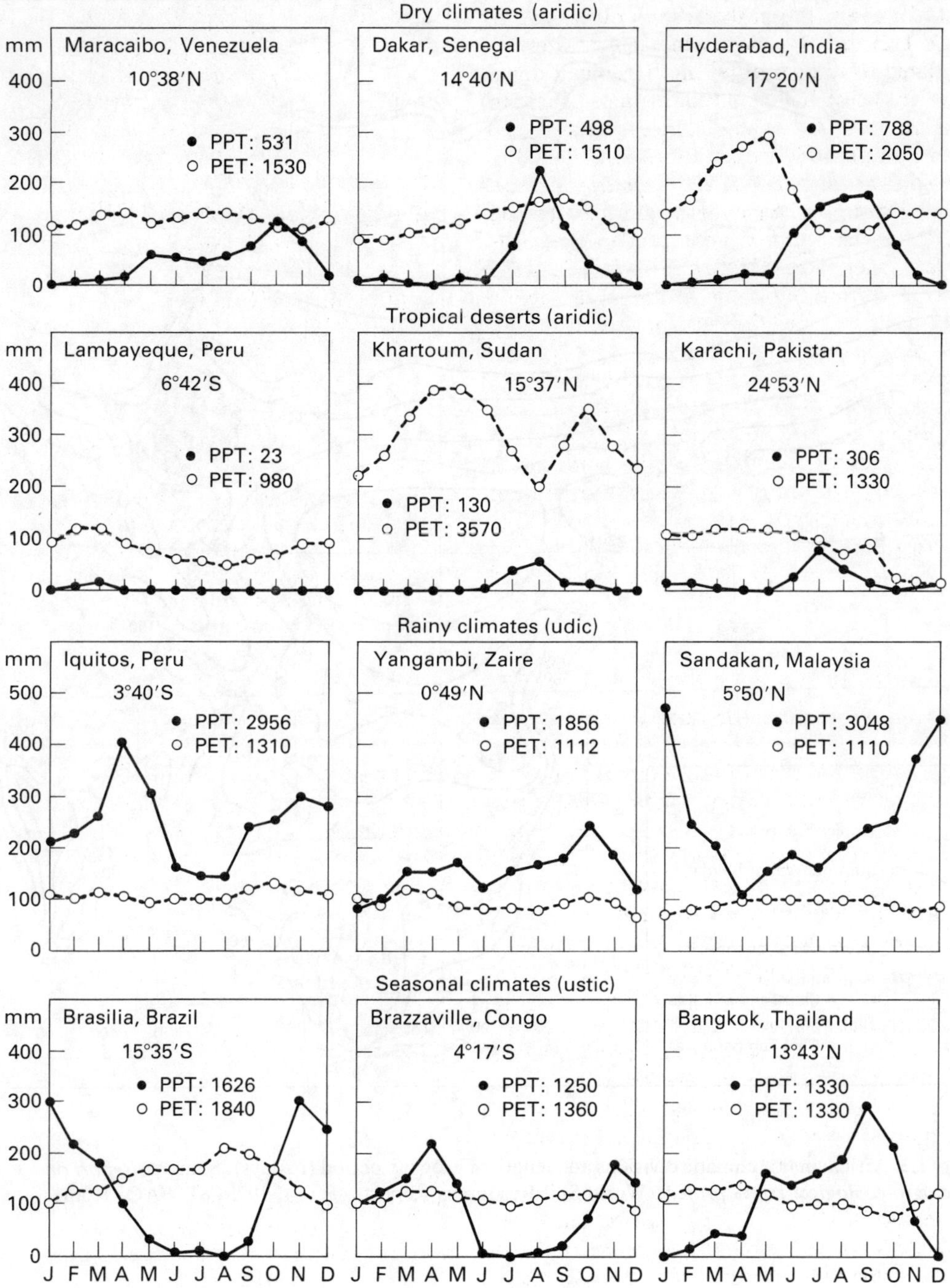
Dry climates (aridic)
mm
Maracaibo, Venezuela
10°38′N
PPT: 531
PET: 1530
Dakar, Senegal
14°40′N
PPT: 498
PET: 1510
Hyderabad, India
17°20′N
PPT: 788
PET: 2050
Tropical deserts (aridic)
Lambayeque, Peru
6°42′S
PPT: 23
PET: 980
Khartoum, Sudan
15°37′N
PPT: 130
PET: 3570
Karachi, Pakistan
24°53′N
PPT: 306
PET: 1330
Rainy climates (udic)
Iquitos, Peru
3°40′S
PPT: 2956
PET: 1310
Yangambi, Zaire
0°49′N
PPT: 1856
PET: 1112
Sandakan, Malaysia
5°50′N
PPT: 3048
PET: 1110
Seasonal climates (ustic)
Brasilia, Brazil
15°35′S
PPT: 1626
PET: 1840
Brazzaville, Congo
4°17′S
PPT: 1250
PET: 1360
Bangkok, Thailand
13°43′N
PPT: 1330
PET: 1330
0
100
200
300
400
500
J F M A M J J A S O N D

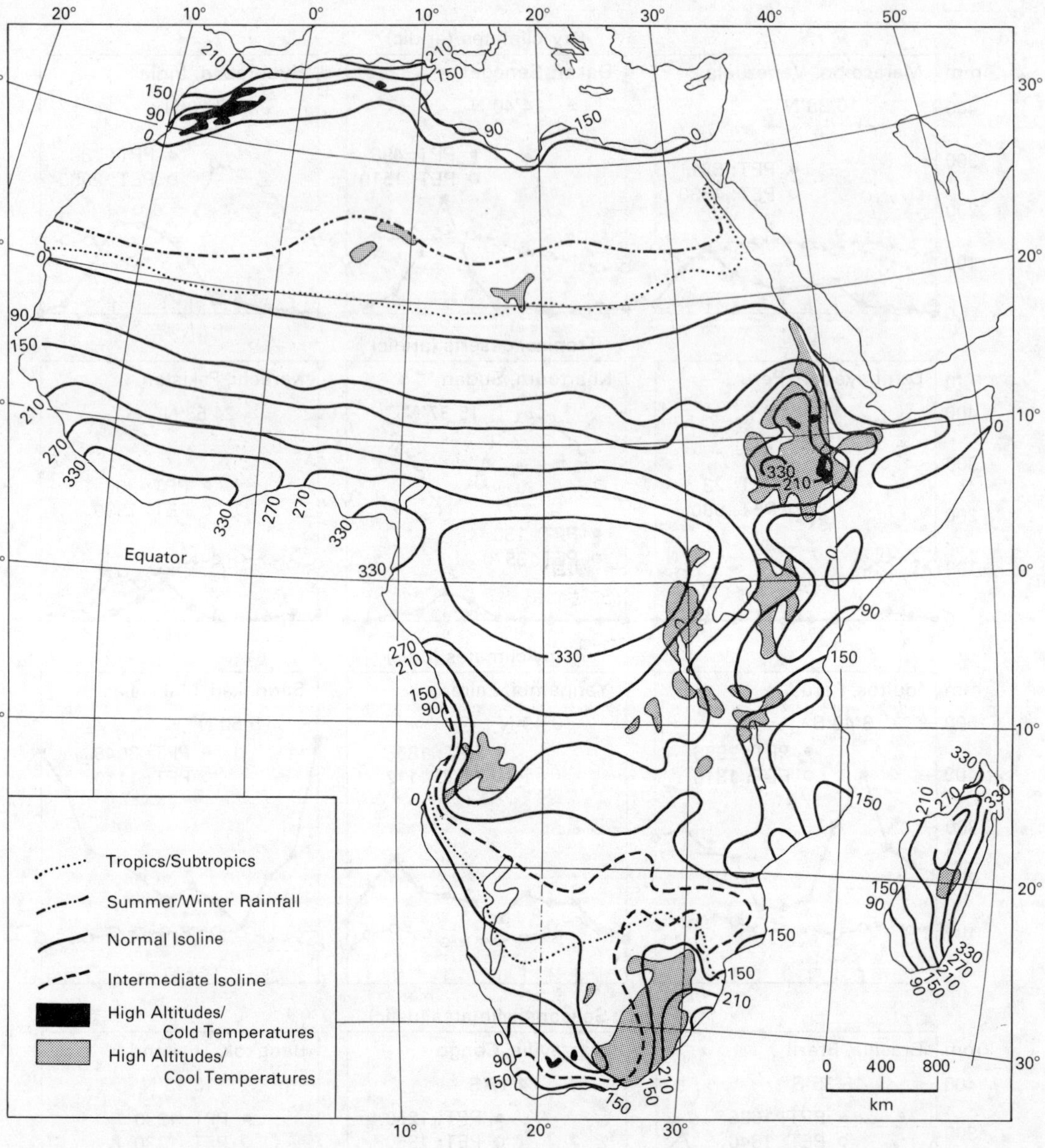

Fig. 1.3 Africa: major climatic divisions and length of growing period (in days). Source: *Report on the agro-ecological zones project.* World Soil Resources Reports 48/1–4, 1978–81. FAO: Rome.

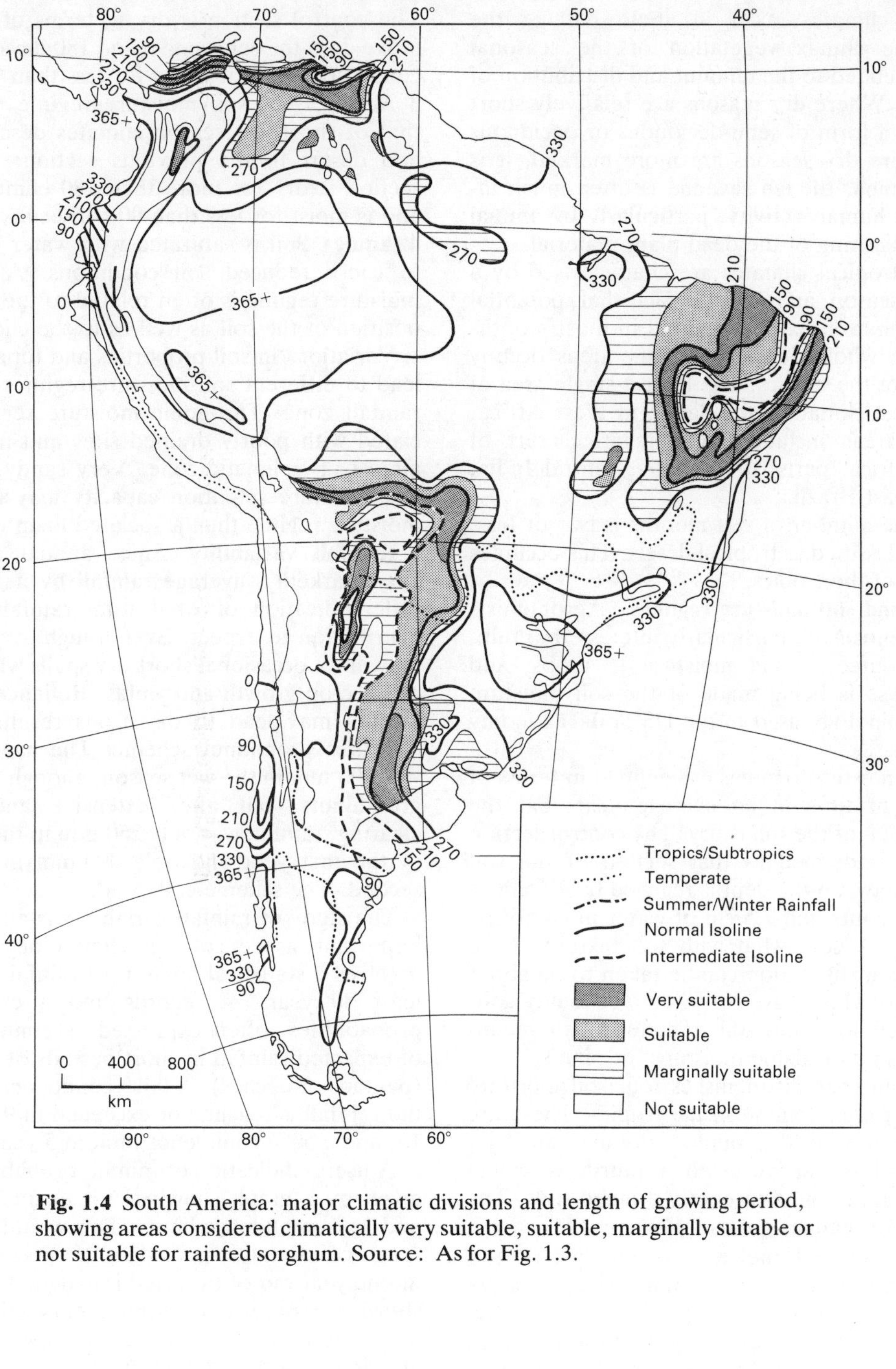

Fig. 1.4 South America: major climatic divisions and length of growing period, showing areas considered climatically very suitable, suitable, marginally suitable or not suitable for rainfed sorghum. Source: As for Fig. 1.3.

Seasonal climates occur in about half of the tropics. The climax vegetation of the seasonal climates is related to the amount and distribution of the rainfall. Where dry seasons are relatively short it is usually a form of semi-deciduous or deciduous forest. Where dry seasons are more marked, it is usually savanna, though savanna is often much influenced by human activity, particularly by annual dry season burning of the dead plant material.

The dry tropical climates are characterised by a short wet season and by the fact that potential evapotranspiration exceeds rainfall in most months and for the whole year. These dry areas occupy some 16% of the tropics. The largest single area of a dry tropical climate is the Sahel, in West Africa, but other areas include the Kalahari Desert of southern Africa, parts of Australia, central India, and north-east Brazil.

Where the number of wet months is two or less, the area is classified as tropical desert. This occupies about 11% of the tropics.

Rainfall and soil moisture regimes Agronomists and soil scientists are particularly interested in rainfall as a source of soil moisture to crops, and increasing use is being made of the soil moisture regime terminology used in the US Soil Taxonomy (see Chapter 8).

The soil moisture regimes are defined in terms of the length of time in an average year that the control section of the soil is dry. The control section of the soil is defined as that section of the soil which lies between the depths reached by 2.5 cm of water in 24 hours and 7.5 cm of water in 48 hours, and these vary somewhat with soil texture. As a guide, the control section can be taken to be about 10–30 cm for clay soils, 20–60 cm for loamy soils and 30–90 cm for sandy soils. The term 'dry' means that there is no available moisture for plants.

The terminology distinguishes four soil moisture regimes that are common in the tropics. These are broadly related to the rainy, seasonal, and dry climates outlined above, with a fourth very wet moisture regime found in poorly drained soils. The four moisture regimes are as follows:

1 **Udic** A wet soil moisture regime where the control section is dry for less than 90 cumulative days.

2 **Ustic** This is essentially a seasonal soil moisture regime with a marked dry season of 3–6 months. The control section is dry in terms of total cumulative days for between 90 and 180 days. In terms of consecutive days it is dry for less than 90 days.

3 **Aridic** This soil moisture regime is related to that of short wet season climates described as dry and desert in the previous section. The control section is dry for more than 180 cumulative days, and is moist for less than 90 consecutive days.

4 **Aquic** Soil is saturated with water long enough to cause reduced soil conditions. An aquic soil moisture regime is often related to the topographic position of the soil as well as to the climatic zone.

Variations in soil properties and topography may lead to different soil moisture regimes in the same rainfall zone. The aquic moisture regime is associated with poorly drained sites and may occur in almost any climatic zone. Very sandy soils with a low moisture retention capacity may have a drier moisture regime than associated loam or clay soils.

Rainfall variability Since annual rainfall may vary markedly, average rainfall by itself may give little indication of what total rainfall or rainfall distribution to expect. Even a high average rainfall may mask occasional short dry spells which severely affect crop growth and yields. Reliance on average rainfall may lead to disastrous results, as in the Tanzania groundnut scheme. The average rainfall was 700 mm in the wet season, enough for a crop of groundnuts, but after extensive and expensive clearing, rainfall was only 250 mm in the wet season of the first year and only 300 mm in that of the second. The scheme collapsed.

The study of rainfall variability is an increasingly important aspect of agroclimatology. It usually involves a statistical analysis of rainfall based on at least 30 years of records and a calculation of probabilities, often expressed as confidence limits of expected rainfall in short periods of 5 or 10 days (pentads or decads). A 90% confidence limit means the rainfall is reached or exceeded in 9 years out of 10, and a 50% confidence limit in 5 years out of 10.

A useful indication of rainfall probabilities can be given by a graph showing, for example, the 25% and 75% confidence limits. This would indicate the range of rainfall expected in the periods indicated in one year out of two, and if matched with a graph showing crop water requirements might usefully draw attention to possible dry spells within the growing season of that crop.

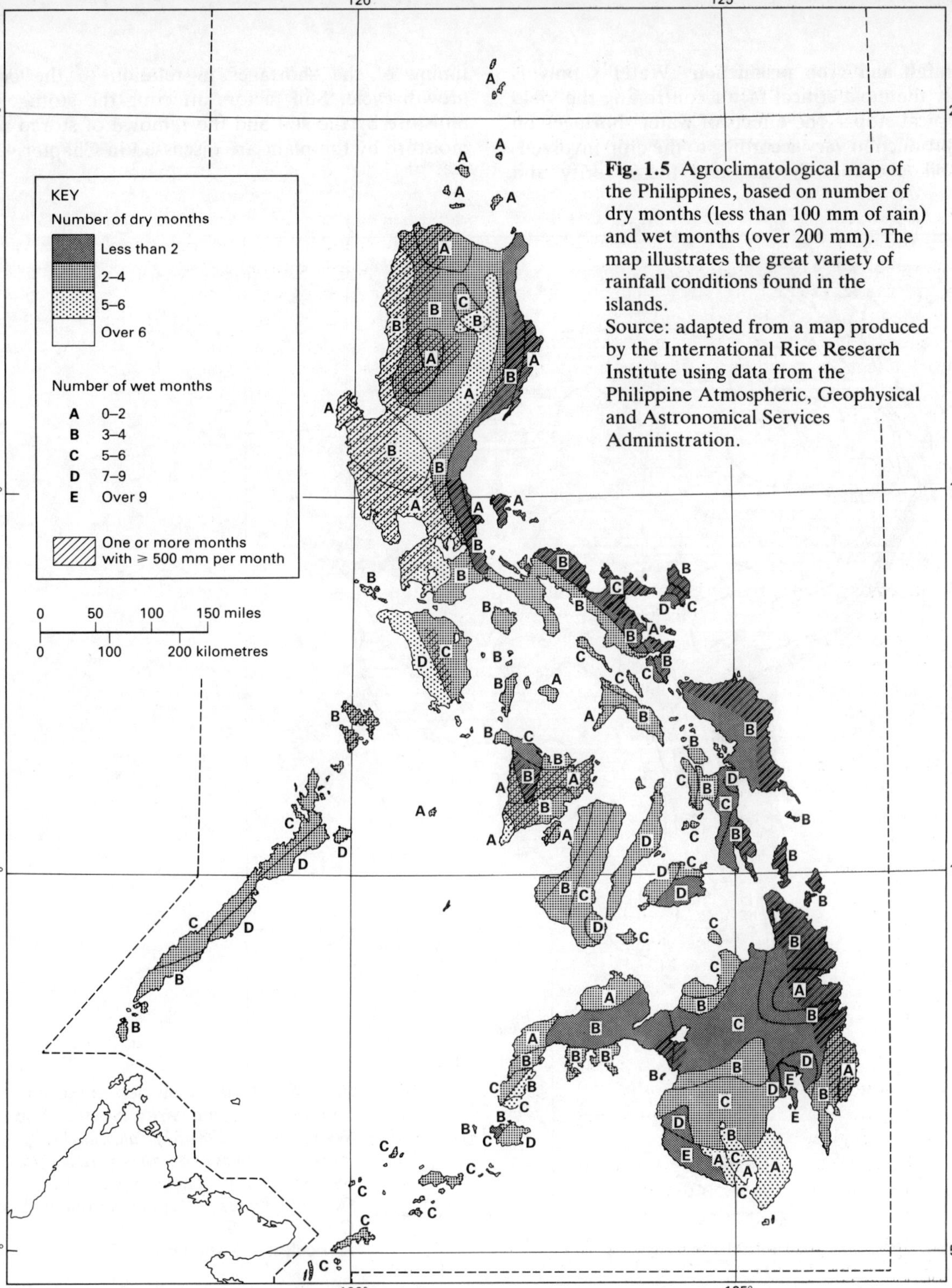

Fig. 1.5 Agroclimatological map of the Philippines, based on number of dry months (less than 100 mm of rain) and wet months (over 200 mm). The map illustrates the great variety of rainfall conditions found in the islands.
Source: adapted from a map produced by the International Rice Research Institute using data from the Philippine Atmospheric, Geophysical and Astronomical Services Administration.

Rainfall and crop production Water supply is usually the most critical factor controlling the yield of tropical crops. The effects of water shortages on crop production vary according to the crop involved, the soil, the root system, and the severity and timing of the shortages in relation to the crop growth cycle. Soil factors affecting the storage of moisture by the soil and the removal of stored soil moisture by the plant are discussed in Chapter 4.

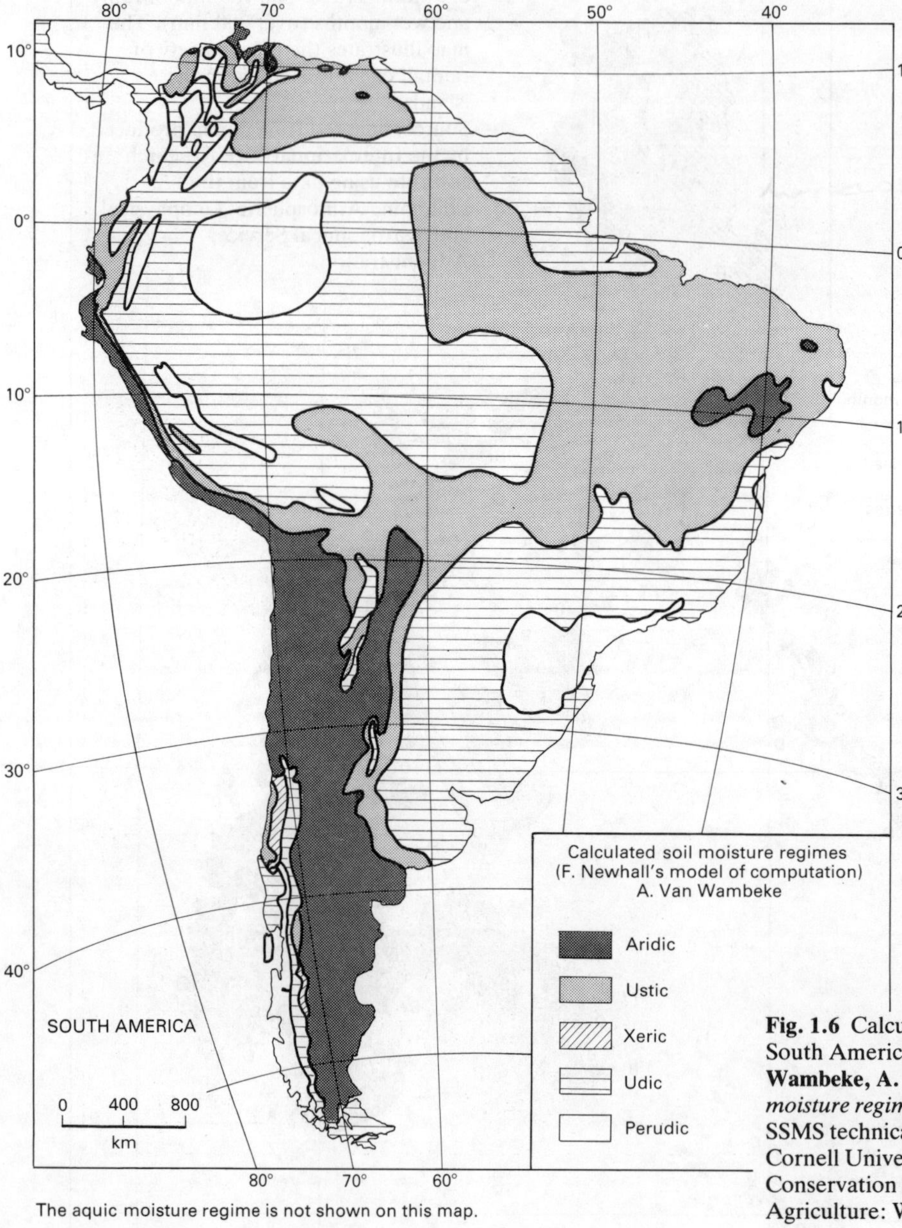

Fig. 1.6 Calculated soil moisture regimes for South America and Africa. Source: **Van Wambeke, A.** (1981/82) *Calculated soil moisture regimes of South America/Africa*. SSMS technical monographs 2 and 3. Cornell University Dept of Agronomy/Soil Conservation Services, US Dept of Agriculture: Washington DC.

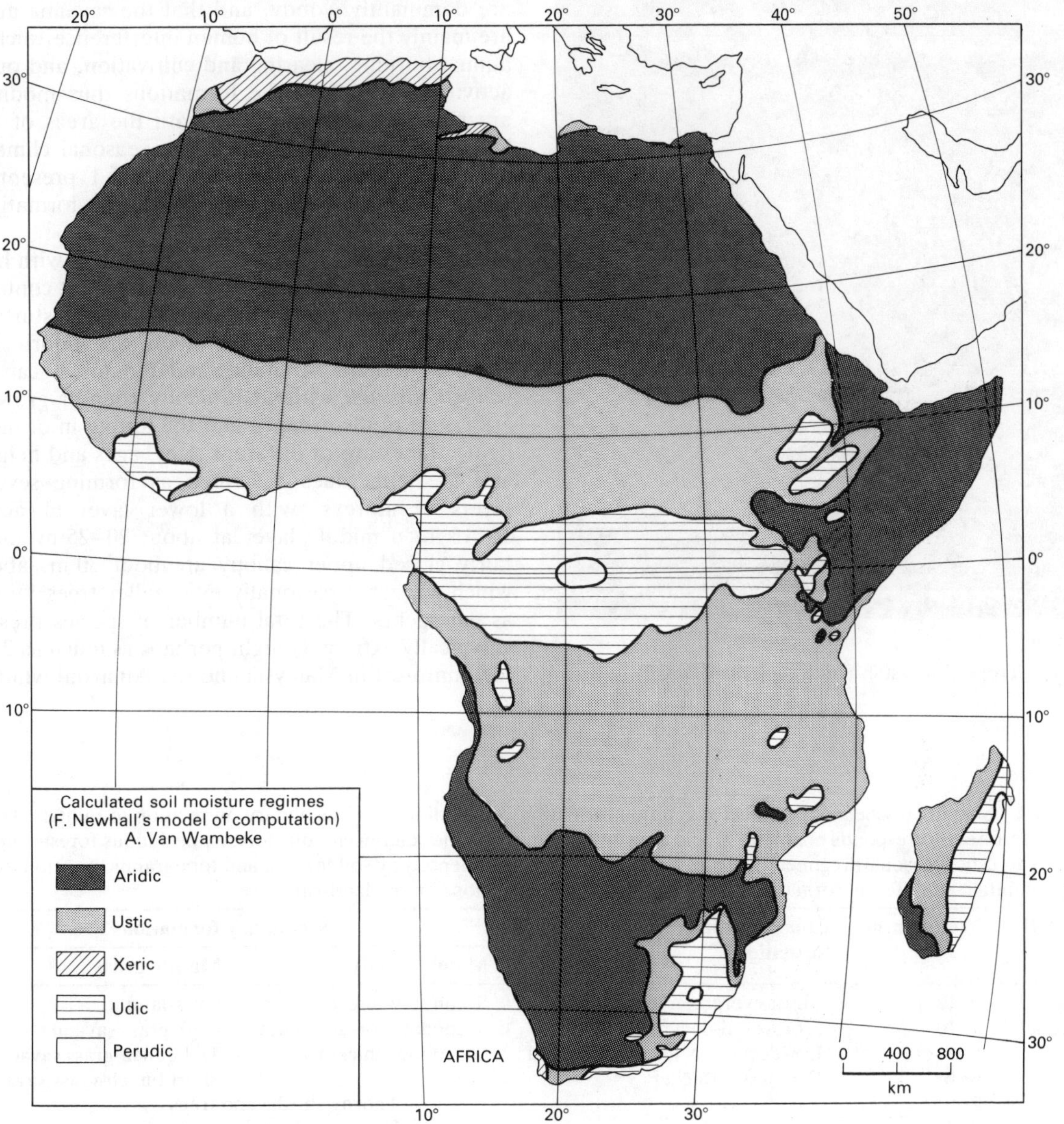

The aquic moisture regime is not shown on this map.

Fig. 1.7 Tropical forest in South America (Brazil).

Vegetation

Primary formations

Natural vegetation in the tropics is closely correlated with climate. It is thought by botanists that all the climatic climax vegetations on free-draining soils are dominantly woody, and that the savanna areas are mainly the result of human interference, such as annual burning, clearing and cultivation, and other activities. However, the formations thus modified are mainly those associated with the areas of less heavy rainfall, or with the more seasonal climates with a marked dry season. Table 1.1 presents a simplified pattern of tropical vegetation formations in relation to rainfall.

The **moist evergreen forest** is associated with high rainfall without pronounced dry seasons. It contains a very large number of species. When undisturbed, the herb and under-shrub layers are poorly developed because of shade, and the forest can be walked through without difficulty (denser growths are a sign of disturbance and the letting in of more light). Trees are of different sizes, ages and heights and are sometimes described as forming several layers or storeys, with a lower layer at about 10–15 m, a middle layer at about 20–25 m and a fairly closed upper canopy at about 30 m, above which appear occasionally even taller trees known as emergents. The total number of species present is typically extremely high, perhaps as much as 2500 for rainforest in Malaysia and the Amazon; while in

Table 1.1 Simplified scheme of tropical vegetation formations, both climax and secondary, in relation to rainfall. The evergreen forest corresponds mainly to a udic moisture regime, and the semi-deciduous forests, deciduous forests and thickets to ustic (seasonal) regimes. The soil moisture regime is influenced by soil texture and topographic position as well as rainfall, so different formations can occur in different positions on the landscape.

Rainfall	Wet months	Climatic climax formation	Secondary formations	
			Mainly woody	Mainly grass
Rainy	9–12	Moist evergreen forest	Secondary forest	Grassland
Seasonal	7–10	Moist semi-deciduous forest	Secondary forest thicket	High grass savanna
	5–8	Dry deciduous forest woodland	Deciduous thicket	Tall bunch-grass savanna
	4–6	Deciduous thicket		Short bunch-grass savanna
Dry	2–5		Thorny shrubs and trees	
Desert	0–2		Desert and semi-desert scrub	

Africa it has been estimated that an area of about 1–2 km^2 may typically contain 600 different species. Deciduous species form only a small proportion of the total. Other typical features described by botanists are drip tips on the leaves, which help rain to fall from the leaf, accelerate drying and reduce mould infestations; lianes and climbers; and the presence of rootless epiphytes which develop in upper branches. The number of large trees per hectare varies with soil conditions and climate, but may be less than that of a well-developed semi-deciduous forest since the soils associated with the high rainfall of rainforests are often leached and acid. This is contrary to the belief expressed in some literature that the evergreen rainforest represents the densest, heaviest and most luxuriant growth in the tropics. The nutrients contained in the standing forest vegetation are discussed further in Chapter 17.

Evergreen rainforests have been estimated to cover about 30% of the tropics. Their distribution between continents is uneven, being most extensive in tropical South America (where they cover about 52%) and least extensive in Africa (12%). In tropical Asia and the Pacific they have been estimated to cover about 38% of the area.

In South America rainforests occur primarily in the Amazon basin, but also in parts of Central America and along the Pacific coast of Colombia. In Africa rainforests occur mainly in the lower Congo basin and in the wetter parts of West Africa where annual rainfall exceeds about 1750 mm. In tropical Asia and the Pacific, rainforests occur mainly in the Malayan peninsula, Sumatra, Borneo, New Guinea, parts of the Philippines and on many smaller islands.

Fig. 1.8 Tropical forest in Africa (a) from the air (Liberia), and (b) on the ground (south-west Ghana).

Fig. 1.9 Savanna in South America (a) in Venezuela (with *Mauritia flexuosa*, locally called moriche, and (b) in the *Cerrado* of Brazil.

The **moist semi-deciduous forest** replaces evergreen rainforest where the distribution of the rainfall is such that the dry seasons are somewhat more marked. In cases where the evergreen rainforest merges into semi-deciduous forest the transition between the two is gradual, being associated with some change in species, and often with a change in structure and weight of the forest. The moist semi-deciduous forest, though sometimes similar in general appearance to evergreen forests, contains more species that lose their leaves, or some of their leaves, at one season of the year, though not necessarily in the dry season or all at the same time.

In West Africa most of the forest areas are classified as moist semi-deciduous, and evergreen rainforest is found only in the wettest parts of the forest zone, including much of Liberia, the south-east and south-west Ivory Coast and south-west Ghana. The moist semi-deciduous forest has a rainfall of between 1000 and 1750 mm and includes the major cocoa and coffee-growing areas of West Africa. A comparison of tree sizes, as shown by all-species girth class curves, indicates that there are more large trees in the West African semi-deciduous forest zone than in the wetter rainforest areas to the south. This greater weight of vegetation is associated with more fertile, less leached soils. The moist semi-deciduous forest in West Africa has been subdivided by foresters and botanists into dry and wet types largely on floristic grounds, with certain indicator species characteristic of each zone. Other deciduous forests are associated with lower rainfall and with longer dry seasons.

In Asia, some semi-deciduous forests in monsoon areas have a more seasonal character than those of Africa and South America. The dry season may be marked by extensive leaf fall and an increase in solar radiation, with a seasonal lowering of leaf-litter decomposition rates. These forests are variously known as semi-evergreen seasonal forest, monsoon forest and dry evergreen forest, and in some areas may contain up to 10% of teak (*Tectona grandis*).

Dry semi-deciduous forest is lower in height than semi-deciduous forest, with emergents at about 25 m. It is leafless during much of the dry season and has a dense understorey of deciduous woody species. In many areas this and similar formations seem to have given way to derived savanna.

Deciduous thickets include a wide range of often dense growths, including densely coppicing shrubs, with scattered emergents rising above the thicket. In Africa, a common emergent is the baobab (*Adansonia digitata*). Where rainfall is low (less than about 600 mm) the common thicket species are drought-resistant thorny shrubs, including many *Acacia* spp.

Secondary formations

Secondary forest formations grow spontaneously when land cleared of evergreen or semi-deciduous forest is abandoned to natural regrowth vegetation. An initial leafy growth of forbs and coppice shoots merges at about two or three years into a secondary thicket. At 8–10 years, depending on soil and climatic conditions and the history of cultivation, the secondary thicket becomes a secondary forest dominated by quick-growing, light-demanding species which cannot regenerate in their own shade and which are gradually replaced, as the secondary forest becomes older, by slower-growing species. The vegetation succession as it occurs when forest that has been cleared and cultivated is abandoned to a natural fallow, is described in more detail in Chapter 17.

In the dry deciduous forest zone, woodland consisting of small and medium-sized trees forming a closed canopy represents another deflected climax influenced by human activities. The understorey is open and there is a herb layer consisting of short grasses. This formation is particularly extensive in the southern half of Africa, where it is known as *miombo*. *Miombo* woodland consists mainly of trees of *Brachystegia* and *Isoberlinia* spp., and it is in these areas of generally poor soils that the *chitimene* system of shifting cultivation is practised. This consists of lopping branches over a wide area, and then piling and burning them in a much smaller cultivated patch (see Chapter 17).

The **savanna areas** of the tropics are extensive on both sides of the equator, particularly in Africa. It is only relatively recently that it has been widely accepted that most savannas are not a natural climax formation but a result of human activities. Savanna is now thought to be a fire subclimax caused by the fact that in the dry season the dead

Fig. 1.10 Savanna in Africa. Savanna vegetation is particularly extensive in Africa and is mostly now regarded as a fire subclimax influenced by deliberate fires in the dry season. (a) Savanna in Sierra Leone, with numerous Lophira trees and seedlings. (b) Savanna woodland immediately after the annual burn. (c) A typical almost flat savanna landscape in northern Ghana, with occasional shea butter trees (*Butyrospermum parkii*) in the middle distance. These and other useful trees are encouraged by selective weeding.

grass and other material is burned and a curtain of fire sweeps across the landscape.

The evidence of plots protected from fire indicates that once burning is kept out, species change and the vegetation becomes denser and more woody, suggesting that the natural climax of most areas that are now savanna is a deciduous forest or woodland. Protection plots which have included both plots burnt early in the dry season, and those burnt late in the dry season when the vegetation is drier and the burn hotter and fiercer, have shown that an early burn is less damaging than a late one.

Occasionally savannas are **edaphic**; that is, they are caused by soil conditions, as when a shallow or impermeable soil is alternately waterlogged in the wet season and dries out in the dry season, thus resulting in a growth of grasses and sedges, but edaphic savannas are limited in extent.

The term 'savanna' is derived originally from a Carib word meaning 'areas not covered with forest', but it is not now easy to define. It is essentially a vegetation formation in which the ecologically dominant stratum is the herb layer, usually composed mainly of grasses, but it may also contain various amounts of shrubs and trees and sometimes palms. In grass savanna, trees are absent or only occasional, but in tree savanna and savanna woodland they are more numerous. In areas where annual burning is regular, the trees are mainly fire-resistant species, and may show blackening or scorching on the trunks from the burn.

Most classifications of savanna vegetation are

based on the height and floristic composition of the grasses rather than on the tree component. In moister areas, including areas that were once under forest, high grass savanna is found. In progressively less wet areas tall bunch-grass savannas and then short bunch-grass savannas occur.

In Africa, where savanna vegetation has been estimated to cover over half the continent and is more extensive than in America or Asia, high grass savanna is dominated floristically by two tribes of grasses, the Paniceae (including *Pennisetum purpureum* (elephant grass) and *Panicum maximum* (Guinea grass), and the Andropogoneae (which include *Andropogon guyanus, Hyparrhenia* spp. and the undesirable *Imperata cylindrica*). High grass savanna forms the fire climax where the rainfall is more than about 1000 mm and the dry season is less than six months. Within these areas, the Paniceae, including the very productive elephant grass, dominate the wetter areas adjacent to forest, but the Andropogoneae dominate a much wider area.

Less tall than the high grass savanna are the tall and short bunch-grass savannas. The grasses of the tall bunch-grass savannas are about 1 m high at flowering and are dominated by the Andropogoneae. The tree component varies within wide limits from near absence to a closed woodland canopy, depending on climate, soil, and the history of cultivation, burning and grazing. In Africa north of the equator the tall bunch-grass savannas occupy a position intermediate between the high grass savannas of the wetter areas and the short bunch-grass savannas and semi-desert grass and scrub of the fringes of the Sahara. In Africa south of the equator, tall bunch-grass savannas cover much of East Africa, Zambia and Zimbabwe, grading into the short bunch-grass savannas where annual rainfall is about 800–900 mm.

Fig. 1.11 Forest and forest clearing in Asia. An area of little-disturbed forest in hills close to the Thailand–Burma border, near Maesod in western Thailand.

In Africa, where much attention has been given in recent years to drought problems in the Sahel, this region is strictly speaking a botanical subdivision of the African savanna. Although botanists disagree about boundaries, and the vegetation transitions are in any case gradual, the West Africa savannas have long been divided into the moister Guinea savanna in the south, the central Sudan savanna, and the drier northern Sahel savanna.

In South America, savanna that is the result of human activity is less extensive than in Africa. Such activities, aided by fire, have caused less widespread modification of the original vegetation, but savanna has nevertheless been estimated to cover about 28% of tropical America. True high grass savanna of the African type is probably absent, but tall bunch-grass savanna, analogous to that of Africa, is very widespread on the Brazilian plateau, on the *llanos* of Colombia and Venezuela, in the Guiana highlands, on the Orinoco plains and in most of Cuba. The short bunch-grass savannas occur in areas of lower rainfall but are of relatively restricted extent.

In Asia, savanna formations are less important and extensive than in Africa and tropical America, but large parts of the Philippines and of Indonesia that were once under forest have been replaced by a derived savanna type of vegetation known as *cogonales* or *alang-alang*, in which the dominant grass is a variety of Imperata (*Imperata cylindrica* var. major). Unlike the African variety, which is displaced in a few years by other grasses that are easier to clear for cultivation, the Asian Imperata is a very persistent weed which is usually eradicated only by a long fallow.

2 The composition of soils and some important properties

Composition of the soil

The soil consists of solid material and of spaces of various sizes and shapes between it. The solid material of the soil is made up of

1 organic material in various stages of decomposition, including humus, and
2 mineral particles.

In the topsoil in particular, organic material and mineral particles are intimately associated, the first binding together the second to form soil crumbs. The organic material stains the mineral particles a dark colour. If the organic material is removed or destroyed, the underlying mineral particles are revealed. These are conveniently divided up according to their size. The division of soil particles into clay, silt and sand-sized particles is a division into size classes according to the diameter of the particles, and not according to their nature and composition, though some of the properties of these particles are related to their size. Very small particles (the clay fraction) have certain characteristic properties because of their very large surface area.

The size classes are as follows:

Clay – less than 2 μm (0.002 mm) in diameter
Silt – between 2 and 60 μm (0.002–0.06 mm) in diameter
Sand – between 60 and 2000 μm (0.06–2.00 mm) in diameter.

The sand and silt fractions are commonly subdivided further, as follows:

Coarse sand	600–2000 μm (0.6–2.0 mm)
Medium sand	200–600 μm (0.2–0.6 mm)
Fine sand	60–200 μm (0.06–0.2 mm)
Coarse silt	20–60 μm (0.02–0.06 mm)
Medium silt	6–20 μm (0.006–0.02 mm)
Fine silt	2–6 μm (0.002–0.006 mm)

The sizes given are those of the international scale, but others are used in some areas (Fig. 2.1). In the

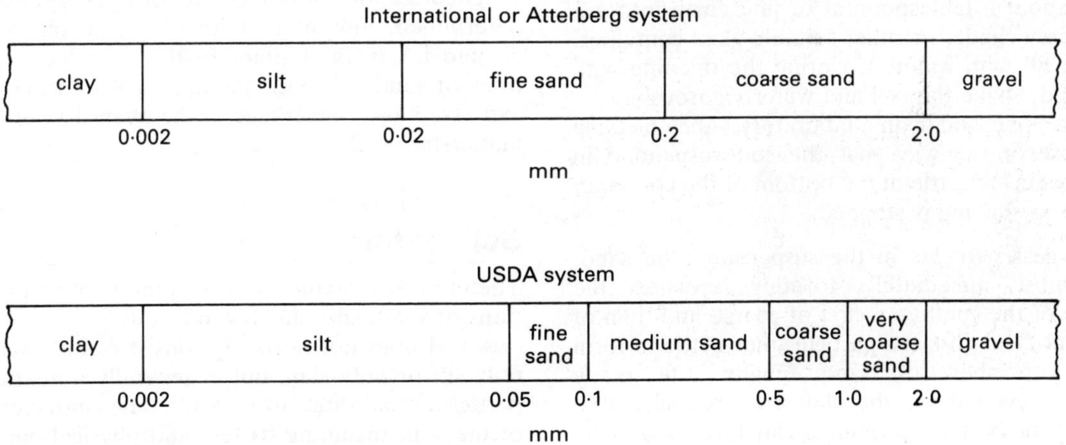

Fig. 2.1 Some different particle size classifications. Source: White, 1979.

US Department of Agriculture system, 50 μm is taken as the upper size limit for silt, and the sand fraction is divided into five sand classes.

The unit of measurement normally employed is the micrometre (μm). This small unit is used because of the very small size of the clay and silt-sized soil particles. The upper size limit for clay-sized particles is 2 μm, but many silicate clays are much smaller than this. For some purposes a distinction is made between fine clay (less than 0.2 μm) and the remainder of the clay fraction.

The soil sample when it reaches the laboratory is normally passed through a 2 mm sieve to separate the coarse fraction, over 2 mm in diameter, from the fine earth. The coarse fraction consists of gravel (2–20 mm), and stones. The fine earth has a maximum diameter of 2 mm which thus forms the upper size limit for sand. It is normally the fine-earth fraction that is analysed in the laboratory, the coarse fraction, if present, usually being described and weighed but not analysed.

The fact that the fine-earth fraction of the soil is composed of particles of different sizes can be conveniently demonstrated by the following simple experiment needing only a 2 mm sieve and a glass measuring cylinder.

1 Take a quantity of soil, preferably a subsoil sample with little humus.
2 Dry the soil, and sieve through a 2 mm sieve to remove any coarse material present.
3 Put about a tablespoonful of fine earth into a 1 litre glass cylinder or other suitable glass container. Almost fill with water. Covering the opening with one hand, shake the soil and water vigorously.
4 Stand the cylinder on a laboratory bench or table and observe the way that the soil suspended in water begins to settle at the bottom of the container as soon as shaking is stopped.

The largest particles in the suspension, the sand, fall almost immediately, forming layers at the bottom of the container, first of coarse and then of fine sand. The silt-sized particles then settle to form finer layers above the sand. Finally, after many hours or even days, the clay-sized particles also settle at the bottom, forming a clay layer above the sand and silt. The water has sorted out the particles according to their size.

The relationship between the settling velocity of particles and their diameter is made use of in the laboratory analysis of soils to determine the proportions in the sample of the various size classes listed above. This procedure is often referred to as a mechanical analysis or particle size analysis of the soil.

The particles in the natural soil adhere to each other to various degrees, to form aggregates. For a mechanical analysis to measure the size of primary particles it is first necessary to separate the individual particles completely. In the simple experiment described above, the clay and silt particles were probably adhering to each other to some extent and settled more quickly than if they had been completely dispersed. The binding agents are, in the topsoil, mainly humus, and in the subsoil various sesquioxides or, in alkaline soils, carbonates. These have first to be destroyed if the soil is to be dispersed. Hydrogen peroxide is normally used to oxidise and thus destroy the humus, and the soil is then shaken up with a dispersing agent (usually sodium hexametaphosphate) for several hours to completely disperse the aggregates into their constituent particles. The sand fractions are then separated with sieves of various diameters. The remaining silt and clay particles are measured by using the relationship between diameter and settling velocity. Samples in a cylinder are either pipetted off at a standard depth at various intervals, dried and weighed, or the specific gravity of the suspension is determined, also at a standard depth and at calculated intervals of time. In this way the proportions of sand, silt and clay are measured and then expressed as a percentage of the oven-dry, organic-matter-free soil.

Soil texture

The term 'soil texture' refers to the relative proportions of sand, silt and clay in a soil.

A soil does not normally consist of only sand or only silt or only clay, but is generally a mixture of particles of all three size classes. The characteristics of the soil, including its feel and physical handling properties, depend greatly on the relative proportions of the different size classes. The feel of the soil gives an indication of these different percentages, and soil surveyors and agronomists sometimes estimate texture by feeling the soil between the thumb and the first finger.

A soil with a high proportion of clay in it is described as a clay soil or, more concisely, as a clay. A very sandy soil may be termed a sand. A soil intermediate between the two in composition and characteristics, with more balanced amounts of sand, silt and clay, is referred to as a loam. In these cases, the terms 'clay' and 'sand' apply to the whole soil, implying the dominance of one size fraction, and must not be confused with the more specific use of the same words applied to a strictly defined particle-size class.

The terms used to describe soil texture are precisely defined according to the range of sand, silt and clay-sized particles that they include. The most widely used set of textural terms in the tropics is that of the US Department of Agriculture (Fig. 2.3). This divides soils into 12 textural classes:

Clay, sandy clay and silty clay
Loam, sandy clay loam, sandy loam, clay loam, silty clay loam and silt loam
Loamy sand and sand
Silt

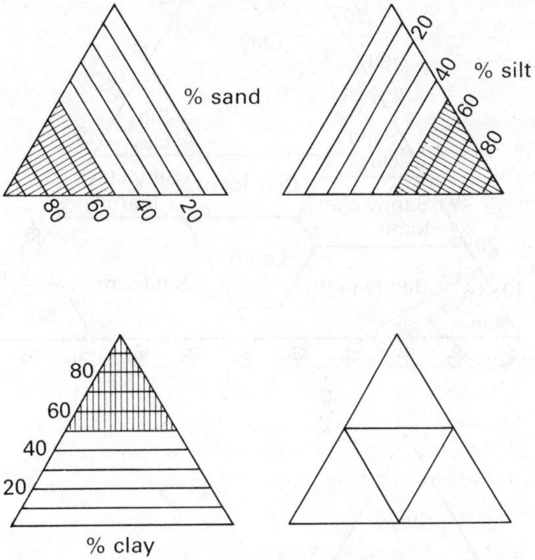

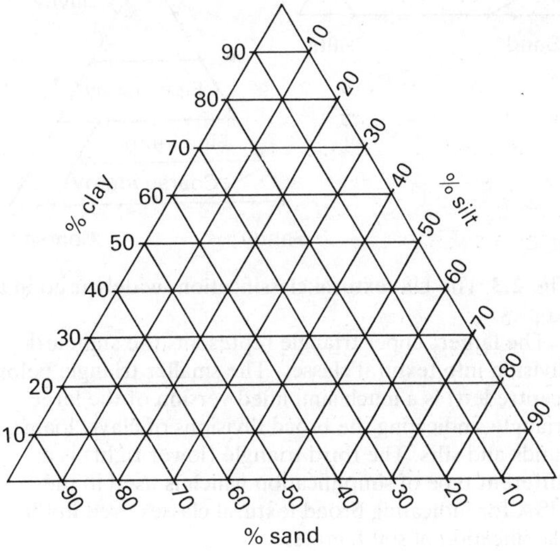

Fig. 2.2 The textural triangle.
The textural triangle really consists of three separate triangles (above) superimposed on each other, one for sand (top left), one for silt (top right) and one for clay (centre left). In these triangles 100 per cent of sand, silt or clay is at the left, right or upper apex respectively. The area shaded in each triangle represents over 50 per cent. In the fourth small triangle (centre right) areas with over 50 per cent sand, silt or clay form the outer triangles, while that in the centre represents soils with less than 50 per cent of each of these.
The full textural triangle (below) is so arranged that any point within it indicates three percentages, i.e. of sand, silt and clay. Thus the texture of a soil or soil horizon can be indicated by a point in the triangle.

A soil described as a clay, for example, can be seen from the textural triangle (Fig. 2.3) to have at least 40% clay-sized particles by weight, 0–40% silt and 0–45% sand.

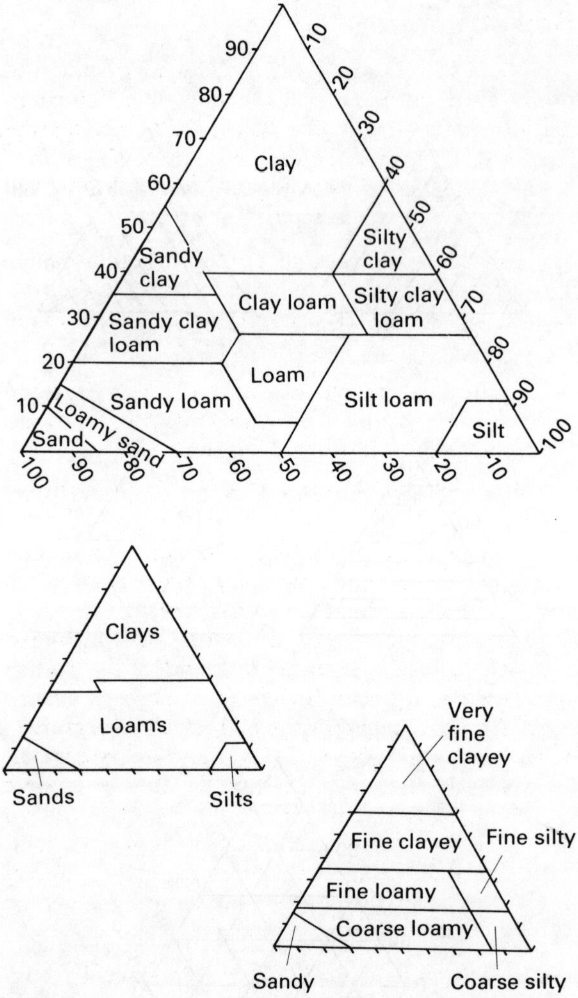

Fig. 2.3 The US textural classification, widely used in the tropics.

The larger, upper triangle represents the standard division into textural classes. The smaller triangle below it (centre left) is a much simplified version of the large triangle, indicating the broad divisions of clays, loams, sands and silts. The third triangle (lower right) is a different type of simplification which is used in the USA for indicating broad textural classes used in the classification of soil families.

Soil texture is a fundamental soil property which in practice the farmer can do little to modify. If a soil is acid he can lime it, and if it is low in nutrients he can use fertilisers, but the physical properties of the soil, including texture, are less easily changed.

Soil texture is related to soil consistency and to soil structure, considered further in following sections. Soil texture is also closely related to the water-holding capacity of soils (Chapter 4), since loams and clays hold more water than do sandy soils.

Farmers often describe clay soils as 'heavy', and sandy soils as 'light'. This refers not to their weight but to their ease of working. A heavy soil requires much greater effort to cultivate than a light, sandy soil, and this is true whether the farmer is using a hand-held hoe or a tractor. Peasant cultivators in the tropics armed with little other than a hoe or matchet usually prefer light soils, except for swamp rice production, and may avoid heavy clays such as the Vertisols (Chapter 8).

In practice the assessment of soil texture by feel, and its consistency and handling properties, is much influenced by the extent to which the primary particles of the soil are bound together to form larger aggregates. In the mechanical analysis of soils, described above, it was emphasised that a correct textural definition depends on complete dispersal of the soil particles. In some tropical soils high in iron and aluminium oxides which bind together the clay particles into larger aggregates, complete dispersal is sometimes not easy to achieve. In soils of this type, including, typically, the highly weathered soils classified as Oxisols (Chapter 8), many soils high or even very high in clay sometimes feel like less heavy soils, often like loams or sandy loams. This is due to the fact that the clay particles in the natural soil form micro-aggregates of silt or fine sand size (2–200 μm in diameter). The feel and handling properties are thus similar to soils containing primary particles of these size classes. This phenomenon of micro-aggregation into water-stable silt and sand-sized particles is considered further below (see 'Soil structure'). As regards soil texture, it is sometimes useful to distinguish between the 'true' texture, which reflects the percentages of primary particles as measured after complete dispersion, and the 'natural' texture, which is what the farmer encounters in the field where the finer particles of many highly weathered tropical soils in particular form micro-aggregates (see Table 2.2).

Soil structure

The term 'soil structure' refers to the aggregation of primary soil particles to larger units. Careful examination of the soil in the field usually reveals the presence of natural aggregates separated by pores, cracks or planes of weakness. A description of soil structure is concerned with

1 the shape and arrangement of the structural units,
2 their size, and
3 their degree of development, distinctness and durability (referred to as their 'grade').

A natural aggregate or structural unit is often referred to as a 'ped'. Structural units or peds are fundamentally of four basic shapes:

1 Those that are more or less flat or platy (platy structure).
2 Those that are prism-like with a predominantly vertical direction, and bounded by fairly flat vertical surfaces (prismatic and columnar structures).
3 Those that are blocky, or many-sided, arranged round a central point, and bounded by flattish surfaces that fit more or less into those of similar blocky units surrounding the ped (blocky structure).
4 Those that are more or less round or spheroidal, or many-sided, which do not fit closely against adjoining peds but merely touch them at certain points so as to leave some space between them. These include crumb structures, in which aggregates are normally porous and bound mainly by humus, and granular structures.

The size of structural units is expressed by conventional, defined terms such as very fine, fine, medium, coarse and very coarse. The actual sizes corresponding to these terms vary according to the type of structure concerned, being smallest for granular and crumb structures and largest for prismatic and columnar structures, as set out in Table 2.1.

The degree of development (grade) of the structure is indicated by the terms weak (barely observable), moderate, and strong (durable peds

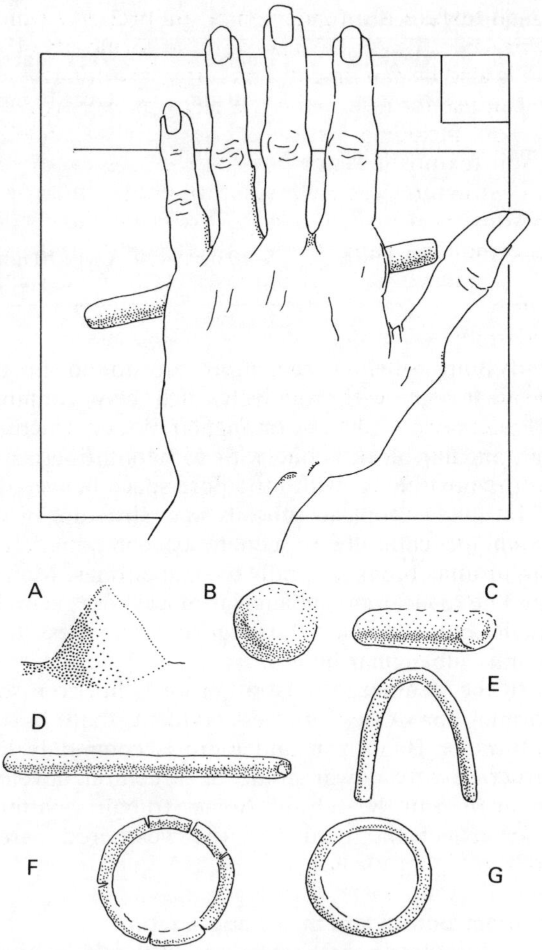

Fig. 2.4 A manipulative texture text which gives a practical indication of soil texture and consistency.

Enough fine earth is taken to make a ball of soil about 2.5 cm across and water is dripped onto the soil until it reaches the sticky point, the point at which the soil adheres to itself but not to the hand. The extent to which the moist soil can be worked is an indication of texture. A sand can only be heaped into a pyramid A, but a loamy sand makes a ball B. If the soil can be rolled out to a short cylinder C it is a light loam. The remaining drawings indicate a loam D, a heavy loam E, a light clay – a circle with cracks F, and a clay – a circle without cracks G.

The cylinder when fully rolled out, as in D, should be about 15 cm long. A board, shown below the hand, is marked with 2.5 cm square and a line 15 cm in long in order to standardise the text.

Table 2.1 Sizes, in millimetres (mm), of structural units as conventionally described

	Very fine	Fine	Medium	Coarse	Very coarse
Granular and crumb structures (diameter)	Less than 1 mm	1–2 mm	2–5 mm	5–10 mm	Over 10 mm
Angular and subangular blocky structures (diameter)	Less than 5 mm	5–10 mm	10–20 mm	20–50 mm	–
Prismatic and columnar structures (diameter)	Less than 10 mm	10–20 mm	20–50 mm	50–100 mm	–
Platy structures (thickness)	Less than 1 mm	1–2 mm	2–5 mm	5–10 mm	Over 10 mm

evident in undisplaced soil). A soil or soil horizon without any regular observable structure is described as structureless, or as massive if there is some degree of cohesion. In very sandy soils in which there is insufficient humus, silt or clay to bind the particles together so that the sand grains remain separate, the soil may be described as 'single-grained'.

Many tropical soils have relatively little regular natural structure. Most soils with some cohesion when dropped will break into fragments, but these fragments are not structural units or peds unless they possess a certain regularity and are bound by natural faces or planes of weakness and are not simply irregular fragments broken by force. Normally what structure there is is best seen by digging a soil pit and letting the exposed soil face dry out for a few days. A structure that is obvious in the exposed surface is usually of moderate to strong grade. Those structures not observed in the soil face but seen by gently picking apart a soil sample in the hand are of weak to moderate grade.

In topsoils with a moderate to good content of humus there is usually a fairly well-developed crumb structure in which soil particles are bound by humus into porous and more or less spheroidal aggregates. Cultivation over a period may have the effect of reducing the topsoil humus content, bringing about a loss of structure and with it a reduction in total porosity of the topsoil. As the crumb structure degrades, the crumbs become both smaller and less well developed, e.g. of weak instead of moderate grade.

In subsoils the structure is usually more related to the type and nature of the clay minerals in the soil than to its humus content. The better developed and more obvious structures are found in the heavier soils with relatively high clay contents. These may have blocky or angular blocky structures (an angular block is one with corners fitting neatly to the next block, with little pore space between).

In many tropical subsoils the structure when examined carefully is seen to be compound, i.e. larger units break naturally to smaller ones. Moderate fine (5–10 mm) subangular blocks, for example, might break down to strong very fine (less than 5 mm) subangular blocks.

In the weathered substratum, or C horizon, soil-forming processes are less evident than in the subsoil or B horizon and there is commonly less structure or a lower grade of structural development than in the subsoil. Some structural patterns may reflect the nature of the weathered parent rock.

Microstructure and micro-aggregation

The preceding paragraphs refer to the macrostructure of the soil, i.e. to the structural units easily seen in the soil with the naked eye. In practice these are from about 1 mm in diameter upwards. Less obvious but sometimes of very great agricultural importance is the microstructure of the soil. Here we are referring to aggregates of silt and fine sand size which, though they may not be immediately recognised as structural units, affect the handling properties of the soil and its apparent texture.

The micro-aggregation of the soil, to form micro-aggregates that are water-stable, is mainly due to the binding effects of iron and aluminium oxides. Pronounced micro-aggregation commonly occurs in those soils of the Alfisol, Ultisol and Oxisol orders of the US Soil Taxonomy that are relatively highly weathered and have high contents of clay, and of iron and aluminium oxides.

One evidence of micro-aggregation is that the texture of the soil as assessed by feeling between the thumb and first finger appears to be lighter than the true texture as determined in a laboratory after the soil has been thoroughly dispersed to separate all aggregates into their primary particles. The difference between the 'apparent texture' as assessed by feel and the 'true texture' as determined in the laboratory thus gives some indication of both the extent and the effects of micro-aggregation.

An example of the effects of micro-aggregation is provided by a volcanic soil in Kenya that is widely used for coffee because of its depth, good drainage and porosity, and is known locally as a 'Kikuyu friable clay'. Texturally this soil is a heavy clay, since it typically contains 60–80% clay throughout the A and deep B horizons. However, it feels and handles like a loam or sandy loam because in the soil in the field the clay particles are mostly aggregated to medium and coarse silt and fine sand-sized aggregates, i.e. to aggregates that are 6–200 μm in diameter (Table 2.2).

Since the binding agents of the micro-aggregates discussed here are mainly iron and aluminium oxides, and not humus, this micro-aggregation is less likely to be lost as humus levels decline in cultivation than is the topsoil macrostructure which is due to organic matter binding.

Table 2.2 A mechanical analysis of a Kenya coffee soil with and without dispersing agent, showing the extent of micro-aggregation in the natural soil. The true texture, using a dispersing agent to separate the soil into its ultimate particles, contrasts with the 'natural texture' of the soil subjected only to eight hours of end-over-end shaking in water, in which water-stable micro-aggregates survive. A comparison of the two shows that whereas clay is over 80% in the B horizon from 90 to 190 cm, in its natural state most of the clay is aggregated to medium and coarse silt and fine sand sized particles, i.e. to particles with diameters of 6–200 μm. This water-stable micro-aggregation makes the soil feel and handle more like a sandy loam than a clay, and contributes to its good water acceptance and aeration.

Depth (cm)	Horizon	SAND			SILT			CLAY
		Coarse 2000–600	Medium 600–200	Fine 200–60	Coarse 60–20	Medium 20–6	Fine 6–2	Below 2
With dispersing agent								
8	A11	1.0	1.9	1.2	14.7	15.1	7.8	58.4
15	A12	1.0	2.0	1.6	21.3	13.3	9.4	51.5
30	B2	0.8	1.5	1.4	8.2	9.9	8.3	69.8
50	B2	0.5	1.5	2.2	4.4	8.3	6.1	77.0
90	B2	0.6	1.1	1.4	3.7	4.8	5.5	82.8
140	B2	0.4	1.4	2.9	1.7	6.2	5.2	82.2
190	B2	0.3	1.1	2.6	4.0	3.0	5.6	83.4
240	B2	0.6	1.2	3.6	16.0	15.5	10.3	52.9
290	B2	3.6	4.0	15.3	15.7	15.7	3.0	42.8
Without dispersing agent								
8	A11	0.3	5.1	14.2	17.8	28.1	14.0	20.5
15	A12	1.2	3.4	30.6	21.2	16.3	8.4	18.9
30	B2	1.0	2.4	36.4	25.6	15.2	15.2	4.2
50	B2	0.5	1.6	34.8	23.9	34.5	3.1	1.6
90	B2	0.4	2.0	43.0	13.8	29.3	9.7	1.8
140	B2	0.5	1.4	27.1	28.6	29.9	11.1	1.4
190	B2	0.3	1.8	33.1	23.5	21.3	19.2	0.8
240	B2	0.4	1.6	24.2	21.6	27.9	21.4	2.9
290	B2	4.6	5.7	44.4	14.2	15.0	18.0	8.1

The practical importance of structure

The structure or absence of structure greatly affects the agricultural productivity of the soil. A well-developed crumb structure in the topsoil increases the total porosity of the horizon, since there are pores within the crumbs as well as generally larger spaces between them, and the greater porosity improves aeration and water infiltration. It makes cultivation easier and favours seedling emergence and root penetration. One of the major agricultural consequences of the lowering of organic matter levels during cultivation is a decline in the grade and size of structural units and consequent reduction of topsoil porosity, a lowering of water infiltration rates, and an increase in runoff which lowers the effective rainfall available to plants and may lead to sheet erosion.

In many subsoils, structure is defined in terms of cracking patterns. These patterns are of importance to water movement down the profile, and to the penetration of roots and rootlets.

The stable structure (both microstructure and macrostructure) associated with many of the more highly weathered tropical soils high in iron and aluminium oxides provides them with generally good and stable physical properties which constitute one of their main agricultural advantages. However, even stable structures can be lost or modified by careless handling and misuse of the soil, as when the soil is cultivated when too wet or subjected to the stresses and compaction of vehicles and heavy machinery.

Soil consistency

The term 'consistence', or 'consistency', refers to the handling properties of the soil in the field, and particularly to its resistance to the insertion of an implement such as a hoe or plough, its resistance to deformation, and its cohesive and adhesive properties. Consistence is an important factor affecting the ease of cultivation of soils and its behaviour when cultivated.

Soil consistency depends very much on whether the soil is wet, moist or dry, and the different sets of descriptive terms that are used to describe the soil in each of these states are given in Table 2.3.

Wet soil is described in terms of its stickiness, i.e. the extent to which it sticks to other objects, and its plasticity – the ability to be pressed into any shape without breaking and to retain that shape. Both of these properties are related to the type and the amount of clay present.

Table 2.3 Terms used to describe consistency at different moisture contents

Wet soil		Moist soil	Dry soil
(a) Stickiness	*(b) Plasticity*		
non-sticky	non-plastic	loose	loose
slightly sticky	slightly plastic	very friable	soft
sticky	plastic	friable	slightly hard
very sticky	very plastic	firm	hard
		very firm	very hard
		extremely firm	extremely hard

Moist soils, because of their lower moisture content, break into small aggregates more easily than do wet soils, even though these aggregates may cohere again when pressed against each other. Consistency of moist soils is estimated by trying to crush a piece of soil in the hand between thumb and forefinger, and estimating its resistance. A very friable soil is easily crushed with gentle pressure, but a very firm soil is barely crushable between thumb and finger and an extremely firm one cannot be crushed in this way.

Dry soils are often relatively rigid, brittle and hard. They may break into angular fragments which do not adhere to each other again when pressed together. A mass of soil should be held in the hand: if it is easily broken between thumb and forefinger it is no more than slightly hard. A very hard soil when dry cannot be broken between thumb and forefinger and can be broken in the hands only with difficulty.

Colour and consistency as described in routine soil descriptions refer, unless otherwise stated, to the soil in its moist state. A more complete description might use such phrases as 'sticky and plastic when wet, very hard when dry', thus emphasising the fact that some soils exhibit very different properties according to their moisture content.

Soil porosity and bulk density

In a good humous topsoil with a well-developed crumb structure, about 50% of the soil is space. In many ways the size, shape and arrangement of these spaces or pores are as important and as deserving of study as the solid soil particles, for they are essential for the circulation of air and water, and for providing spaces where soil organisms live and function.

Pores and spaces in the soil occur both between structural units such as crumbs or blocks, and within those units. The total pore space, or porosity of the soil, is thus related to the type and degree of development (grade) of the structure, as well as to related factors such as texture and organic matter content.

The total porosity of the soil is measured as the pore space ratio (PSR), which is obtained by expressing the volume of all pores as a percentage of the total volume of the soil. Pores are classified according to their size. Those that can be seen with the naked eye, i.e. larger than about 60 μm in diameter, are termed 'macropores'. These often form a large proportion of the total space between soil peds, including that created by faunal activity. Pores smaller than about 60 μm, too small to be easily distinguished with the naked eye, are termed 'micropores'. Water moves fairly freely through the macropores under the action of gravity, but is retained by micropores owing to capillary attraction by the pore walls, as discussed further in Chapter 4.

The bulk density of the soil refers to its weight in relation to its volume. The more compact the soil and the lower the porosity, the higher will be its bulk density. Bulk density thus affords a convenient way of estimating total porosity (Table 2.4).

Table 2.4 Relationship between bulk density (BD) and total porosity (rounded to the nearest 1%) for a soil with an average particle density (PD) of 2.65

BD	1.0	1.1	1.2	1.3	1.4	1.5	1.6	1.7	1.8
% porosity	62	58	55	50	47	43	40	36	32

The percentage by volume of solids in the soil (% solid) is $\frac{BD}{PD} \times 100$. The percentage pore space is 100 – % solid. The two relationships can be combined to give the formula

% pore space $= 100 - \frac{BD}{PD} \times 100$. Thus for an assumed specific gravity of 2.65 g/cm^3, % pore space $= 1 - \frac{BD}{2.55} \times 100$. For a BD of 1.5 this becomes $1 - \frac{1.5}{2.65} \times 100$, or 43.4%, which is rounded in Table 2.4 to 43%.

For soils high in organic matter, the average PD is less than the 2.65 assumed in Table 2.4, so that the porosity at any given bulk density is correspondingly lower.

In compacted soils, including those which have been compacted by the use of heavy machinery, with bulk densities in excess of 1.7, corresponding to porosities of less than 36%, root growth normally becomes slower or ceases. In very gravelly soils, bulk density is increased by the high proportion of solid gravel, and the overall porosity reduced accordingly, but the porosity in the soil material between the gravel may be much higher than the overall porosity, and may thus nevertheless be high enough to allow roots to grow and penetrate the soil.

The bulk density of a soil depends on the specific gravity of the particles it contains and on their packing arrangement, and is conventionally expressed in terms of g/cm^3. The specific gravity of the mineral particles of the soil is often about the same as that of quartz, i.e. 2.65, meaning that they weigh 2.65 times the same volume of water.

For organic soils the bulk density might be less than 1.0. For topsoils with a good humus content and crumb structure, a bulk density of about 1.2 is normal. This means that the topsoil weighs about 1.2 times the same volume of water, and that it has a porosity of about 50% or less (see caption to Table 2.4). In subsoils there is typically less humus, which has a lower specific gravity than the mineral particles of the soil, and a lower porosity than in a well-structured topsoil, so that bulk densities are higher than in the topsoil and might be of the order of 1.5–1.6 (see Fig. 2.10). They may be even higher than this in compacted or indurated horizons and in those containing large amounts of stones or gravel.

Our knowledge of the shape and distribution of small soil pores and of the microstructure of the soil has been greatly increased in recent years by the use of soil thin sections. Undisturbed soil samples are impregnated and hardened with a colourless plastic resin, and then ground to form thin sections for examination under the microscope. Even without a microscope much can be learnt by using a hand-lens and gently breaking open a fragment of soil with a needle and examining the fresh surface, pores and channels exposed.

The clay fraction

The clay fraction of the soil has been defined above as consisting of those mineral soil particles that have a diameter of less than 2 μm. Because of the small size of these particles, little was known about them until the advent of techniques such as X-ray diffraction and electron microscopy. The material in this size class is now known to consist mainly of crystalline clay minerals, together with the weathered residues of primary minerals and various amorphous materials which include silica, iron oxides and aluminium oxides. These amorphous materials may occur as coatings on the surfaces of the clay minerals. In soils over chalk and limestone and in high pH soils of arid regions, calcite (calcium carbonate) may also occur in the clay fraction.

Most of the crystalline clay minerals were found to have a platy structure similar to that of the micas. They are hydrous aluminium silicates, mostly with a layer-lattice structure. Some of them are formed by relatively slight alteration of material in the parent rock such as mica, but most are synthesised in the soil during the weathering of primary silicate minerals, so that the type of clay formed reflects the nature of the parent rock and the immediate soil environment in the weathering rock zone.

The most important and widespread clay minerals in tropical soils are of three broad types, belonging to the kaolin, illite and smectite groups. All are characterised by being built up of layers of silica and alumina sheets. In the kaolin group, belonging to the so-called 1:1 lattice clays, a sheet in the crystal consists of one silica layer bonded to one of alumina. In the remaining groups, belonging to the 2:1 lattice clays, the sheets consist of two silica layers sandwiching an alumina layer between them. The structure and properties of the clays of these three groups are briefly discussed in the following paragraphs.

Kaolin or kaolinite, the dominant clay mineral in

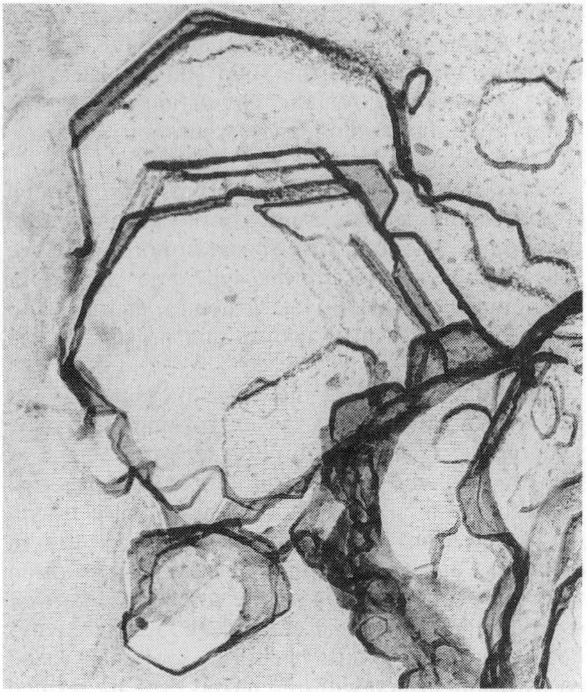

Fig. 2.5 Electron microscope photograph of kaolinite crystals. The hexagonal crystals shown here are about 1 micron across.

most highly weathered tropical soils, has relatively large crystal flakes about 0.2–2.0 μm in diameter, often more or less hexagonal and made up of a large number of adjacent sheets (Figs 2.5, 2.6). The sheets consist of a double layer, one of a lattice of silica tetrahedra and the other of a lattice of aluminium octahedra (gibbsite), the two layers being bonded together by shared oxygen atoms. Each of these double-layer crystal units is held to the next double layer by a hydrogen bond between the oxygen and hydroxyl of adjacent sheets, and the distance between these double layers is both small and fairly constant. Kaolin therefore has a one-to-one (1:1) layer lattice, i.e. one layer of silica to one of alumina, and is described as a non-expanding clay since the distance between the double-layer sheets remains about the same.

Illites, or hydrous micas, are examples of 2:1 lattice clays in which each sheet consists of a triple layer with two outer silica tetrahedra layers enclosing an alumina octahedral layer between them

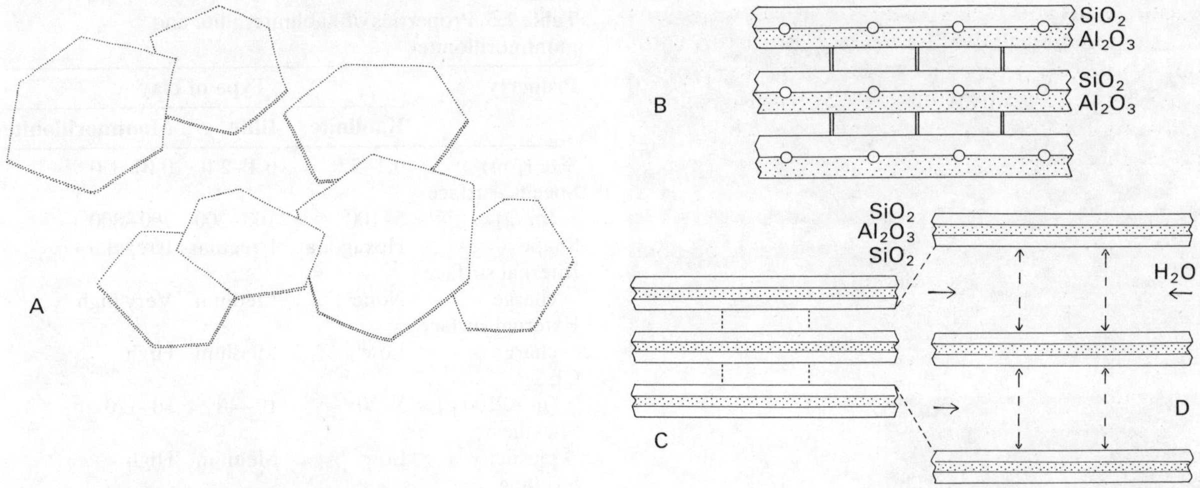

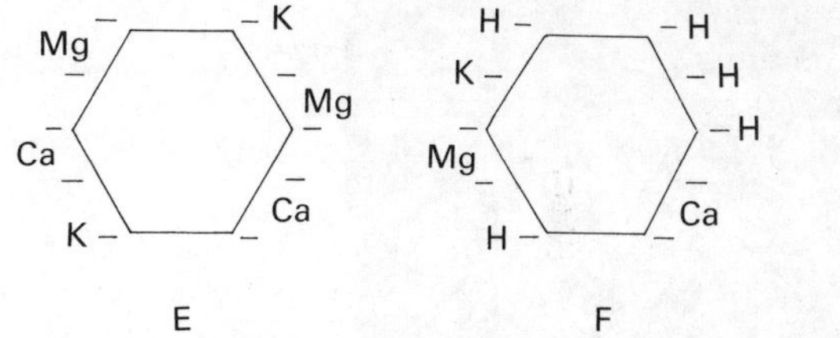

Fig. 2.6 Clay minerals and their structure.

The drawing A indicates kaolin flakes as they appear in an electron microscope photograph. The magnification is about 30 000×.

B indicates the structure of a kaolin flake, composed of many double layers, each of silica and alumina sheets bonded by shared oxygen atoms (circles). Each double layer is bound to the double layer above and below by oxygen–hydroxyl bonds (double vertical lines) which allow very little expansion between the layers, so that ions and water molecules cannot penetrate easily. Because of the very limited amount of expansion possible, kaolin is described as a non-expanding lattice clay.

C represents a montmorillonite flake in which each layer-unit consists of three layers, two layers of silica with a layer of alumina sandwiched between them. The bonds between the layer-units are relatively weak and the lattice can expand considerably, as represented in D. When wetted, water enters between the units of the lattice and the lattice expands.

Diagrams E and F represent clay flakes and their associated ions attracted to the negative charges on the clay. The flake E represents a soil which is saturated with metallic cations. All the negative exchange sites are occupied by calcium, magnesium, and potassium ions, and the soil pH would be about 7.0 or slightly above. Diagram F represents what occurs when some of the cations are removed, as when soils are leached. The place of the cations is usually assumed to be taken by hydrogen ions and these give an acid reaction to the soil, but in acid soils with a pH of below 5.0 free aluminium ions can be expected in increasing amounts as the soil becomes more acid.

bonded by oxygen atoms. Adjacent sheets are bonded by potassium ions and the distance between them remains, as in kaolin, fairly constant, to give a non-expanding lattice which does not swell or shrink appreciably on wetting and drying. If the potassium bonds are removed during weathering, illite changes to an expanding lattice clay known as vermiculite. As considered further in the next section, the cation exchange capacity of illite is greater than that of kaolinite owing to isomorphous substitution within the crystal lattices.

Smectites, of which the most important is montmorillonite, are also 2:1 type clays in which the basic crystal unit is a triple layer of two silica tetrahedral layers enclosing an alumina octahedral layer, but they differ from illite inasmuch as the crystal sheets are only weakly bonded to each other by oxygen–oxygen linkages and the distance between them can expand. They are thus expanding lattice clays which change their volume considerably when wetting or drying. As in the case of illites, their cation exchange capacity (see below) is

Fig. 2.7 A well-structured clay soil, near Bogotá, Colombia. The dark topsoil has a well-developed, fine angular blocky structure. The pale subsoil has an angular blocky to columnar structure.

relatively high owing to isomorphous substitution in the crystal lattice.

The three major types of clays outlined differ considerably in their properties and these in turn affect the agricultural characteristics of the soils in which they occur. The differences include differences relating to

1 the size of the clay particles and their surface area,
2 the extent to which they shrink or expand with changes in moisture,
3 their consistency properties of cohesion and plasticity, and
4 their cation exchange capacity.

Table 2.5 Properties of kaolinite, illite and montmorillonite

Property	Type of clay		
	Kaolinite	**Illite**	**Montmorillonite**
Size (μm)	0.1–5.0	0.1–2.0	0.01–1.0
Specific surface (m^2/g)	5–100	100–200	700–800
Shape	Hexagonal	Irregular	Irregular
Internal surface charge	None	Medium	Very high
External surface charge	Low	Medium	High
CEC (m.e./100 g)	3–20	10–40	80–120
Stickiness, plasticity	Low	Medium	High
Swelling capacity	Low	Medium	High

These differences are summarised, for kaolinite, illite and montmorillonite, in Table 2.5. Kaolinite has the largest crystals, the lowest specific surface area and the lowest cation exchange capacity, gives a relatively low plasticity and cohesion to the soil, and swells and shrinks to only a low degree on wetting and drying. In contrast, montmorrillonite has very much smaller flakes of more irregular shape and very much larger surface area. It also has a very much higher cation exchange capacity owing partly to the fact that internal surfaces contribute to exchange reactions, as considered further in the next section. Montmorillonite gives a much higher degree of cohesion and plasticity to soils than do kaolinite and illite, and swells and shrinks much more than they do on wetting and drying because of its expanding clay lattice which expands when water molecules enter between the sheets. Illite is in many respects intermediate between kaolinite and montmorillonite in the properties discussed here.

Cation exchange capacity

The properties of the clays discussed above are related both to their very small size and hence very great surface area, and to the internal molecular structure of their crystal lattices. The smaller a particle of matter, the greater is its surface area in relation to its weight and the greater the proportion of its molecules which are at the surface and able to participate in surface reactions. A molecule at the interior of a particle is connected to other molecules and is usually electrically satisfied, but one at the edge may not be. At the edges of the clay flakes, and in some cases on the surfaces between the sheets which make up the flakes, are molecules with unsatisfied electrical charges. These unsatisfied charges are both positive and negative, but the negative charges are usually much more numerous than the positive. These negative positions attract positively charged ions, the soil cations, to them, and these may be regarded as being loosely adsorbed on to the clay surface or as being attracted to positions close to it. A cation thus adsorbed or attracted can easily be displaced by another cation or by a hydrogen ion or taken up by the plant. The negative charges on the clays thus give to them a cation exchange capacity that is equivalent to the sum total of exposed negative charges.

The cation exchange capacity of a clay or a soil is measured in the laboratory by first saturating the sample with a cation such as ammonium in such a way as to displace all other cations previously held and to occupy all the exchange positions, and then measuring the amount of cation retained. The cation exchange capacity thus determined is conventionally expressed as the weight of cations which can be adsorbed by 100 g of the substance. It is convenient to express this in terms of the equivalent weight of cations, rather than the actual weight which would vary with the species of cation concerned. Since the quantities involved in clay exchanges are relatively small, the standard unit of measurement is one-thousandth part of the equivalent weight, the milliequivalent (m.e.). Cation exchange capacity is thus conventionally expressed as milliequivalents per 100 g (m.e./100 g) of the clay, soil or other substance referred to.

The cation exchange capacity of kaolin, with its non-expanding lattice, is due almost entirely to negative charges at the edges of the sheets only, and it is relatively low. It is normally regarded as having a cation exchange capacity in the range 3–20 m.e./100 g and in practice it is found that the kaolinitic clay of many tropical soils has a cation exchange capacity of about 8–10 m.e./100 g. For rough and ready calculations a figure of about 10 m.e. is often used.

The cation exchange capacity of montmorillonite is due partly to isomorphous substitution in the crystal lattices, which increases the net negative charge, and to the very small size of the clay flakes. Isomorphous substitution involves the replacement of a constituent silicon ion in the tetrahedral layer by aluminium, or of a constituent aluminium ion in the octahedral layer by magnesium. Since both these substitutions involve another cation of lesser valency, this gives rise to unsatisfied negative charges within the crystal lattices both at the external surfaces of the sheets and at internal surfaces. Montmorillonite normally exhibits a very high cation exchange capacity of the order of 10–15 times that of kaolinite. The usual range of cation exchange capacity for montmorillonite is 80–120 m.e./100 g, and for rough calculations a figure of 100 m.e./100 g is often assumed.

As with many other properties (Table 2.5), illite is intermediate in cation exchange capacity between kaolinite and montmorillonite. Its usual range is 10–40 m.e./100 g.

The cation exchange capacity of a soil is due both to the clay and to the organic matter it contains. The cation exchange capacity of organic matter (Chapter 3) is normally higher than that of any of the clays discussed above. For purposes of rough calculation, a figure of 200 m.e./100 g humus can be assumed.

If a soil sample is analysed so as to give the organic matter content and the percentage clay, then approximate calculations can be made to give an indication of the main type or types of clay mineral which appear to be present, remembering that in many soils more than one type of clay mineral is often present. Examples of this sort of calculation are now given to illustrate their ease and usefulness.

Example 1 A topsoil has 3% organic matter, 40% clay and a cation exchange capacity (CEC) of 10 m.e./100 g. Assuming a CEC of the organic matter of 200 m.e./100 g, the humus accounts for 6 m.e. of the CEC, leaving 4 m.e. assumed due to the clay fraction. The apparent CEC of the clay fraction is $\frac{100}{40}$ times 4, or 10 m.e./100 g. This low apparent cation exchange capacity of the clay suggests that it is kaolinitic.

Example 2 A subsoil sample with negligible organic matter has 50% clay and a CEC of 45 m.e./100 g. Assuming that the CEC is due to the clay fraction alone, its apparent cation exchange capacity is $\frac{100}{50}$ times 45, or 90 m.e./100 g. This high cation exchange capacity suggests that the clay mineral involved is dominantly of the montmorillonite type.

pH dependent cation exchange capacity

The cation exchange capacity of soils is not a constant property but varies with the pH of the soil and the concentration of ions in the soil solution. In normal laboratory analyses it is measured at a standard pH 7.0 and the cation exchange capacity given thus refers, strictly speaking, only to the soil or clay mineral when at that pH.

The cation exchange capacity of a clay mineral is due partly to a permanent negative charge which remains the same irrespective of soil pH, and partly to additional negative charges which open up and become active only at certain pH values. The latter is termed the pH dependent charge or cation exchange capacity.

The permanent negative charge on the clays is that due to isomorphous substitution within the crystal lattices. The pH dependent charge is due to the dissociation of carboxyl and phenolic (COOH) groups and the development of charges on oxygens and hydroxyls at the edges of clay or sesquioxide crystals. These phenomena depend in turn on the solution concentration of H^+ or OH^- ions which can be reversibly absorbed at such sites, and thus on the pH of the soil. Of the clays, edge charges are relatively important for kaolin, which shows a considerable pH dependent charge. Free oxides in the soil, particularly hydrated aluminium and iron oxides, exhibit pH dependent charges due to the reversible absorption of H^+ ions. Since free oxides are widespread in the soils of the tropics, the cation exchange capacity of many tropical soils is not a simple concept but must be thought of as subject to change when the reaction of the soil changes. The negative charges on organic matter, and hence its cation exchange capacity, are also pH dependent, so that organic matter has a higher cation exchange capacity at alkaline pH values than it does at an acid pH.

The effect of pH dependent charges is that to some extent, depending on the type of clay mineral involved, the cation exchange capacity of the clay fraction of the soil is, like that of organic matter, higher at high pH values and lower in acid soils. At a very low pH value only the so-called 'permanent' charges on the clays and some of those on the organic matter are available for cation exchange. As the pH is raised, additional exchange sites on the mineral and organic colloids are released. At pH 7.0, at which cation exchange capacity is conventionally determined, most of the pH dependent charges are active as well as the more permanent ones.

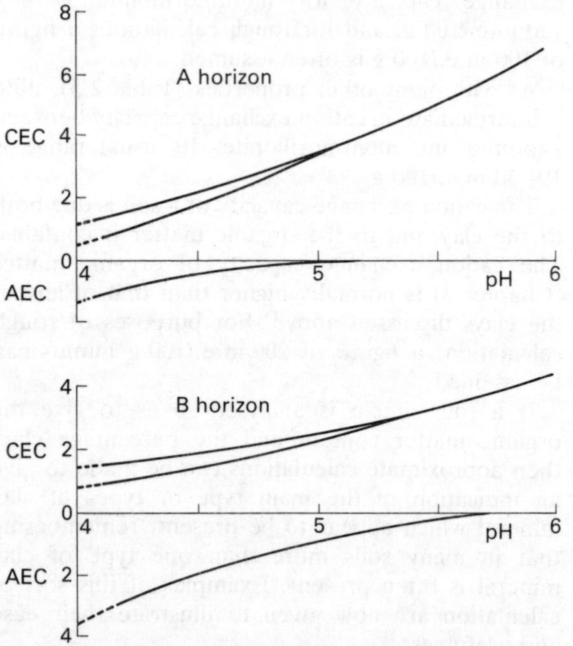

Fig. 2.8 Cation exchange capacities (CEC) and anion exchange capacities (AEC) of the A and B horizons of a Brazilian Oxisol. Of the two lines showing CEC, the upper one represents adsorbed Ca^{2+} and Al^{3+}, and the lower only adsorbed Ca^{2+}. The difference between the two represents adsorbed Al^{3+}, found in increasing amounts below about pH 5.0. Source: **Gillman, G. P.** (1987) IBSRAM Newsletter 5.

The practical implications of pH dependent charges are first that the cation exchange capacity of the soil can be modified by changes in reaction, however induced, becoming less as the soil becomes more acid, or more as it becomes less acid, and secondly that the cation exchange capacity of the soil as conventionally measured in routine soil analyses at pH 7.0 does not necessarily reflect the cation exchange capacity of the soil in the field if it is at a different pH.

The dominant soil cations

The dominant metallic cations in most soils except extremely acid and extremely alkaline ones are calcium (Ca^{2+}), magnesium (Mg^{2+}), and potassium (K^{+}), often in that order, and these are the cations normally determined in routine soil analyses. Sodium (Na^{+}) may be present, particularly in some alkaline soils, as well as minor amounts of manganese (Mn^{2+}), and very small quantities of trace elements such as copper (Cu^{2+}) and zinc (Zn^{2+}). In addition, the soil normally contains the ammonium cation (NH^{4+}) in amounts that vary according to the extent of microbial nitrification (Chapter 9).

In acid soils the dominant cation may be aluminium. Exchangeable aluminium in the soil may have important adverse effects on root growth; it is considered in more detail in Chapter 11.

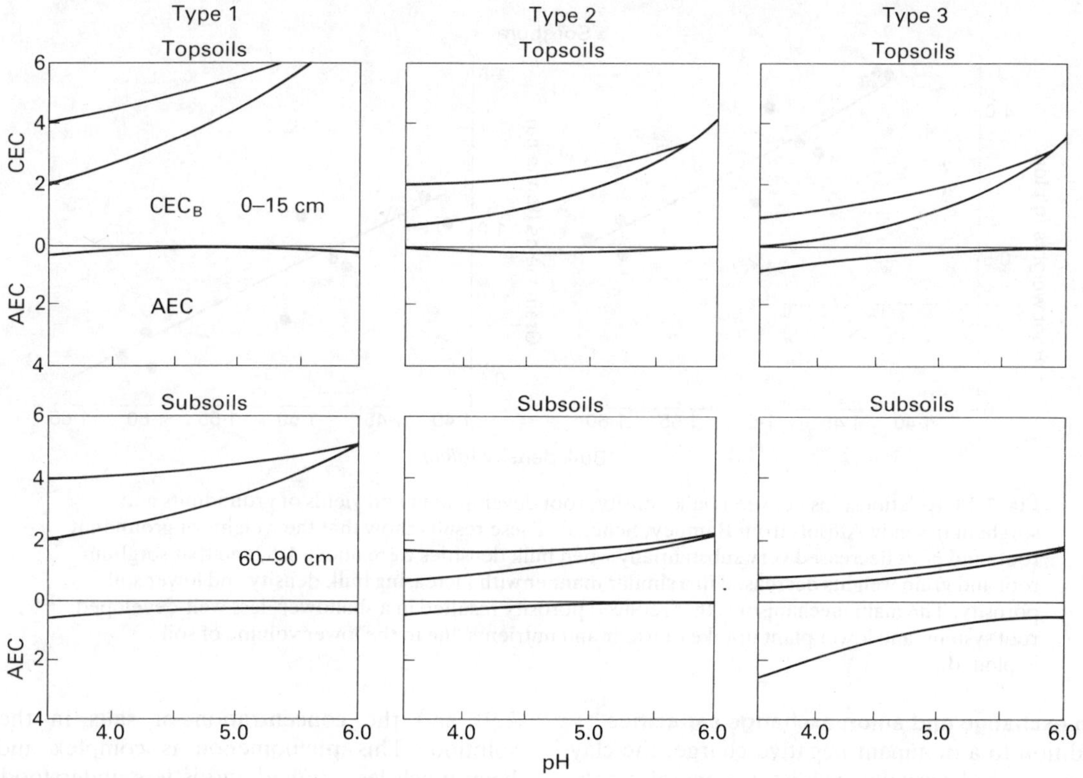

Fig. 2.9 Mean CEC and AEC curves for topsoils and subsoils of three groups of soils from Queensland, Australia. Type 1 with high CEC and low AEC consists of soils developed in alluvium. Type 2 with both CEC and AEC low consists of soils developed from granite and metamorphic rocks. Type 3 with low CEC and high AEC consists of soils derived from basaltic rocks. Source: As for Fig. 2.8.

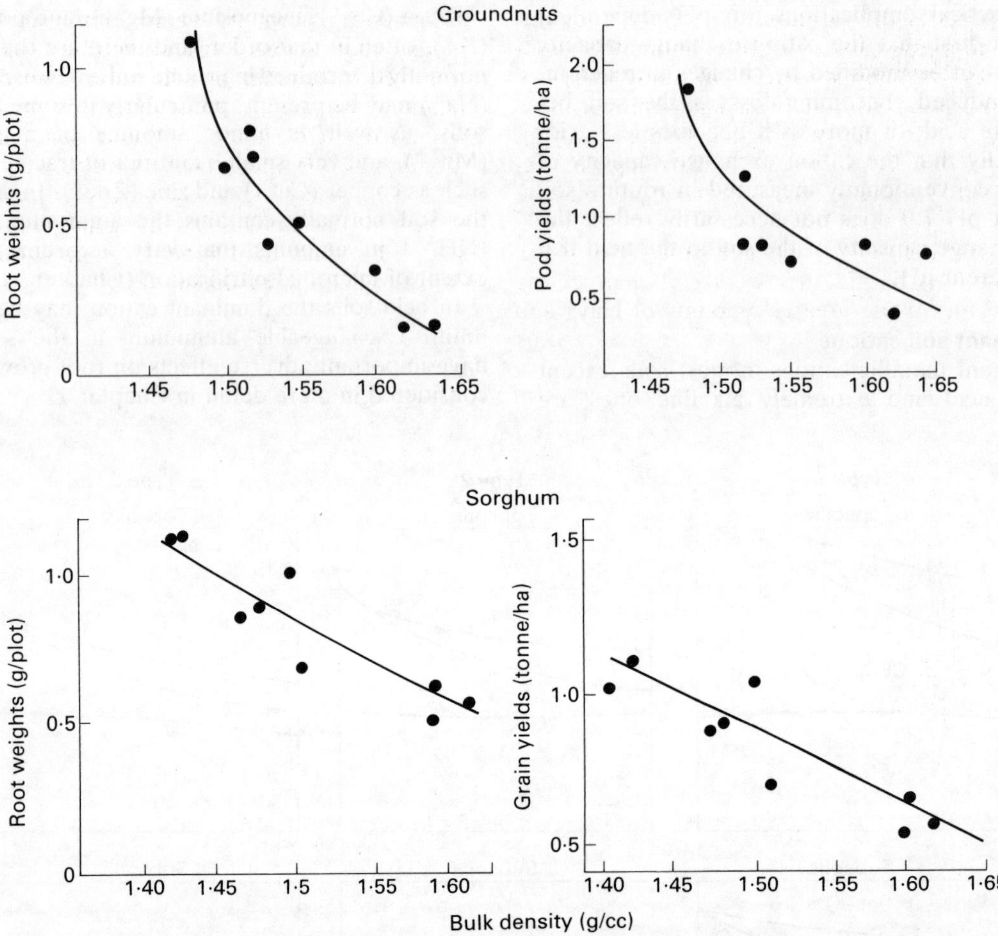

Fig. 2.10 Relationships between bulk density, root development and yields of groundnuts and sorghum in sandy Alfisols from Bambey, Senegal. These results show that the weights of groundnut roots and pods decreased very substantially when bulk densities were above 1.5, and that sorghum root and grain weights decreased in a similar manner with increasing bulk density and lower soil porosity. The main mechanism is that reduced porosity resulted in a shallower, less well-developed root system, and lower plant uptake of water and nutrients due to the lower volume of soil exploited.

Anion exchange and anion exchange capacities

In addition to a dominant negative charge, the clay fractions of soils usually possess a generally subordinate positive charge, particularly under acid conditions, which serves to attract negatively charged ions, including important soil anions such as phosphate, nitrate and sulphate. This anion exchange capacity varies with the soil pH (Figs 2.8, 2.9) and the concentration of salts in the soil solution. This phenomenon is complex and has been much less studied and is less understood than cation exchange capacity, and receives little attention in many soil science textbooks.

In tropical soils an important source of anion exchange capacity is provided by the iron and aluminium oxides and hydrated oxides (hydroxides in various stages of dehydration). These are frequently found in appreciable quantities intimately associated with the clay minerals. In some cases of highly weathered soils in which kaolinitic clay is associated with particularly high amounts of sesquioxides, particularly aluminium hydroxides, the net charge when the soil is below a certain pH may be positive.

In practice the anion exchange capacity of many tropical soils is higher in the subsoil than in the topsoil owing to the greater amounts of clays and sesquioxides normally found in the subsoil. This means that anions such as nitrate and sulphate removed by percolating soil water from the topsoil might be retained to some extent by the increased anion exchange capacity of the subsoil. This phenomenon has been held to explain the fact that seedlings and young plants showing sulphur deficiency symptoms (Chapter 9) may grow out of these symptoms when their roots are long enough to tap sulphate held in the deeper layers of the soil.

3 Soil organic matter

Formation and properties of soil humus

The organic matter in the soil is formed from the various organic plant and animal residues it receives. Dead organic matter is colonised and attacked by a wide range of soil organisms. During the initial stages of decomposition of these fresh residues some of the essential elements are converted from organic combinations to inorganic forms such as carbon dioxide (CO_2), ammonium (NH_4^+), phosphate ($H_2PO_4^-$) and sulphate (SO_4^{2-}) and released into the soil where they are available to growing plants and other organisms.

The term 'humus' refers to that more or less stable part of the soil organic matter which remains in the soil after the initial decomposition processes have decomposed the major portion of the plant and animal residues added to the soil. The original organic structures are completely obliterated, and, when separated, humus appears as a dark-coloured, glue-like substance. The processes involved in the production of humus are very complex and involve the oxidation of aromatic compounds which condense with other oxidation products and amino acids to form humic polymers containing 2–5% nitrogen. The oxidation process is catalysed by enzymes secreted by soil fungi, bacteria and actinomycetes.

It is necessary to distinguish clearly between the initial relatively rapid decomposition of raw, freshly added organic material to the soil, which stimulates microbial activity and results in the release of nutrient ions as stated above, and the subsequent very much slower mineralisation of the resultant humus which has been added to the soil as the result of the initial decomposition. The rate of mineralisation of the humus in the soil depends on its amount and composition, and on environmental conditions, including moisture supply and temperature. In many conditions in the tropics it is in the range 1–5% a year (see Table 9.4 p. 122). This mineralisation of humus involves its complete decomposition into its constituent elements and ions, and forms a major source of nutrient elements to plants in balanced amounts.

Humus normally becomes intimately associated with the mineral particles of the soil, coating them and binding them into crumb-like aggregates (Chapter 2). In the topsoil, where organic matter in the soil profile is normally at its maximum, humus typically coats the sand, silt and clay particles, staining them a dark colour, thus giving the topsoil its characteristic dark greyish-brown to almost black colour. If the organic matter in the topsoil sample is removed, as by treatment with hydrogen peroxide, the crumb structure produced by the binding effect of the humus is lost and the underlying colour of the sand, silt and clay particles is exposed.

Humus is an organic colloid whose important properties are related to its great surface area and high cation exchange capacity (Chapter 2). The humus in the soil normally makes a substantial contribution to the total cation exchange capacity, particularly in soils that are low in clay. Humus affects the physical as well as the chemical properties of the soil because of its role in the formation and stabilisation of soil structure. Humus is absorbed at positive sites on the edges of clay minerals and on sesquioxide films associated with them, and forms an intimate clay–humus complex. Its general binding effect results in the formation of crumb-like aggregates typical of topsoils, and, since humus is water-insoluble, these aggregates are fairly stable as long as the humus level in the soil remains high enough. The lowering of soil organic matter levels by cultivation, however, brings with it a loss of topsoil structure, and hence of soil porosity.

Humus also has effects on soil consistency and handling properties. It holds together the more sandy soils and gives them body, but has the effect of lightening the heavier, clay soils, and making them more porous and easier to work.

The colloidal properties of humus resemble those

Fig. 3.1 A well-structured humus topsoil about 12 cm thick, showing numerous feeding roots. The topsoil has a crumb structure, the subsoil below it an angular blocky structure. The measuring tape at the right is divided into 10 cm units.

of clays. Humus can absorb many times its own weight of water and it also increases soil water storage capacity (Chapter 4) by improving soil structure and hence increasing the volume of soil pores.

The composition of humus varies both from place to place according to local conditions of climate, vegetation and soil, and also in time. On average it contains something like 60% of carbon. In laboratory soil analyses the content of soil organic matter as such is not usually determined, but is calculated by determining the organic carbon content of the soil by a simple analysis involving the oxidation of the carbon by potassium dichromate in the presence of a standard amount of sulphuric acid, and by multiplying the result by a factor of 1.724, which reflects the assumed average carbon content of soil organic matter, to give the organic matter content of the sample.

The proportion of carbon to nitrogen in the soil organic matter, the C/N ratio, is also frequently determined. It reflects environmental conditions and the composition of the residue sources that have contributed to the humus. Normally it is about 10–12 in agricultural topsoils but falls to lower levels in the subsoil where humus amounts are typically very much less than in the topsoil (Fig. 3.1).

The amount of organic matter in a soil is related to environmental conditions, particularly vegetation and climate, to the history of cultivation, and to soil texture and drainage. Other factors being equal, clay soils have higher organic matter contents than sandy soils because of slower humus decomposition rates when the humus is intimately associated with the clay fraction. In poorly drained soils decomposition is also impeded, and organic matter may accumulate.

Typical organic matter contents for some tropical soils in relation to their vegetation and history of cultivation are given and discussed in Chapter 7.

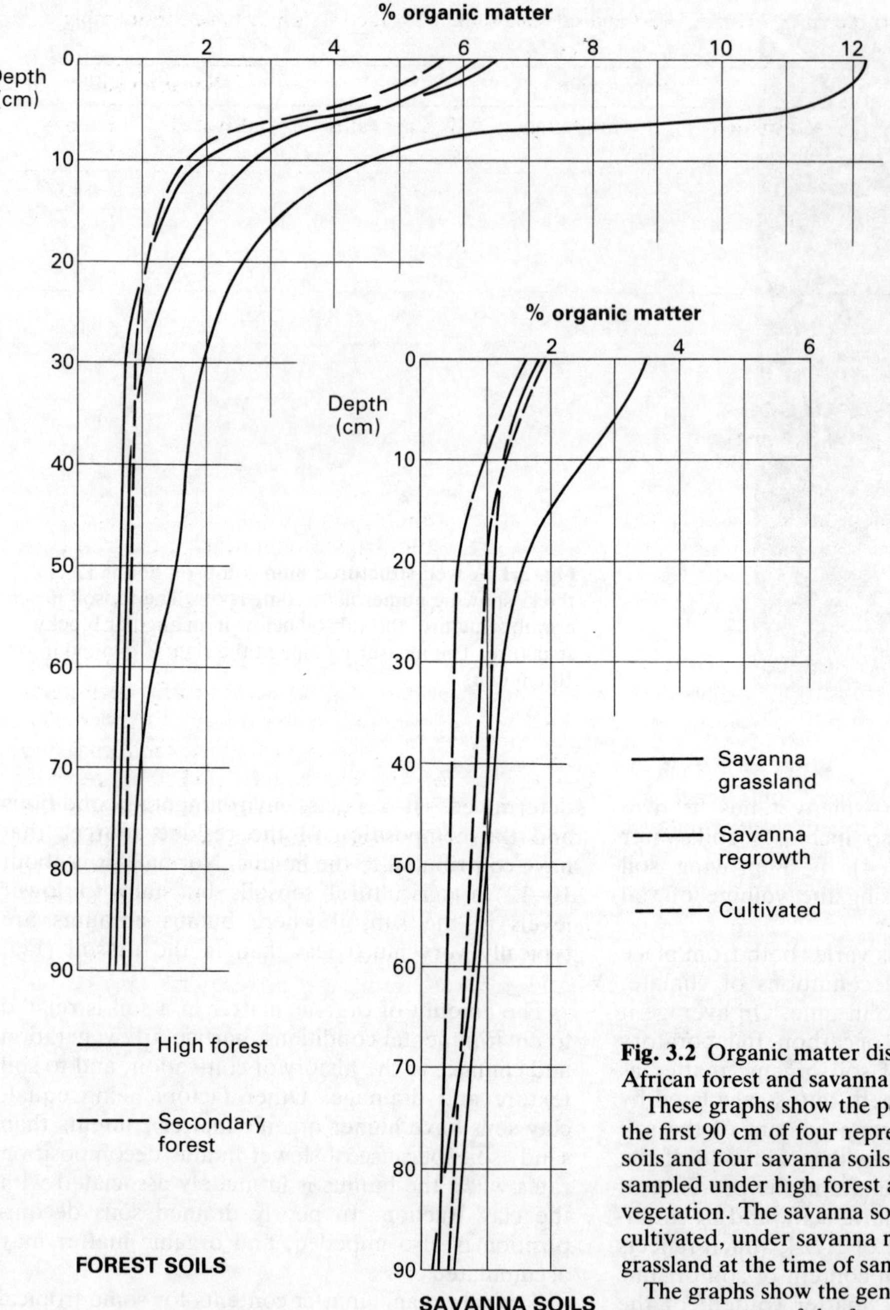

Fig. 3.2 Organic matter distribution in typical West African forest and savanna profiles.

These graphs show the percentage of organic matter in the first 90 cm of four representative West African forest soils and four savanna soils. The forest profiles were sampled under high forest and under secondary forest vegetation. The savanna soils include profiles which were cultivated, under savanna regrowth, and under savanna grassland at the time of sampling.

The graphs show the generally relatively high organic matter contents of the forest soils in the upper 10–15 cm, but indicate that there is less difference between the organic matter contents of the subsoils of forest and savanna soils. Source: Ahn, 1970.

Table 3.1 Typical organic matter contents of African forest and savanna soils. Degraded soils may have considerably lower contents

Horizon	Forest soils				Savanna soils	
	Virgin %	Cultivated %	3 years fallow %	10 years fallow %	Cultivated %	Fallow %
Upper topsoil	5.0–12.0	2.0–4.0	3.0–5.0	3.0–7.0	1.0–1.5	1.0–3.0
Lower topsoil	1.0–3.0	1.0	1.0–1.5	1.0–2.0	1.0	1.0–2.0
Subsoil	0.5–1.5	0.5	0.5	0.5–1.0	0.5	0.5–1.0
Weathered substratum	0.1–0.2	0.1	0.1	0.1	0.1	0.1

The practical importance of soil organic matter levels and humus mineralisation

The contribution of soil organic matter to soil productivity has been much discussed, particularly in the light of the fact that in many tropical areas cultivation is progressively reducing the levels of soil organic matter in the soil (Chapter 17).

The effects of a reduction in the amount of soil humus, either by humus mineralisation, or by the physical removal of all or part of the topsoil due to accelerated erosion, include

1 a deterioration of soil structure, and hence a lowering of topsoil porosity, aeration, and rainfall acceptance;
2 a lowering of the cation exchange capacity of the soil;
3 a reduction in its water-holding capacity; and
4 the loss of the potential nutrients in the humus itself, and, if mineralisation rates remain constant, a decline in the amounts of nutrients released to plants by humus mineralisation.

The amount of humus in the soil also has important effects on the biological population of the soil (see Chapter 7).

Reductions in humus levels may therefore lead to spectacular declines in soil productivity. In practice, humus levels in tropical soils can only be built up again rather slowly, owing to generally high temperatures and mineralisation rates (Chapter 17). Once the topsoil is lost or partly lost, the remaining productivity of the soil and its ability to recover depends in part on the nutrients remaining in the subsoil and weathered substratum, and their physical properties. In some tropical soils, particularly those that are highly weathered and leached, the subsoil may be very much poorer in nutrients than the original topsoil, and if the soil is very deep the weathered substratum and its nutrients may be out of reach of most crop roots.

4 Soil air, soil water and soil temperature

The study of soil texture, structure, and porosity, the circulation of air in the soil, and the movement and retention of soil water, are often referred to by the collective term, 'soil physics'. In many ways the physics of the soil are at least as important as its chemistry, partly because shortage of either air or water can cause rapid death of the plant, and partly because physical properties are less easily altered than are chemical ones. Many chemical deficiencies can be made good by adding the necessary fertilisers, but adding nutrients will not make up for poor physical properties.

Soil porosity and aeration

Plant roots require both air and water and normally the soil has to be able to supply the plant with both. If all the pores are filled with water, as in a waterlogged soil, air cannot circulate in the soil and the roots of most plants are unable to live. If, on the other hand, there is insufficient water in the soil pores, aeration may be adequate but the plant dies for lack of water. Thus the soil pores normally have the dual task both of holding water in some of the pores, and allowing air to circulate and gases to diffuse in others. These two different functions are possible because soil pores are of very different sizes and shapes and in general the larger pores are those in which air circulates and those which allow the rapid drainage of excess water, while the smaller pores hold water by capillary attraction.

The total pore space in the soil depends on texture, structure, organic matter content, faunal activity and other factors. It is increased by working and cultivating the soil, but reduced by compaction, such as that caused by heavy machinery. Total porosity is reflected in the bulk density of the soil (Chapter 2) which becomes higher as porosity gets less. Sandy surface soils may have a total porosity of 35–50%, whereas that of medium and fine-textured topsoils is somewhat higher than this, perhaps 40–60%. Porosity is much influenced by organic matter content and the degree to which the soil has a crumb structure, and may in some cases

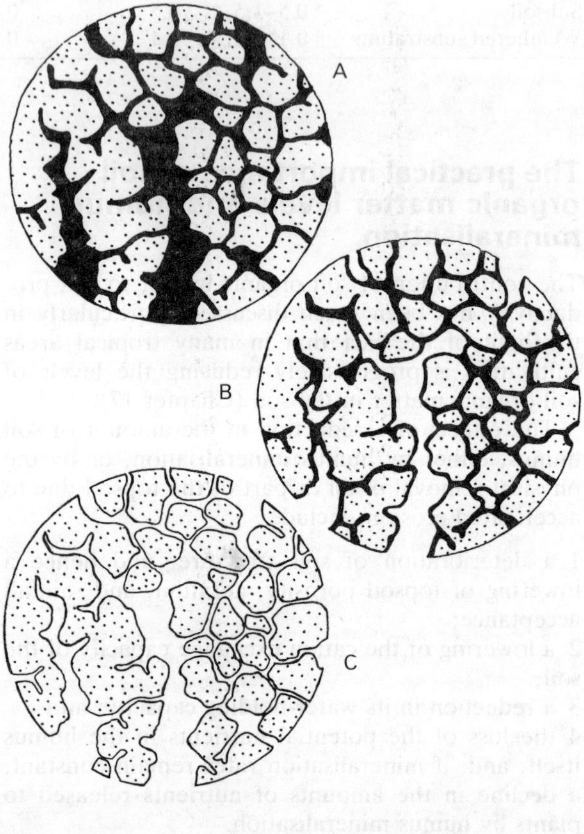

Fig. 4.1 Drawing representing the soil when saturated, at field capacity, and at wilting point. The upper drawing A represents the soil when saturated with water. Both the capillary and non-capillary pores are filled with water.

The centre drawing B shows the soil after gravitational water has drained out of the non-capillary pores. The soil is at field capacity. Water is retained in the capillary pores.

In the lower drawing C all the water available to plants has been removed. The plant wilts and dies. The soil is at wilting point. The soil still contains some water but this is held too tightly to be available to plants.

be higher than 60%. The crumb structure increases porosity mainly by providing extra macropores in the spaces between the crumbs. Total pore space is commonly less in the subsoil than in the topsoil, particularly where the subsoil is compacted or contains substantial amounts of gravel.

Three broad classes of pores are commonly recognised: macropores, capillary pores and micropores. The macropores are the larger pores which serve to drain away excess water percolating through them and which allow the circulation of air. The capillary pores are smaller pores from which the moisture can be emptied by plant roots. They are small enough to hold water by capillary attraction to the pore walls. They remain filled with water when excess water has drained out of the larger soil pores. The micropores are very small pores which can be emptied of water only under strong drying conditions. Pore sizes are often defined by the suction needed to empty them.

In sandy soils total porosity is less than in clay soils, but the movement of air and water is rapid because of the dominance of macropores, and these soils are normally well aerated. Clay soils have a higher total porosity but most of this consists of small micropores, so that gas and water movement is relatively slow and aeration may be poorer than in sandy soils.

The term soil aeration refers to the movement and diffusion of gases in the soil, allowing oxygen to move in and carbon dioxide to move out. Oxygen is constantly used for respiration by roots, by aerobic micro-organisms and by soil fauna. Aerobic respiration by micro-organisms involves the breakdown by the micro-organisms of complex carbon molecules, a process which requires oxygen and releases carbon dioxide, water and energy. When the oxygen supply is not limiting, the volume of CO_2 released is the same as the volume of oxygen consumed.

When oxygen is limiting and respiration becomes anaerobic, then oxygen is no longer consumed but carbon dioxide continues to be produced. The respiration of micro-organisms is augmented by the respiration of plant roots. The roots themselves stimulate bacterial activity and respiration in the immediate root zone by the carbonaceous material they exude or slough off.

Air normally contains approximately 79% nitrogen (N_2), 21% oxygen (O_2), and 0.03% carbon dioxide (CO_2) by volume. The composition of soil air, i.e. the air in soil pores, may differ from that of the air of the atmosphere owing to the respiration of soil organisms and of roots which use up oxygen but return at least an equal volume of CO_2. The diffusion of CO_2 out of the soil and O_2 into it tends to maintain an equilibrium between soil air and the atmosphere, but the degree to which this equilibrium is maintained depends on the soil respiration rate and on the factors affecting the difficulty of gas movement within the soil. The greater the respiration and the slower the diffusion, the more the soil air will diverge in composition from the atmosphere, resulting in it having less oxygen and more carbon dioxide and water vapour than the air outside the soil.

Gaseous diffusion in soils is a linear function of the effective pore space. The effective pore space is that part of the total pore space that permits free passage of gases, i.e. the pore space not filled with water and not blocked off from other pores by small water-filled pores. The effective pore space and the diffusion of gases thus depend not only on the shape, size and arrangement of the soil pores, but on the extent to which those pores are filled with water. The diameter of pores that are filled with water at any one time is directly related to the tension at which the water is held by the soil, as considered further below.

When all or most of the pores are filled with water, aeration is eliminated or severely reduced. The soil becomes deficient in oxygen, and anaerobic. The partial pressure of oxygen may fall to nothing, while that of carbon dioxide rises. Under these conditions anaerobic micro-organisms grow rapidly in numbers, producing both carbon dioxide and various reduced forms of nitrogen, manganese, iron and sulphur. Many of these products of anaerobic decomposition are toxic to most plant roots. The intensity of the ensuing reducing conditions is reflected in the limitation of various oxidation–reduction systems usually present in well-aerated soils.

Some of the chemical changes that occur in a waterlogged soil are considered in more detail in Chapter 20.

The visible signs of poor drainage and of reducing conditions in most soils are the progressive disappearance of the bright red and brown colours that are indicative of good drainage and aeration, the presence of mottles of a different colour from the soil matrix, and the appearance and spread of greyish, blueish and greenish colours owing to the presence of reduced compounds of iron and aluminium. The appearance of colours reflecting reducing conditions are the external signs of the process known as 'gleying' (Chapter 6).

Soil water

The entry of water into the soil

Rain falling onto the surface of the soil may evaporate again almost immediately, it may infiltrate into the soil, or it may run off the surface down the slope without penetrating the soil. Evaporation may be relatively important when a light shower of rain falls on plants or ground heated by the sun. If the topsoil has a good rainfall acceptance capacity, then perhaps much of the rain will infiltrate the soil, particularly if the rain is not too intense. Since rainfall acceptance is related to the number of coarse pores in the soil, it is increased when the topsoil has a good humus content and well-developed crumb structure, though the surface tilth has to be maintained by protecting the topsoil from the mechanical force of falling raindrops. Runoff is increased when the rainfall is intense, i.e. when a large amount falls in a relatively short time, and when the soil is already wet.

Sandy soils, particularly those containing a high proportion of medium and coarse sand, are usually well supplied with macropores which allow water to infiltrate rapidly and move downwards into the soil.

Loamy soils with adequate structure normally have satisfactory rainfall acceptance properties and balance between macropores and capillary pores.

Clay soils have a high total porosity, but most of this is due to very small pores, and rainfall acceptance is influenced both by the structure and the existing moisture content when rain falls. A particularly striking example of the wide range of rainfall acceptance properties exhibited by the same soil at different moisture contents is provided by the black cracking clays, or Vertisols (Chapter 8). During dry periods, drying of these montmorillonitic soils causes the formation of soil cracks which may penetrate the soil for depths of 1 m or more. A fall of rain when the cracks are open, as at the beginning of the rains, may penetrate the soil easily and deeply, but as the soil is progressively wetted it expands again and the cracks close up and disappear, thus making further rainfall infiltration very difficult. A wet, heavy clay of this type may be relatively impermeable so that water infiltrates only extremely slowly, and a large proportion of the total rain received may then be lost by evaporation or as runoff.

Infiltration is affected by the following:

1 The macroporosity of the topsoil, which is related to its structure and humus content, and which is improved by cultivation.

2 The nature of the plant cover, since plants break the fall of the rain and allow it to drop more slowly onto the soil.

3 The slope of the ground, since runoff is least likely on flat sites and most likely on steep ones.

4 The presence of a layer of plant litter or mulch, which slows down water movement over the soil surface and thus gives the water more time to penetrate the soil and become available to plants.

By paying attention to these factors, the farmer can ensure that as much as possible of the rain that falls onto his land enters the soil and is available to plants. This is particularly important in areas of low or marginal rainfall. A rough cloddy soil surface may not be as ideal a seedbed as a finely tilled one, but experiments in marginal rainfall areas of Kenya showed that a rough surface improved rainfall acceptance by increasing macroporosity and providing small surface depressions in which water collects, thus slowing down runoff. As far as practicable the farmer should avoid leaving a bare soil surface, particularly at the beginning of the rains when rainfall intensities might be particularly high. However, in seasonal rainfall areas of the tropics where all the litter that has been used by people has been burnt, or eaten by animals, this is often very difficult to achieve. Other measures that can be taken to slow down runoff and allow water time to infiltrate include the construction of small ridges and ditches at intervals along the contour. Runoff is to be avoided both because it represents a loss of water to the crop, and because of the dangers of accelerated surface erosion and topsoil removal.

Rainfall acceptance in some tropical soils may be severely reduced by phenomena known as crusting and sealing, both of which are most likely to occur on the surface of soils with poor topsoil structures. The formation of a thin crust on the surface of the soil, often typically only 3–10 mm thick, appears to be connected with the beating action of rain and the sorting of soil particles to give an arrangement in which smaller particles occupy the spaces between the larger ones so that the crust is relatively compact and may show fine bands or layers. Crusting occurs on a range of soils, even fairly sandy ones, provided that they contain enough fine sand and silt to fill the spaces between larger particles to form a compact layer. Crusts have been studied in various parts of the West African Sahel, where they are broken by cultivation but in some cases form again quite quickly after a few falls of rain. In north-western Nigeria some soils that have developed in so-called coversands, seal and crust so easily that they are generally avoided for cultivation. These include the Sangiwa coversand deposits which reseal with the first rains after the surface has been disturbed by cultivation.

In some cases of soils on flattish sites, temporary waterlogging of the soil surface and the growth of algae may contribute an algal layer which reinforces the sealing effect of the crust.

Storage of water by the soil

Every soil and soil horizon can hold a certain maximum amount of water, beyond which any additional water received is lost by downward percolation. After a heavy fall of rain the upper horizons of the soil may be saturated with water, and all or most of the soil pores temporarily filled with water, but water then drains out of the larger pores, the macropores, owing to gravity, and moves to lower horizons. The speed with which this excess water drains downwards is related to soil texture, to soil structure and to other factors that affect the amount and size of soil pores, and in some heavy clay soils may be very slow. For most soils the excess water has usually drained downwards after 24–48 hours, leaving the wetted layer now holding as much water as it can against the force of gravity. The excess water that has drained down is termed 'gravitational water'. When the gravitational water has drained downwards, leaving the soil holding as much as it can against gravity and the smaller, capillary pores full of water, the soil is said to be 'at field capacity'. In some tropical soils, including Vertisols, however, downward drainage continues at a slow rate for a considerable period, so that field capacity becomes more difficult to define.

The amount of water held at field capacity varies from soil to soil and from horizon to horizon within a soil profile, depending on the content of soil colloids, both clays and humus, and the number and size of small pores. These smaller pores, or capillary pores, hold water by capillary attraction, the natural attraction between a solid and a liquid, in this case the attraction between the water and the walls of the pore. For small pores this attraction is enough to hold the water against gravity and keep the pore full of water. A productive soil should have a balance between capillary and non-capillary pores.

The water held at field capacity depends not only on the total capillary porosity, but also on the content of clay and humus colloids which attract and absorb water. This absorption by clay and humus contributes to the fact that clay soils hold more water at field capacity than sandy soils, and that topsoils high in humus often store more water in relation to their weight than underlying horizons or otherwise similar horizons with lower humus contents.

Water losses by plant removal

Since a soil horizon at field capacity has already been drained of any surplus gravitational water, it will not lose any further water by downward seepage unless additional water is received. Subsequent loss of water from a horizon at field capacity can take place in two ways. Water can be lost by evaporation, mainly from the surface soil, and water can be lost through removal by plant roots.

Not all the water held by the soil when it is at field capacity is available to plants. Plants take in water through very small root hairs, and soil water enters these root hairs through the force of osmosis.

The water moves from the weak soil solution to join the denser solution within the plant tissues, the cell sap. The water that is more loosely held by the soil is given up most easily and will move first, but as the soil water is progressively removed the root will have to extract water further away from the root hair and water that is more tightly held by the soil. The extraction of this more tightly held water will require progressively more force as the soil water gets less. Finally a point comes when the suctional forces exerted by the plant are not enough to give it the water it needs. The plant then wilts and dies. At this point there is still some water left in the soil but it is held with a greater force than the plant can exert to remove it, though some of it could be driven off by evaporation and all of it by heating the soil in an oven. A soil from which all the water available to plants has been removed is said to be 'at wilting point'.

Water contained by the soil when it is at field capacity can be divided into that which is available to plants – the available moisture – and that which is not. The available water stored by the soil is conventionally taken as the difference between the water held at field capacity and that still held at wilting point – the point at which the remaining water is held at too high tensions for the plant to be able to extract it.

The force attracting the soil water to the soil solids is known as 'the matric potential', the total force holding the water to the soil matrix. The matric potential is made up of the capillary attraction of water for soil solids plus the forces of adsorption onto the colloids. It may be thought of in terms of the suction or tension needed to remove the water. This suction or tension can be measured in terms of the height in centimetres of a column of water whose weight just equals the suction. The higher the column, the greater the suction. The force or tenacity with which the soil moisture is held can thus be conveniently expressed in terms of a height in centimetres of water, or, alternatively, in terms of the logarithm of the height in centimetres, the pF (Table 4.1). The pF system has the advantage that the units used are smaller than centimetres. Alternatively, the suction can be expressed in terms of bars, a bar being 1 kg/cm^2, or atmospheres (there are 1033 mb to the standard atmosphere).

Table 4.1 Approximate equivalents of three different ways of expressing the suction or force with which water is held by the soil: in terms of cm of water, in terms of the pF scale, and in terms of atmospheres.

Height of column of water in cm		**Logarithm of (pF)**	**Atmospheres**
	10	1	0.01
	100	2	0.1
FC	346	2.53	one-third
	1 000	3	1
	10 000	4	10
WP	15 849	4.18	15
	31 623	4.5	31
	100 000	5	100

FC = field capacity
WP = wilting point

The term 'millibar' (mb) is used for one-thousandth part of a bar, and is the same as the weight of a column of water 1 cm in height, so that a force or suction of 10 cm can also be expressed as 10 mb. Publications on soil water commonly refer to the force with which water is held either in terms of atmospheres, or of bars, or of pF, so that a familiarity with all three sets of terms is desirable (Table 4.1).

Although the term 'available water' is conventionally applied to that water held between field capacity and wilting point, and it was once assumed by some soil scientists that all this water was equally available to plants, it is now known that this assumption is incorrect. The water held with lower tensions is more easily extracted by the plant than that held with higher tensions. Thus the water held with tensions of between one-third and one atmosphere is very easily available to plants, and that held at tensions approaching the 15 atmospheres limit is less available, so that this residual moisture requires progressively more force (or suction) to extract it. This is particularly important for some vegetable crops which can exert only relatively low forces to extract soil moisture.

Soil moisture storage capacities

The available water that a soil can store is often determined in the laboratory. If a saturated sample of soil is subjected to a suction of about one-third of an atmosphere, equivalent to the weight of 346 cm of water (Table 4.1), this is sufficient to remove the gravitational water from the sample and leave it at field capacity. If a saturated sample is subjected to the much higher suction of 15 atmospheres, the available moisture will be removed and the sample will contain an amount of water corresponding to that left in the soil when most plants wilt, the wilting point. The water held by the soil at field capacity (one-third of an atmosphere) and at wilting point (15 atmospheres) is then calculated as a percentage by weight of the weight of the oven-dry soil. The differences between the two percentages gives the water available to plants, the available moisture. At field capacity, for example, a soil may hold 32%, and at wilting point 15% of the oven-dry weight. In this case the weight of the water available to plants is 32–15%, or 17% of the oven-dry weight of the soil. Sometimes the amount of water is expressed as grams of water per 100 g oven-dry soil instead of as a percentage.

To complete the calculation it is convenient to use the bulk density of the soil (Chapter 2) instead of the actual weight of the oven-dry soil. If we assume that a soil horizon 50 cm thick has a bulk density of 1.5 and has an available moisture capacity of 20%, then the actual amount of available water held by that horizon when at field capacity can be calculated as follows:

1 Since the soil has a bulk density of 1.5, i.e. the oven-dry soil weighs 1.5 times the same volume of water, the weight of 50 cm of oven-dry soil is equivalent to 50×1.5 cm = 75 cm of water.
2 Since the available water is 20% of the oven-dry weight of the soil, the available water is 20% of 75 cm = 15 cm.

This simple calculation avoids the need to know the actual weight of the oven-dry soil, since the ratio of the weight of the oven-dry soil to the weight of the same volume of water is given by the bulk density. In the above case the available water, 15 cm, was calculated in respect to a horizon 50 cm thick. An alternative way is to express the available water in terms of centimetres of water per metre of soil (cm/m). For the horizon considered in the above example, a metre of soil would have held 30 cm, and the available moisture is 30 cm/m.

The water storage capacity of a soil profile often varies from horizon to horizon, depending mainly on texture and organic matter content. It is sometimes useful to know how much water a soil profile holds at field capacity and how much of this is available to the plant. To measure these, it is necessary to sample the soil horizon by horizon, preferably by taking undisturbed core samples in tubes, and to determine the water held at field capacity and at wilting point in the laboratory in order to calculate the available water storage capacity of the soil for each horizon separately. This is then added up to the depth required (e.g. to the depth of the roots of the crop in question).

The approximate relationship between soil texture and the amount of available water held by a soil is shown in Fig. 4.2. The water held at field capacity

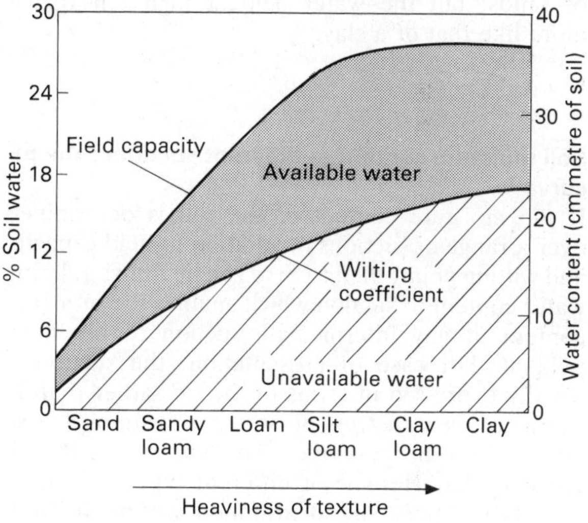

Fig. 4.2 Typical amounts of water held at field capacity and at wilting point for soils of different textures. The area between the two lines indicates water conventionally considered available to plants, and that below the wilting point line that water which is held by the soil which is not available. Sandy soils hold the least moisture and clay soils the most, but in clay soils much of the water held at field capacity is not available to plants. The greatest amount of available moisture is often held by soils with intermediate textures. In practice water storage is influenced by soil structure and organic matter content as well as by texture.

increases from a minimum in sands to a maximum in clays. However, the water still held by the soil at wilting point follows a somewhat different curve, since it rises sharply for clay soils. The available moisture – the difference between the two – is often relatively low for sands, moderate for clays, but highest for the soils with intermediate textures, the loams. Available water is increased in soils with high organic matter contents.

Tropical soils with a high degree of micro-aggregation (Chapter 2) may have characteristic moisture retention, aeration and drainage properties because they combine a high clay content, and therefore a high content of water still in the soil at wilting point, with a pore size distribution related to the size of the water-stable micro-aggregates which are commonly of silt and sand size. The water held and released at low tensions may be comparable to that of sands, but the water held at high tensions is more like that of a clay.

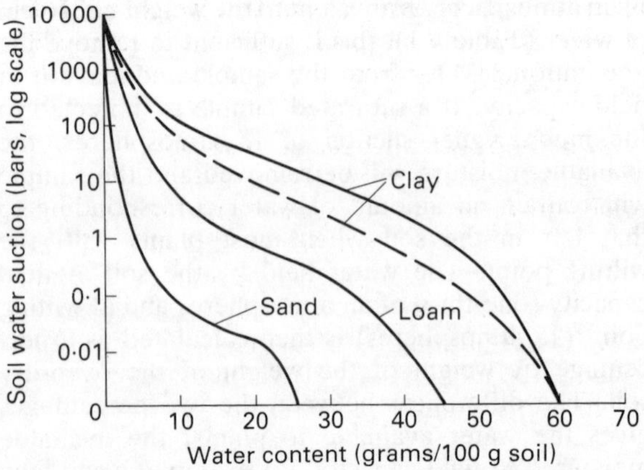

Fig. 4.3 Typical soil water suction curves for a sand, a loam and a clay, showing the percentage water in the soil at different tensions. At all suctions most water is held by the clay and least by the sand. The soil water content in relation to suction is slightly different for wet soils being dried (continuous lines) than it is for dry soils being wetted (shown, for the clay soil only, as a broken line). The difference is known as the *hysteresis effect*.

Soil moisture content at different suctions: the pF curve

If the moisture content of the soil is determined over a range of suctions in addition to field capacity and wilting point, then a curve can be drawn relating water content to suction which enables the moisture content at any intermediate suction to be interpolated. For ease of presentation, the suction is usually expressed in terms of the pF rather than in terms of centimetres of water or atmospheres, hence the term 'pF curve'. To some extent the pF curve is characteristic of different types of soil. It reflects the pore size distribution within the soil, since with increasing suction, water is removed successively from pores of a specific equivalent diameter. At field capacity, pores up to about 50 μm may remain filled with water, but as suctions increase, all the remaining silt-sized pores and then the clay-sized pores will be progressively emptied. At wilting point, only very very small pores of 0.2 μm or less equivalent diameter are still filled with water. The curve showing moisture held at different suctions is therefore also the curve showing the pores still filled with water at those suctions and the pore size distribution in the soil sample.

Water losses from the soil by evaporation

If no plants are growing in a soil, water will still be lost by evaporation from the surface. Evaporation can dry out soils more than can plants, and thus reduce the soil water content below wilting point. However, evaporation in practice can dry out only the upper part of the profile, the maximum depth that can be affected depending mainly on soil texture. It is common experience to find that even if the upper layers of a soil without plants are very dry, lower horizons may still be moist, even in a dry season after considerable periods without rain.

Evaporation from the soil surface might be quite rapid immediately after rain, being faster when temperatures are high, when the air is dry, and when there is a wind blowing. When the first few centimetres of the soil have dried out, evaporation rates may be sharply reduced, and further evaporation may take place only at a very slow rate because of the protecting effect of the dry surface layer of soil. When this layer is thick enough, further evaporation virtually stops. The thickness of the layer necessary to virtually prevent further losses depends mainly on soil texture, being least for sandy soils and most for clays, but is commonly in the range of 40–80 cm.

Water carry-over from season to season

It is possible to store water in the soil for long periods if there are no plants growing in it by allowing a dry upper layer to protect water in the subsoil. This principle is made use of in dry farming methods which store the moisture of one year or season for use during the following year or season.

Where the amount of water is marginal for crop growth, the amount of stored water carried forward in the soil profile from one year to another may critically affect crop growth and yields. This is a function of the rainfall received and of losses by evaporation and crop removal, but since different crops have different water demands and remove water to different depths, they have different effects on the extent to which a profile is dried out and the amounts of water still left at the end of the cropping

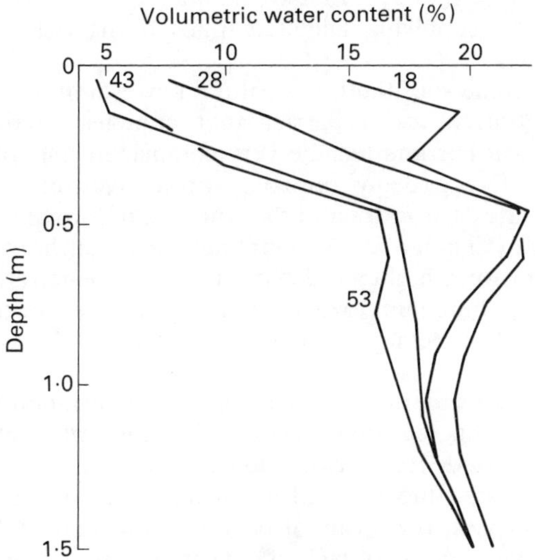

Fig. 4.4 Changes in water content of the soil beneath a crop of millet grown on stored soil water. The numbers on the graph lines show days after sowing. Very little water was removed below a depth of 1 m. The root system of most annual cereals does not normally grow very much after flowering, and moisture may remain in the soil after harvest. Source: **Squire G. R.** *et al.* 1984 *Experimental Agriculture* **20**, 135.

season to be carried over to the next. Shallow-rooted annual crops such as beans may dry out a profile much less deeply than deep-rooted crops such as maize. A crop of cotton may dry out a profile more thoroughly than a cereal crop, partly because its root system is relatively deep and partly because its growing season is indeterminate and it may continue to flower until all the soil moisture is used, leaving little carry-over to the next season. In many areas of seasonal rainfall in the tropics, the deep-rooted natural fallow vegetation dries out the soil much more than do most agricultural crops.

Measuring the water content of soils in the field

The measurement of soil moisture storage capacity in the laboratory depends on weighing wet samples subjected to various tensions. To know how much water is present in a soil in the field at any time, four main methods can be used:

1 In **gravimetric analysis** of soil moisture, a soil sample is taken, put at once into a tin or bag so that it loses no moisture, taken to the laboratory and weighed, and then oven dried and weighed again. The moisture lost on drying as a percentage by weight of the oven-dry soil represents the percentage (or grams per 100 g) of water in the soil sample when it was taken in the field. The method accurately shows the moisture content in the sample but is laborious and requires frequent sampling if moisture changes are to be monitored over a period.

2 **Gypsum resistance blocks**, in which electrodes are embedded, are inserted in the soil at various desired depths so as to make good contact with the surrounding soil, and left there until an equilibrium is established between water in the soil and water in the blocks. The water content of the block, assumed to reflect that of the surrounding soil, is then obtained by measuring the resistance to a current across the electrodes. A dry soil will show a high resistance, a wet soil a low one. Resistances recorded have to be calibrated with moisture contents in the block in order to arrive at a reliable soil moisture content. This method is not highly accurate but allows continuous monitoring and shows changes at various soil depths without taking soil samples.

3 Instead of resistance blocks, **tensiometers** of various lengths are inserted into the soil so that a porous lower end through which water can move

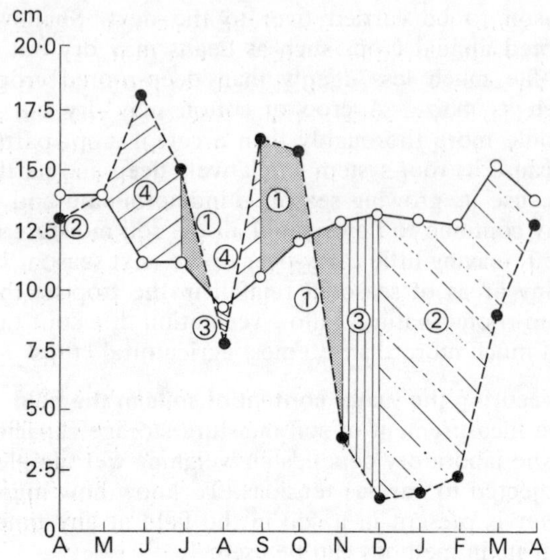

Fig. 4.5 Soil water at Moor Plantation, Ibadan, Nigeria. The diagram shows average annual rainfall (broken line) in relation to potential evapotranspiration (solid line joining small circles).

There are two wet seasons: May to July, and September to October. The rains of the first season cause soil water recharge (4) culminating in a water surplus (1). This is followed by soil water utilisation (3) in the August break in the rains, and and a second period of water recharge (4) at the beginning of the second wet season.

There is a water surplus (1) during most of the second rains and into the beginning of the main dry season. After this soil water utilisation (3) results in a deficit (2) for most of the main dry season. Source: Smyth and Montgomery, 1962.

out into the soil is positioned at the required depths. It is assumed that the suction in the tensiometer, measured on a suction gauge at the top, is the same as that in the soil. This method is best used only for low suction values of less than 1 atmosphere, e.g. for monitoring the need for irrigation water.

4 A **neutron probe** containing a source of fast neutrons is lowered into the soil to the desired depth through a previously inserted tube. When the fast neutrons collide with the hydrogen atoms in soil water, they are slowed down. Slowed neutrons are measured by a detector in the probe connected to a scaler held above the ground. The reading on the scaler can be calibrated with the moisture content of the soil. The method measures hydrogen in the soil and reflects the water content only in soils in which other sources of hydrogen are not important. It does not work so well for organic soils which have combined hydrogen in the organic matter, nor can it be used for the surface horizon. With these limitations it can, if well calibrated with the soil in question, give a reliable indication of moisture contents and changes with comparative ease and speed.

Root depth

Both the amount of water available to a crop and its nutrient supply is related to its depth of rooting. If moisture is in short supply, the deeper-rooted plants may continue to obtain enough water while shallower-rooted ones wilt and die. The ability of a crop to survive a dry period might depend on the depth to which it has been able to send its roots before water shortages become severe, and in such cases it might be profitable to stimulate deep rooting, e.g. by supplying adequate phosphorus (Chapter 10).

In some soils there is a natural lower limit to root penetration, or a barrier that impedes rooting. Physical barriers include very compacted horizons, extremely gravelly horizons, or a layer of hard plinthite or ironstone in the soil within normal root depth (Chapter 7). An extremely acid soil horizon would form a chemical barrier to root penetration due to aluminium toxicity, and in such cases rooting depth can be increased by deep liming (Chapter 11).

A high water table also stops root penetration for most plants, because the soil below the water table is saturated. Except for a few adapted plants, such as rice and those found in swamps and mangrove areas, plant roots cannot penetrate the water table very far because of lack of oxygen and are confined to the soil above. In such cases, root depth can be increased by draining the soil in such a way as to lower the water table.

If plants grow with a water table near the surface and their roots are confined to a shallow soil zone, a subsequent sudden drop in the water table followed by drying out of the upper layers of the soil might result in their dying for lack of water, but if

Fig. 4.6 Crop mounds. Mounds such as those shown here are widespread in the West African savanna and forest zones, as well as in some other parts of the tropics. In the example shown, near the Black Volta river, north-western Ghana, the thin sandy soil, developed over pale granites, has been scraped up with a hoe into cultivation mounds. These are used for maize, sorghum and millet, often planted in that order, as well as for yams (*Dioscorea* sp.). Mounding increases root depth and minimises the harmful effects of waterlogging and poor aeration after heavy falls of rain.

the water table falls slowly the roots might be able to grow downwards to keep pace with it.

The moisture storage capacity of a soil has to be considered for agricultural purposes in relation to the local climate. In the case of perennial crops, it should be assessed in relation to the length and severity of the dry seasons. In the same climatic area, a perennial tree crop might survive on the deeper soils but be eliminated by dry-season water shortages on shallower soils with lower moisture storage capacities.

Making the best use of low and erratic rainfall

Much of the tropics suffers from seasonal water shortages. In general the lower the rainfall, the less reliable it is. Making the best use of low and erratic rainfall requires that the farmer applies some of the principles of soil physics outlined above. The farmer's aims should be

1 to get as much of the total rainfall into the soil as possible, by reducing the proportions lost by evaporation and runoff;
2 to reduce unnecessary losses from the soil so as to increase the proportion available to the crop as against that which is taken up by weeds or lost by evaporation from the soil surface;
3 to make the best use of the wet season expected by planting at the right time crops that have suitable water requirements and length of growing season, and which are otherwise adapted to local conditions.

Runoff can be reduced by ensuring that the macroporosity of the topsoil remains high, so that it remains open and permeable and has good rainfall acceptance properties. These will be enhanced by a good crumb structure. The beating action of intense tropical rains may destroy topsoil crumb structures, reducing macroporosity and permeability, unless the soil is protected. A plant cover breaks the fall of the rain but cannot be maintained in most seasonal cropping conditions, though in many shifting cultivation situations farmers leave a few trees during clearing which provide shade, and break the force of the rain.

Anything that slows down the movement of water at the soil surface and gives it more time to infiltrate the soil has the effect of reducing runoff. A layer of mulch or plant residues has this effect, as

Fig. 4.7 Constructing broad beds on Vertisols, ICRISAT station, Hyderabad, central India. An ox-drawn cultivator equipped with ridging tools on either side, with a chain between the two, prepares the broad bed ready for dry planting of seeds immediately prior to the expected monsoon rains.

does planting along the contour, and contour ridges and ditches at intervals can intercept runoff and increase infiltration. Alternatively the whole crop can be planted on ridges more or less on the contour, a method that has proved very effective in some of the drier regions of the tropics. It can be even more effective when small cross-ridges, or ties, are built across the ridges at intervals, so limiting water movement along the furrows and creating a series of small rectangular basins which retain water long enough to allow it to sink into the soil. This method is known as tie or basin ridging. However, one of the problems of low rainfall areas in the tropics is that when rain comes it may fall in very heavy, intense downpours, and may cause temporary waterlogging and even flooding. At such times the basins may fill with water and plants die because of shortage of air. To prevent this it has been found helpful to make the cross-ridges, or ties, across only alternate pairs of furrows, or alternatively to dig the ridges later in the growing season when the heavy downpours associated with the beginning of the rains are less likely.

Cultivation in many areas of the tropics is not on the flat but on ridges or mounds. Cultivation mounds are formed by scraping up the topsoil into heaps, usually with a hoe. This is traditional practice for yams in West Africa, but a range of grain and other crops, including maize, sorghum and millet, are also planted on mounds. This is in part a measure designed to avoid temporary flooding and waterlogging, and to prevent crops dying from poor aeration, during the usually short periods of heavy rain. Thus the farmer may have to cope with water shortages most of the year but at the same time be ready for occasional heavy storms.

The second aspect of making the best use of low or erratic rainfall concerns measures aimed at reducing water losses from the soil other than those due to crop uptake. Weeds compete for water, but can be cut down, pulled out or destroyed with weedicides. Cover crops, such as *Pueraria* and *Stylosanthes*, though desirable because they protect the soil and may fix nitrogen, also compete for water with the crop and their advantages and disadvantages must be carefully assessed where water is likely to be limiting. Surface evaporation losses are effectively reduced with a grass mulch, which

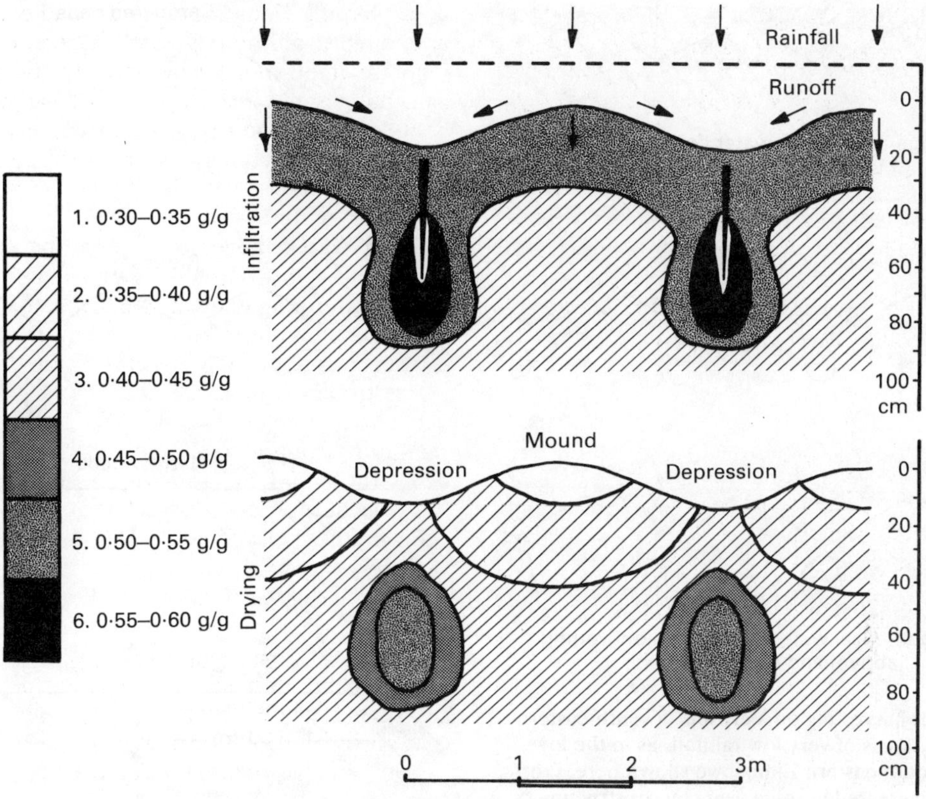

Fig. 4.8 Water distribution in a Guadeloupe Vertisol during wetting (above) and drying (below), showing the uneven distribution of water in relation to the microrelief.
Source: **Jaillard, B. and Cabidoche**, Y. M. (1984). *Science du Sol* 3: 239–251.

also reduces runoff and has the longer-term effect of adding humus to the soil.

The third approach is to assess the rainfall probability of the wet season, and choose or breed crops that are adapted to the local growing season. It might, for example, be better to plant a four-month maize with a moderate yield that is well adapted to the local wet season rather than a higher-yielding five-month maize that can be expected to fail because of inadequate rain about one year in three or four. Crop varieties can be bred with increased resistance to dry spells, drought resistance being shown by such qualities as the capacity to recover from a partial wilt, or with deeper rooting systems. Enormous improvements have been achieved in tropical agriculture in recent years by the selection and breeding of crop varieties for specific soil conditions, including erratic moisture supply, soil acidity and low phosphorus levels.

Where rainfall and soil moisture are likely to be inadequate, a simple and very long-established method of increasing the water available is to increase the spacing between plants, thus giving a lower population density and allowing more soil per plant.

Soil temperature

Soil temperature is of interest and great importance because it directly affects the rates of plant growth, of microbial activity and of the decomposition both of fresh residues added to the soil and of the soil

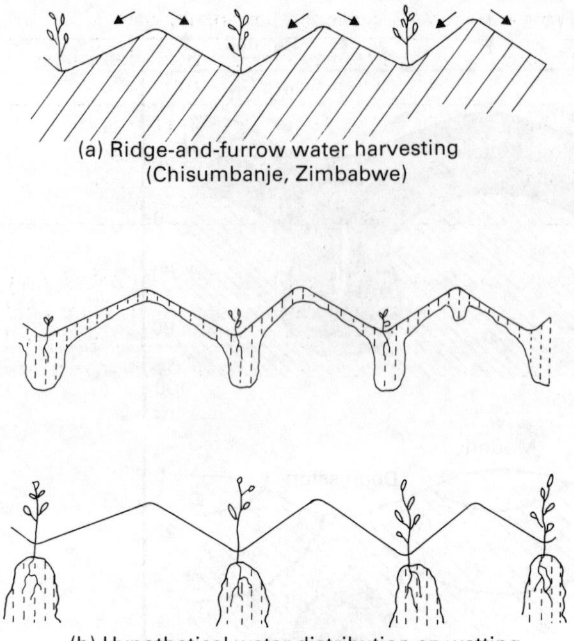

Fig. 4.9 Water harvesting on Vertisols in south-eastern Zimbabwe. In areas of very low rainfall, as in the low veld area of south-eastern Zimbabwe shown here, crop yields can be increased in some years by constructing ridges at 1.5 m intervals and planting the crop in the furrow between the ridge. Water moves down the side of the ridge and collects in the furrow.

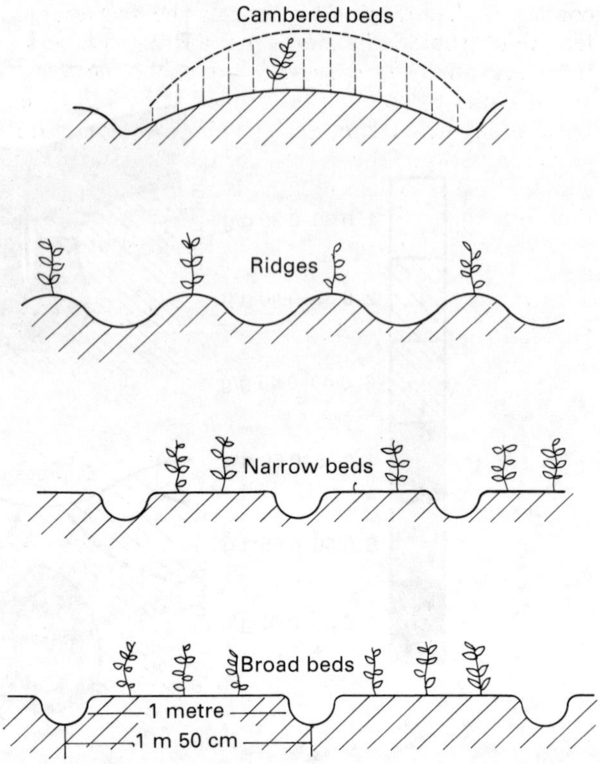

Fig. 4.10 Some artificial microrelief patterns designed to improve surface drainage: cambered beds, ridges, narrow beds and broad beds.

humus. The rate of chemical reactions in the soil doubles with every rise of about 10 °C in temperature. Not only the macroclimate but also local conditions affecting soil temperatures, such as the type and amount of plant cover and the amount of shade, affect the rate of chemical processes in the soil, and plant growth.

The temperature of the soil at the surface fluctuates during the day in relation to air temperature and insolation. Dry soils change temperature more easily than wet or poorly drained ones because of the extra heat needed to raise the temperature of the water as compared to the mineral soil which has a lower specific heat. The highest soil temperatures recorded, over 80 °C, have been for dry, bare soils exposed to the sun, but normally unexposed surface soils do not greatly exceed air temperatures.

Below the surface, soil temperatures change much less, and more slowly. At a depth of 30 cm, daily temperature changes rarely exceed 3 °C. At a depth of 60 cm they may reach 1 °C, while at 1 m they are virtually absent.

Soil temperature regimes as defined and used in the US Soil Taxonomy (Chapter 8) reflect subsoil temperatures, and in terms of these the tropics experience isotemperature regimes. A soil isotemperature regime is one that shows little change throughout the year, being defined as having 'less than 5 °C difference between the mean summer and mean winter temperatures at 50 cm or at a lithic contact if shallower'. This reflects the fact that a feature of tropical climates is that average temperatures are relatively constant throughout the year, and that the tropics can be defined (Chapter 1) as those areas where the mean monthly temperature variation between the average of the three warmest and the three coolest months is 5 °C or less. The four isotemperature soil regimes characteristic of the tropics, and the approximate elevations at which they are found, are as in Table 4.2.

Soil surface temperatures can be controlled and fluctuations dampened to some extent by a mulch. This might consist of a layer of dead grasses or other plants, or of a sheet of plastic, or even of stones or dry soil. In cool climates, mulches may raise soil temperatures and protect them from cold spells, but in hot climates they may lower soil temperature by shading the soil from direct insolation.

Table 4.2 Tropical isotemperature soil regimes

Regime	Mean annual temperature (°C)	Elevation range (m)
Isohyperthermic	Over 22	0–600
Isothermic	15–22	600–1800
Isomesic	8–15	1800–3000
Isofrigid	Less than 8	Over 3000

5 Looking at soils in the field

The soil profile: the major soil horizons and their properties

To examine a soil in the field it is not usually sufficient to look only at the top few centimetres of the soil, even though the surface horizon is in some ways the layer of soil that most affects plant growth. The whole soil profile should be examined.

A soil profile is a vertical cross-section through the soil showing the horizons of which it is composed. These horizons are the different soil layers which together form the soil profile, each horizon differing in some respects from the horizons above and below.

In the hot, wet tropics, rock weathering is sometimes deep and intensive. Those profiles developed from an underlying rock may be several metres or even tens of metres deep before they reach consolidated rock below. Deep profiles are not confined to wet areas. They may occur in areas that are now relatively dry but are due to deep weathering during wetter climatic periods in the past. Other profiles, in contrast, may be so shallow that plants do not have enough root room, with rock or ironstone at less than 1 m. Some soil profiles developed in very deep alluvium deposits may not go down to hard rock at all, but simply be developed by soil-forming processes in the upper few metres of the alluvium, with little changed alluvium below.

Soil profiles that show at least the upper horizons of the soil can be examined in road cuttings, in ditches if they are deep enough, in quarries and other holes dug in the ground, including wells, or in pits specially dug for the purpose and known as 'profile pits'.

Typical local soils should be examined carefully and in as much detail as possible. Soils that are examined should preferably include some that have not been disturbed or cultivated for some time.

The topsoil

The uppermost 5–10 cm or more of most soils that have not been disturbed by cultivation are somewhat stained by humus, and darker than the soil below, forming a fairly well-defined uppermost horizon, the topsoil.

The amount of humus in the topsoil, and hence the colour, varies considerably with the vegetation, history of cultivation and other factors. A humous topsoil is typically a dark greyish-brown in colour, even though often more loosely described as 'black'. Humus itself, when separated from the soil, is seen to be a black, sticky, amorphous, glue-like substance (Chapter 3). It is derived from plant remains that fall on to the surface of the soil, from roots that die and decompose within the soil, and from the bodies of soil microbes and other soil fauna (Chapter 7). The humus coats and impregnates the soil particles, staining them a dark colour, and also helps to bind them together so that in topsoils they typically form crumb-like aggregates (Chapter 2).

Humus has colloidal properties that profoundly influence the soil. It affects soil structure, porosity and consistency, and because it absorbs many times its own weight of water, it increases the water storage capacity of the soil. A small proportion of the humus in the soil, depending on environmental conditions, is mineralised each year to provide a range of plant nutrients. Thus humus affects both physical and chemical soil properties, and hence soil fertility and productivity.

Leaves and other plant parts may be found on the surface of the soil, both fresh and partly decomposed. This layer of organic debris is the leaf litter layer. In savanna areas and where the vegetation cover is sparse, the leaf litter layer may be thin and discontinuous, with areas with little dead material on the soil surface. In forest areas the quantity of dead leaves, flowers, twigs, branches and other plant parts may be relatively large, and the leaf litter layer continuous, though still normally fairly thin because the organic material falling onto the soil surface decomposes rapidly under hot wet conditions and does not accumulate. Leaf litter layers may also be observed under perennial tree

On the surface: leaf litter layer

Topsoil

0–8 cm Very dark greyish-brown humous loam, crumbly and loose.

8–15 cm Dark greyish-brown loam, less humous than above, weakly crumbly.

Subsoil

15–120 cm Reddish-brown light clay, weak fine subangular blocky structure, containing frequent to abundant ironstone concretions and some quartz gravel and stones.

129–180 cm Reddish-brown light clay to heavy loam, transitional to the horizon below, usually coarsely mottled and sometimes indurated.

Weathered substratum

Below 180 cm Brown or red silty loam, usually slightly mottled, with traces of soft weathered parent rock (phyllite or schist).

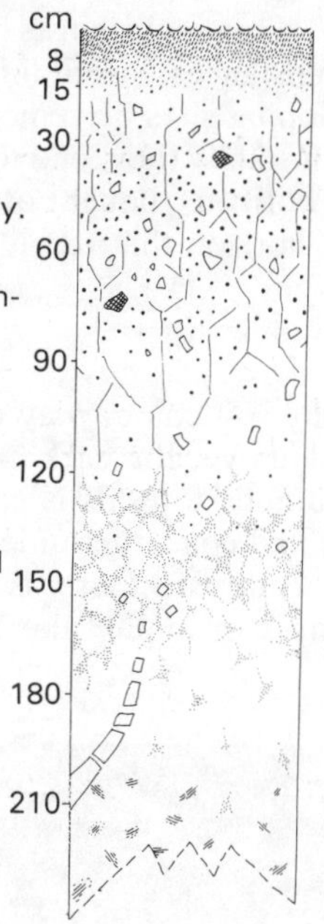

Fig. 5.1 Diagram of a typical well-drained West African upland sedentary soil developed over phyllite or schist. A short, simple description of the soil is given next to the profile. For more detailed work a fuller description would be necessary, further subdividing the horizons shown. Figures on the left of the profile diagram refer to the depth in cm.

crops, such as rubber and coffee, and are sometimes particularly well developed under tea which provides a dense cover, much leafy material, and shades the ground.

In undisturbed soils under long-established natural vegetation, the topsoil is not uniform but may consist of a darker upper topsoil 4–8 cm thick over a less dark, less humous lower topsoil. In cultivated areas the upper 15–20 cm of soil is frequently mixed by cultivation and the topsoil made more homogenous. In temperate areas this layer mixed by cultivation is often referred to as a 'plough layer'. In those areas of the tropics where cultivation is mainly by hand, it would be better called a 'hoe layer', or a 'cultivation layer'.

Samples of the topsoil should be examined in detail, preferably with a hand-lens. Soils are normally examined both by eye and by feel. The soil should be felt by rubbing a little between thumb and forefinger. The experienced soil surveyor or agronomist can assess the texture of the soil in this way.

Taking the soil apart enables the structure to be examined and the pores and natural spaces in the soil observed. The soil may contain gravel or stones, and there are probably many small roots which may form a dense root-mat, for the topsoil is the main feeding zone of plants, with fewer roots in the subsoil.

The above paragraphs describe some of the common topsoil characteristics in the field. Pedologically a topsoil that meets certain criteria is termed an 'A horizon', but a precise definition of an A horizon is not easy to formulate. Normally the essential characteristics are that it is the horizon of maximum biological activity, and it is also a horizon of eluviation in the sense that it loses fine material in suspension and some substances in solution. These removed substances, including clay-sized particles, may on sloping sites be removed laterally to some extent but normally they tend to accumulate in the subsoil, or B horizon, thought of as a horizon of illuviation, i.e. a horizon that receives material.

The subsoil

Below the topsoil the soil colour may change quite markedly. As noted above, the dark greyish-brown colour typical of humous topsoils is caused by the humus which stains and coats the mineral particles. In the subsoil, in contrast, there is usually not enough humus to mask or modify the colour of the mineral soil. As a result, subsoil colours are often more striking or varied than the more uniform colour of humus-stained topsoils. When a soil is described as being brown, yellowish-brown, yellow or red it is usually the colour of the subsoil that is being referred to, though some soils are red to the surface.

Fig. 5.2 A soil pit dug in a gravelly forest soil similar to that shown in Fig. 5.1.

The warm red and brown colours of soils are mainly due to the presence of iron componds. If these are removed from a subsoil sample in the laboratory the resulting soil material may be very pale or almost white. The colour depends partly on the amount of iron present but mainly on the form in which it occurs. In free-draining soils, i.e. in soils in which soil water moves down fairly rapidly and oxygen circulates adequately, the iron is oxidised to reddish ferric oxide (as in the rusting of iron), which occurs as the mineral haematite, and even very small quantities of iron III oxide coating the fine soil particles or adsorbed onto them can give the soil a striking reddish colour. A red colour does not necessarily indicate a fertile soil, but it does normally indicate a well-drained soil in which water drains down rapidly enough for the air to move freely in the larger soil pores.

On middle and lower slope sites where the soils remain wetter somewhat longer than on the reddish summits and upper slopes, and sometimes on all slopes of very high rainfall areas, the soil iron may be hydrated, i.e. combined with water. The hydrated forms of iron include the mineral goethite, which is brown or yellow in colour, and for this reason the soils containing hydrated iron oxides range in colour from warm brown to yellow.

On poorly drained sites such as are found in some valley bottoms, swamps and other areas of impeded drainage, the shortage of air oxygen brings about reducing conditions (Chapter 4) and the formation of compounds which give predominantly grey, often mottled, colours to soils.

The porosity and structure of the subsoil should be examined carefully. Subsoil structures of loams and clays are normally blocky, sometimes prismatic or columnar, and may be compound, i.e. with aggregates breaking to smaller ones (Chapter 2). In some well-weathered tropical soils, macrostructure may be weak but microstructure may be well developed, with the formation of sand and silt-sized aggregates. In sandy soils structure may be weaker and more difficult to discern.

The consistency can be estimated by feeling the soil to establish whether it is friable or firm when moist, or loose or hard when dry. The wet soil should be worked in the hand and between the fingers to establish to what extent it is plastic and sticky (Chapter 2).

In many tropical soils the subsoil is the horizon with the highest clay content. It is thus common to find that the subsoil has a heavier texture than the topsoil. This is often due to the loss of some clay from the topsoil and its accumulation in the subsoil, described above as a horizon of illuviation. Evidence of clay movement can be seen in some subsoils in the form of thin clay skins, sometimes shiny, coating the outsides of structural units and some soil pores and channels. Clay skins are discussed further in Chapter 7.

The subsoils of some tropical soils contain stones and gravel. The gravel fraction may consist simply of quartz gravel but in some areas soils contain large amounts of ironstone nodules and concretions, sometimes referred to as 'pea gravel'. Soils developed in alluvium may contain rounded, water-worn pebbles.

The subsoil is often described as the 'B horizon'. The characteristic properties of B horizons are first that there has been some degree of alteration and weathering of the parent material, as shown by the formation of structure, liberation of oxides, formation of clays and other processes, and secondly that there has been some illuviation from the horizon above.

The weathered substratum

The boundary between the topsoil and the subsoil below is often easy to see because of the darker colour of the topsoil, caused by humus, and the fact that the topsoil is often more porous and easy to work, whereas the subsoil may be more compact. The boundary between the subsoil and the weathered substratum may however be much more diffuse and gradual, and difficult to see. In some cases one can discern a transition zone between the two horizons. The differences between the two vary according to the type of soil and the nature of the underlying weathered substratum, which may or may not be similar to the material from which the subsoil is assumed to have developed.

Fig. 5.3 A soil scientist looking at a soil pit dug in a black cracking clay (Vertisol).

The weathered substratum, often also called the 'C horizon', is the zone where the soil material has been less changed and modified by soil-forming processes than it has in the subsoil. If the soil is a sedentary soil derived from the rock below, then the weathered substratum contains the weathered products of that rock and perhaps also some relatively unweathered fragments or core stones, or some rock that has been softened but still retains something of its original structure and arrangement.

In the field, the weathered substratum or C horizon can be distinguished by changes in soil texture, structure and consistency (Chapter 2). It may be lighter coloured, or more varied in colour because of mottles or traces or fragments of soft weathered rock. Sometimes the C horizon is less compact than the subsoil, and has a lower bulk density.

It is difficult to generalise about the nature of the weathered substratum because it varies very greatly with the nature of the soil material, while in some cases the C horizon is not the parent material from which the solum (the topsoil and subsoil) is assumed to have developed. In such cases there may sometimes be a relatively abrupt transition between the solum and the material below it.

Types of soil horizons and their symbols

In the above description, the general terms of topsoil, subsoil and weathered substratum were used in conjunction with the pedological terms A horizon, B horizon, and C horizon. The A B C horizon designations are convenient to indicate the morphology of a profile and its assumed genesis. An A B C profile is a normal profile in which all three major horizons are represented, but an A C profile is a young profile that has not developed a B horizon (Chapter 7), while a truncated profile that has lost its A horizon owing to accelerated erosion might be referred to as a B C profile.

Pedologists frequently subdivide the main horizons indicated above according to subordinate characteristics and may also add suffixes to designate certain properties. A B2t horizon, for example, is a horizon in which B qualities are well developed (B2) with illuvial clay (t).

The technical definitions of the various pedological horizons recognised and the commonly employed suffixes are as follows.

O horizons organic horizons formed or forming at the surface of a mineral soils characterised by fresh or decomposing organic matter. These include what is often termed the leaf litter layer, and are subdivided:

O1 organic horizons in which the forms of the organic matter are visible to the naked eye;

O2 organic horizons in which the original form of the organic matter is no longer discernible.

A horizons mineral horizons including (1) horizons of organic matter accumulation at or near the soil surface, and (2) horizons that have lost clay, iron, aluminium or humus with a relative enrichment of quartz silt and sand. The latter include the pale, bleached lower topsoils of podsols. A horizons are also subdivided:

A1 horizons formed or forming at or near the soil surface, characterised mainly by an accumulation of humified organic matter more or less intimately associated with the mineral fraction of the soil;

A2 mineral soil horizons characterised by losses of clay, iron, aluminium or humus with a relative enrichment of quartz or other resistant minerals of silt and sand size;

A3 transitional horizons between the A and B horizons, dominated by the characteristics of an overlying A1 or A2 but possessing some properties of an underlying B.

B horizons horizons in which the dominant characteristics related to one or more of the following: (1) an enrichment, due to illuviation, of clay, iron, aluminium or humus, alone or in combination; (2) a relative enrichment in clay or in sesquioxides, alone or mixed, resulting from processes other than the solution or removal of carbonates or of more soluble salts; (3) the formation of sesquioxide coatings giving colours markedly more deep, bright or red than adjacent horizons in the same sequence but without apparent iron illuviation; (4) in cases where processes (1) to (3) cannot be recognised, an

Fig. 5.4 A shallow soil profile over base-rich rock, ICRISAT station, Hyderabad, India. A shallow, dark, clay-textured topsoil overlies a C horizon containing weathered rock material and boulders of relatively unweathered rock.

Fig. 5.5 Examining a shallow Vertisol profile developed in clay alluvium near Lake Chad, West Africa. The heavy textured alluvium in which the profile is developed is here less than 1.5 m thick and there is a very sharp transition to contrasting light-coloured sand below. In the lower photograph deeper soil observations are being made with a soil auger.

alteration of original material destroying the original rock structure, forming silicate clay minerals and/or liberating oxides, and giving rise to granular, blocky or prismatic structures if soil textures are such that moisture changes produce a change in soil volume.
B1 horizons transitional between the B and the overlying A1 or A2 but showing dominantly B horizon properties.
B2 that part of the B horizon in which B horizon properties are best developed, and lacking transitional properties to the A above or the C or R horizons below.
B3 transition horizon to the C or R horizon below, clearly showing the distinctive properties of the B above but also exhibiting some characteristics of the C or R horizon below.

C horizons mineral horizons or layers, excluding the underlying rock, which are either similar to, or different from, the material from which the B horizon is assumed to have formed, relatively little affected by pedogenetic processes and lacking the distinctive properties of the A and B horizons, but including material modified by any of the following: (1) weathering outside the zone of maximum biological activity, (2) reversible cementation, (3) gleying, (4) accumulation of calcium or magnesium carbonates or of salts more soluble than these, or cementation by them, or (4) cementation by siliceous substances soluble in alkali, or by iron or by silica.
IIC the use of the prefix Roman II indicates that the C horizon is not thought to be genetically related to the B horizon above. Roman numerals can also be used for B horizons (e.g. IIB, IIIB) to indicate an apparent discontinuity and change of assumed parent material within the solum.

R a continuous rocky substratum, such as granite, schist, sandstone or limestone, assumed related to the horizons above.
IIR an R horizon assumed to be unrelated to the soil horizons found above it.

Suffixes

The horizon designations given above reflect both the observations made on the soil and the opinion of the person describing the soil in question on the assumed genesis of the horizons described. Further information as to the dominant characteristics of these horizons is given by suffixes, of which some of the more commonly used are the following:

b buried horizon
ca accumulation of calcium carbonates
cs accumulation of calcium sulphate
cn accumulation of ironstone concretions or nodules ('pea gravel')
g showing considerable gleying
m cemented (massive)
p modified and mixed by cultivation (plough or cultivation layer)
ir with illuvial iron
t with illuvial clay

The soil catena

Once a particular local soil profile has been examined in some detail, it should be compared and contrasted with other soil profiles in different topographic positions, such as those on hilltops, upper slope sites, lower slope sites and valley bottoms, and also on particularly steep sites. It can also be instructive to compare soils over different parent materials and under different types of vegetation and land use, since these factors also influence profile development and soil properties (Chapters 6 and 7).

In many tropical situations there is a sequence of soils related to relief, so that different profiles are encountered in a more or less regular manner as one moves down from a summit site, down the slope, and to the valley bottom. Such a recurring sequence of soils related to relief is known as a 'soil catena', a term coined by a British pedologist, G. Milne, working in East Africa in the 1930s. *Catena* is Latin for chain, and the soils succeed each other like links on a chain.

Soil catenas are very varied, but in essence the sequence of soils they include usually consists of soils which differ

1 in their topographic position, and hence in their drainage and in their relationship with each other; and
2 in their local soil parent material, which is related in turn to geomorphology and relief.

Differences in soil drainage are often reflected in the colour of the subsoil. The subsoil of well-drained upland soils may be red or warm brown, owing to the presence of haematite. Subsoil colours of middle and lower slope soils may change to brown or yellow. This colour change reflects the fact that the soils remain moist longer, and that the iron oxides are more or less hydrated. Finally, in poorly drained valley bottom sites, intermittent or permanent saturation with water results in the reduction of iron III to the iron II form, with the production of soil mottles and grey, blue or green soil colours which reflect the poor drainage.

In addition to differences between the soils of a catena due to drainage, further differences are related to the fact that whereas upland soils are frequently sedentary, i.e. developed in the weathered products of the rock below, soils in lower slopes may be developed in local colluvium, and those on river terraces and in valley bottoms in alluvium. Colluvium is the material that has moved down the slope under the action of gravity, whereas alluvium is material transported, sorted and then deposited by wind or water.

Soil catenas are discussed in more detail in Chapter 6, in relation to relief as a soil-forming factor.

Soil description and soil sampling

Soil description

Detailed descriptions of soils in the field are routinely made by soil surveyors and agronomists, and the methods of detailed soil description are now fairly standardised. They are well set out in such publications as the FAO booklet *Guidelines for Soil Description*, which draws heavily on the older *US Soil Survey Manual (Handbook 18)*. The fact that methods and terminology for detailed soil description have been more or less internationally standardised greatly simplifies the task of using and understanding detailed soil descriptions made world wide, and in correlating and classifying the soils described.

Despite this standardisation of terminology, considerable differences in both the length and content of soil descriptions are found in published reports and articles. These differences are due both to variations in approach and emphasis between countries, and to the amount of detail that a particular worker finds it necessary to go into for a specific purpose. In some cases soil descriptions are much shorter than the manuals referred to suggest, depending partly on the purpose for which the soil is being described. For some purposes very brief descriptions are sufficient. Soil profile descriptions are sometimes accompanied by profile diagrams.

A very full description of a soil profile would ideally include the following:

1 Information on the site (locality, altitude, topographic position, gradient, vegetation or land use, and climate).
2 General information about the soil (parent rock, drainage, moisture status at sampling, depth of water table, presence of stones or rock outcrops, evidence of erosion, presence of salts on the surface).
3 A very brief description of the soil profile, drawing attention to its main features.
4 A detailed horizon-by-horizon description of the soil which gives the following information in the order given:

(a) Horizon symbol, e.g. A1, B1, B2, B3, C, R.
(b) Depth of horizon in centimetres, e.g. 15–33 cm.
(c) Colour according to the standard Munsell colour chart or equivalent for the moist soil and for the air-dry soil. The colour is designated by the use of the standard colour term given in the colour chart, followed by the symbols it gives for hue (colour), value (lightness or darkness) and chroma (colour purity and strength); e.g. light yellowish-brown (10 YR 6/4) moist, very pale brown (10 YR 7/4) dry. When the moisture state of the soil is not specified, then the colour is for the moist soil.
(d) Abundance, size, contrast, prominence and colour of mottles; e.g. with numerous, fine and medium, distinct, strong reddish-brown mottles.
(e) Texture, e.g. clay, sandy loam. Where the field texture as estimated by feel (the apparent texture) is markedly different from the texture as determined in the laboratory (the true texture), then this can be indicated by a phrase such as 'clay that feels like a sandy loam'.
(f) Grade, size and type of structure, e.g. weak, fine, subangular blocky.
(g) Consistency when wet, moist and dry, as indicated in Table 2.1; e.g. very sticky wet, very firm moist, very hard dry. When the moisture state is not specified, then the consistency given is for the moist soil.
(h) Presence of cutans, pressure faces, or slickensides. For cutans the abundance (rare, occasional frequent, very frequent), thickness and location is usually indicated; e.g. frequent thin to moderately thick clay films in pores and on ped faces.
(i) Extent, if any, of cementation or induration.
(j) Porosity (frequency, size, continuity, orientation, distribution and morphology); e.g. occasional very fine, continuous, vertical, internal and external, interstitial and tubular pores, or, for a less detailed description, frequent fine vesicular pores.
(k) Presence of boulders (over 20 cm in diameter), stones (2–20 cm), rock fragments or gravel (2–20 mm), indicating their abundance, shape and nature; e.g. with occasional angular quartz stones and very abundant nodular ironstone gravel.
(l) Presence of mineral nodules, such as calcium carbonate concretions:
(m) Presence of a compacted horizon, or pan.
(n) Presence of carbonates or soluble salts.
(o) Frequency and size of roots and their distribution, e.g. with few fine and very fine roots, mostly between peds.
(p) Nature of the boundary with the horizon below (diffuse, gradual or abrupt, and whether horizontal or wavy).
(q) the pH of the horizon.

It should be emphasised that this list is sometimes shortened when a more concise description is considered adequate.
5 Observations that interpret the above information, suggesting, for example, the agricultural value or limitations of the soil described, and its possible vocation (pasture, tree crops, intensive vegetable production, etc.).

For further information on standard soil description terminology, readers are referred to the two publications mentioned above (p. 56).

Soil sampling

An old saying, 'The analysis is no better than the sample', draws attention to the fact that great care must be taken to sample carefully, and to ensure that the sample taken is typical of what it is meant to represent. Nowadays laboratory procedures for soil analysis are relatively standardised and accurate, and the main source of error often comes from the initial sampling of the soil.

Methods of sampling depend on the purpose for which the sample is required. Broadly speaking there are two main types of sampling:

1 The pedological sampling of a soil profile horizon by horizon to characterise and classify the soil profile.
2 The taking of agricultural samples from the surface layer, or from both the surface and deeper layers, of a field or area considered fairly homogenous in order to make analyses designed to indicate soil fertility and productivity and to lead to fertiliser recommendations.

Profile samples should always be taken from a freshly dug soil face, and not from one that has been exposed for a long period; nor should they be taken from an old road cutting or other such exposure. When a soil profile is sampled it is first necessary to divide it into suitable sampling horizons according to its morphology as seen in the soil profile pit, the soil normally being described at the same time. Beginning at the bottom of the pit, to avoid contamination by soil falling during sampling, each horizon is successively sampled and a sufficient quantity put into a canvas or plastic bag or other container. The container is identified both inside and on the outside by a number, or by details of the soil and the horizon depth. For deep subsoil horizons sampling may be along a vertical band between horizon boundaries taken in such a way as to represent the whole of the horizon more or less equally, while for shallower topsoil horizons a horizontal band may be taken. The quantity taken depends on the analyses to be performed, but normally 1–2 kg is sufficient except for very gravelly soils for which a larger sample might be needed in order to yield enough fine earth after the coarse fraction has been removed by passing through a 2 mm sieve.

For some purposes, such as the laboratory determination of bulk density and of water suction curves (Chapter 4), or the preparation of soil thin sections for microscope investigation, additional undisturbed soil samples are required.

The second type of sampling concerns the agricultural sampling of a field or part of it that is considered fairly uniform as regards soil characteristics and history of cultivation. A pit need not necessarily be dug, since the sampling does not take into account the natural soil horizons. Instead, core samples are taken to a standard depth, commonly to 15 cm, at a number of points scattered more or less at random in the sampling area. The core samples are then thoroughly mixed to form a bulk sample which is subsampled to give a final sample for analysis. It is important to have a sufficient number of core samples to ensure that when mixed they approach the average of the field. It can be shown statistically that 20–40 core samples are needed to do this adequately, and that the number of core samples needed is the same for a small area as a large area if both are seemingly fairly uniform, since the variations averaged usually occur over small distances, depending on the soil property concerned. Many agronomists make the mistake of taking too few core samples for bulking. It is also important to make sure that the individual core samples are taken to exactly the standard depth, and are of uniform width from the surface to the lower limit, so as not to weight one part of the soil layer sampled against another.

In some cases, additional samples might be taken for special purposes or investigations, and the methods of sampling might be different from those outlined above. For example, to assess the extent of nematode infestation in sandy soils in which it is thought that this might be important, the soil has to be sampled moist and received by the laboratory within five days.

6 Soil formation: parent material and relief

Soil-forming factors

A soil-forming factor may be thought of as a factor that influences soil formation and properties, and thus serves to cause differences between soils.

In an area of similar climate and vegetation, differences between soils may be found due to differences in the local geology, according, for example, to whether the soil is developed over sandstone, granite or limestone. Since the parent material causes differences between soils, parent material is regarded as a soil-forming factor. Further differences might be found between well-drained upland soils, and soils of lower slopes and valley bottoms, according to their position in the landscape, so that relief is also a soil-forming factor.

On a broader scale, the soils of dry areas are found to be very different from those of wet areas, and the soils under forest to have different properties from those under savanna. Climate and vegetation, causing differences between soils, are thus also soil-forming factors.

Even where all the soil-forming factors outlined above are broadly similar, differences may be found between young soils that have not developed very much and older soils with a more marked, advanced degree of profile development. In this sense, time, as shown by degree of profile development, is also a soil-forming factor.

The five environmental soil-forming factors commonly recognised and studied by pedologists are therefore as follows:

1 Parent material – the material from or in which the soil is developed.
2 Climate, past and present.
3 Vegetation and soil fauna.
4 Relief.
5 Time.

Our understanding of soil formation has been greatly helped by studies aimed at showing the effects of each of these environmental soil-forming factors on soil-forming processes and properties. Pedologists assume that

1 where two soil profiles are very similar, then all the factors of soil formation are also similar;
2 that where soil profiles differ, this is due to a difference in one or more of the soil-forming factors.

However, two further factors have to be kept in mind. First, the soil-forming factors cannot, strictly speaking, be treated as independent variables. To take an obvious example, vegetation is closely connected with climate, but is also influenced by soil characteristics, including those that modify climatic factors such as available moisture. Thus the reciprocal relations between climate, vegetation and soil are complex. To take another example, relief is influenced by parent rock, as when resistant rocks stand up as hills.

Secondly, the environmental soil-forming factors resulted in the formation of natural, virgin soils, but in practice most agricultural soils have been influenced to various extents by human activity, both directly and also indirectly by influencing the soil-forming factors. Human activity may thus be considered a sixth soil-forming factor. People influence soils directly by cultivating them, draining or flooding them, and by adding amendments and fertilisers. They influence them indirectly by removing or modifying the natural vegetation, and by changing microclimates and the water regimes of catchments.

An important further consideration is that even the soil that has been modified by human activities continues to be influenced by the environmental factors of soil formation. In heavy rainfall areas, for example, soils are strongly leached and acid in reaction (Chapter 7). The acidity can be temporarily corrected by appropriate management practices, but the tendency of the climate to cause leaching remains. Thus an appreciation of how a soil has been formed may help in understanding how it is evolving at present.

Parent material as a soil-forming factor

Parent rock and parent material

Any material from which soil develops, and in which a soil profile begins to form, is parent material. Many soils develop in the products of rock weathering. In these cases the underlying rock from which the soil is presumed to have formed is known as the parent rock, but the weathered rock material in and from which the soil forms is the immediate parent material. Soils can also develop in deposits of alluvium (material transported, sorted and deposited by wind or water), in deposits of colluvium (material that moves downslope under the action of gravity) or even in organic matter accumulations such as peat. The term 'soil parent material' is therefore a wide one, while the term 'parent rock' is more restricted, referring to the rocks in whose decomposition products the soil forms.

A soil that is thought to have derived ultimately from the rock below it is regarded as a 'sedentary' or 'residual' soil, but soils development in transported materials are 'non-sedentary'.

The soil profile shown in Fig. 5.1 is a sedentary one, and there are gradual transitions from topsoil to underlying rock. Each horizon is related to the one below it. In some soil profiles there is a sharp transition or discontinuity between the upper parts of the profile and the weathered substratum below it, or within the subsoil. Such might be the case in a soil developed in shallow alluvium or colluvium: the upper horizons might be developed in shallow transported material, the alluvium or colluvium, but the lower horizon may be derived from the underlying rock. In the case of soils developed in alluvium it may be necessary to distinguish between soils developed in young, recent alluvium, such as that which forms the present river banks and floodplain, and older alluvium, perhaps further away from the river and at a higher level, which has been subjected to a greater degree of weathering and profile development than the young alluvium.

The study of soil formation therefore requires a knowledge both of the rocks that form the earth's crust and which weather to become soil parent materials, and of landscape-forming processes that have carved out the present-day relief and resulted in the formation of secondary deposits which have subsequently also become parent materials. The first is the province of geology, and the second of physical geology or geomorphology.

The main classes of rocks

The rocks that make up the earth's crust have been formed in three main ways. Molten rock material coming up from the interior of the earth cools and solidifies at or near the surface. These primary rocks, from which the other classes of rocks are ultimately derived, are known as **igneous rocks**, from the Latin word for fire, *ignis*, suggesting the heat which kept them molten before they cooled or hardened. When rocks are exposed at the surface they are liable to be weathered and worn down, and the weathered material to be transported and then deposited as a sediment. Rocks derived from these sediments are known as **sedimentary rocks**. In some cases previously existing rocks, either igneous or sedimentary, are fundamentally changed by heat and pressure to form **metamorphic rocks**. These three groups of rocks, and some of the important minerals which they contain, are briefly considered in the following sections.

Igneous rocks

Igneous rocks are derived from molten material, known as magma, pushed up from the lower regions of the earth's crust to positions nearer the surface, or even at the surface, where the magma cools and solidifies. As the magma cools, crystals of different minerals, depending on the composition of the magma, are formed. Most igneous rocks consist of a number of different mineral crystals which interlock to form the rock. The slower the cooling of the magma, the larger are these crystals likely to be, for slow cooling gives them time to grow before the rock sets hard. In contrast, sudden cooling is likely to give a rock with very small crystals of individual minerals, too small perhaps to be seen with the naked eye, so that the rock appears to be uniform and without crystals until it is looked at through a microscope.

Rocks vary greatly in the types of mineral they contain, and in the size, arrangement and chemical composition of the minerals.

A mineral may be defined as a naturally occurring substance with a fairly uniform chemical composition within certain limits, and with a regular, well-defined structure. This structure shows itself in the formation of a crystal, though the exact chemical composition of a mineral can vary somewhat owing to the substitution of one element for another in the crystal structure.

Silicate minerals The majority of rock-forming minerals are silicates, i.e. combinations of silicon and oxygen with other elements. The basic structural unit of silicate minerals is a silica tetrahedron, or pyramid, a four-sided unit in which one relatively small silicon atom at the centre is bonded to four much larger oxygen atoms which surround it and which form the four corners of the regular tetrahedron (Fig. 6.1).

Silicate minerals can be classified according to the way these fundamental building units have linked up to form the mineral. In the first group are minerals containing silica tetrahedra that are not linked to each other directly, but are connected by other elements. In the second group, the silica tetrahedra are linked directly by sharing oxygen atoms to form groups or rings. In a third group they form chains of tetrahedra, and in a fourth they form flat sheets. Finally, the silica tetrahedra may share all their oxygen atoms with adjoining tetrahedra to give a three-dimensional lattice forming a block structure.

Since the structure of silicate minerals helps us to understand how they weather to form soil material, these types of structure are discussed briefly, with examples of common rock-forming minerals found in each structural group.

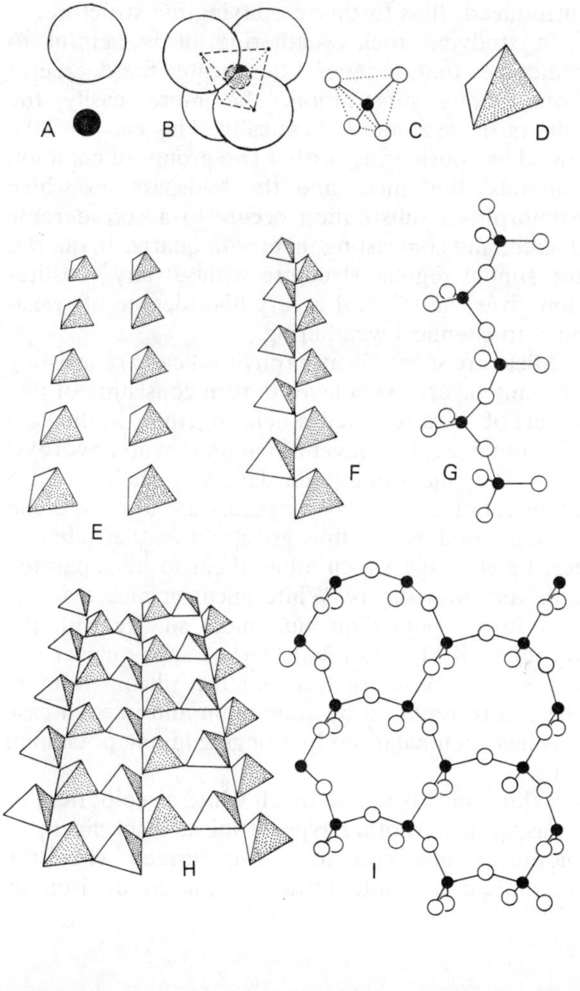

Fig. 6.1 The structure of silicate minerals. At top left A an oxygen atom (circle) is shown in proportion to an atom of silicon (black). B is a diagram representing a silica tetrahedron, showing the four relatively large oxygen atoms almost hiding the smaller central silicon atom. The dashed lines join the centres of the four oxygen atoms, indicating the four-sided pyramid-like figure, the tetrahedron, which they form. C is a structural diagram of the tetrahedron indicating the relative positions of the atoms it contains. At D the silica molecule is represented pictorially as a pyramid.

In the centre the pictorial representation at the left, E, shows isolated silica tetrahedra as they occur in the nesosilicates. In this group of silicates the tetrahedra are linked by metallic cations (not shown). In the central drawing F the silica tetrahedra are linked to each other by sharing oxygen atoms: each tetrahedron shares two of its oxygen atoms with its two neighbours to form a chain. This structure is shown diagrammatically on the right: single and double chains of tetrahedra are characteristic of the inosilicates.

In the phyllosilicates the tetrahedra share three of their four oxygen atoms to link up to form sheets. This is shown pictorially at H and diagrammatically at I. Micas and clays are examples of phyllosilicates.

Nesosilicates (from the Greek word *nesos* island) are those silicates in which the silica tetrahedra remain separate from each other with no shared oxygen atoms. They are linked instead by intermediate cations. The olivine group of minerals form the most important example of nesosilicates. In the olivines the silica tetrahedra are linked by divalent magnesium (Mg^{2+}) and iron (Fe^{2+}) ions. Olivine is relatively easily weathered. The linking ferrous and magnesium cations are exposed at the edge of the crystal and are oxidised or hydrated, and the mineral disintegrates to release iron, magnesium and silica.

Olivine is an example of a group of minerals termed ferromagnesian because of their content of iron and magnesium. They include minerals of other structural groups than the nesosilicates, are usually dark in colour, weather relatively easily, and release iron and magnesium when they do so.

In a second group of silicate minerals the silica tetrahedra form groups or rings, while in an important third group the silica tetrahedra join to form long chains. These are the **inosilicates** (from *inos*, fibre). In the pyroxene family they form a single chain, while in the amphibole family they form double chains which run the length of the crystal. In both families the silica chains are linked to each other by Ca^{2+} and Mg^{2+} ions. Thus pyroxenes and amphiboles (of which the most important mineral is hornblende) are chemically calcium magnesium silicates, but because of isomorphous substitution (discussed further below), other cations, such as Fe^{2+}, Mn^{2+} or Na^{+} can be introduced, giving rise to different minerals within the family. The pyroxenes and amphiboles are often dark-coloured minerals which weather relatively easily and, as would be expected from their composition, release calcium and magnesium to the soil.

Phyllosilicates (from *phyllon*, a leaf) are a very important group of silicate minerals in which the silica tetrahedra share three of their oxygen atoms to form flat sheets of tetrahedra. The sheets are tied to those above and below by linking cations. Phyllosilicates include the micas, discussed below, and the clay minerals (Chapter 2).

Isomorphous substitution is an important process in mineral formation which modifies the basic structural patterns discussed above because of the substitution of one ion in the crystal lattice by another of approximately the same size. In practice the substitute ion may be a little larger or smaller than the one replaced, and have a different valency. When an ion of a slightly different size is fitted into the crystal structure, the structure is no longer as regular as it was, and the extra stresses and strains caused by this irregularity may weaken the lattice and cause it to decompose or weather more rapidly than a more regular one. Secondly, when a cation replaces another it may have a different valency, as when aluminium (three valencies) replaces silicon (four valencies), or magnesium (two valencies) replaces aluminium. In each case there will be a negative charge of the oxygen left unsatisfied. In order to restore electrical neutrality, an additional cation, such as sodium or potassium, has to be introduced, thus further modifying the structure.

In studying rock weathering, it is helpful to remember that in general the greater the degree of isomorphous substitution, the more easily the mineral is weathered chemically. This can be illustrated by considering further two groups of common minerals, the micas and the feldspars, in which isomorphous substitution occurs to a considerable degree, and contrasting them with quartz. In quartz, the simple regular structure without any substitution gives the mineral a very high degree of resistance to chemical weathering.

Micas are sheet silicates (phyllosilicates) consisting of many layers, each layer in turn consisting of two sheets of silica tetrahedra held together, in the case of white mica, by a layer of aluminium and hydroxyl ions. The silica–aluminium layers are held to each other relatively weakly by potassium ions, and can be separated easily, thus giving micas their characteristic cleavage which allow them to be separated into very thin sheets. White mica, or muscovite, is therefore a potassium aluminium silicate with the formula $KAl_2(AlSi_3O_{10})(OH)_2$. One-quarter of the silica tetrahedra has had the silicon atom at the centre replaced by aluminium and in each case this has been balanced by bringing in one potassium ion.

White mica is not so much white as colourless or transparent. Another type of mica, black mica, or biotite, is also common. It is formed when the aluminium of white mica is replaced by iron or magnesium, so that $(Mg)_3$ or $(Fe)_3$ replaces Al_2 in the formula for white mica given above. Both types of mica contain potassium, but biotite also contains magnesium and iron, so that it is the richer of the two as regards plant nutrients. As would be expected from the greater degree of isomorphous substitution, biotite is also considerably more easily weathered than the relatively resistant muscovite. Biotite is one of the ferromagnesian minerals referred to above, and has the dark colour typical of these minerals.

Feldspars are examples of **tectosilicates**. They have a three-dimensional block structure in which all the silica tetrahedra share oxygen atoms with all adjacent tetrahedra. Each oxygen atom is therefore shared by two tetrahedra. A proportion of the central silicon ions (with four valencies) have been replaced by aluminium ions (three valencies), giving an excess negative charge which has then to be satisfied by another cation. When the negative charge is satisfied by an introduced potassium ion, a potassium feldspar is formed. Often the negative charges are satisfied by a combination of potassium and sodium (the alkali feldspars) or of sodium and calcium (the plagioclase feldspars). Of the group, the calcium feldspars, with the greatest degree of isomorphous substitution, are the most easily weathered, and the potassium feldspars the least easily weathered.

Quartz has the same three-dimensional block structure as the feldspars but there is no isomorphous substitution. The mineral consists of silica tetrahedra and nothing else, and has the formula SiO_2. Because there are no basic cations to be attacked by the forces of chemical weathering, and because there has been no isomorphous substitution to weaken the simple regular structure of the mineral, quartz is extremely resistant to chemical weathering. In neutral and acid soils, chemical weathering of quartz is negligible and it merely breaks down physically to smaller particles. Because of its resistant nature, it survives in soils after less resistant minerals have been broken down, and accumulates. It is frequently the main or almost the only constituent of the sand fraction, particularly in highly weathered soils. Quartz is a hard, usually colourless, transparent mineral occurring both as discrete crystals in rocks, as in granite, and also as veins of quartz which may have been injected into previous rock formations.

Minerals and soil formation There are two important properties of a mineral in regard to soil formation:

1 How easily does it weather?
2 What elements does it release when it does weather?

A knowledge of the structure and composition of minerals helps in understanding these two properties. Quartz is the extreme example in each case. It is very resistant to weathering, and liberates nothing except silica on decomposition.

The chemical composition of igneous rocks The structural classification of minerals outlined above is linked to some extent with their chemical composition. Chemical composition can be used by itself, without reference to structure, to classify minerals and rocks in a different way.

Chemically, rocks are divided into the more basic rocks, those containing relatively high proportions of the metallic cations, and the more acid rocks in which an increasing proportion of the total composition is silica. It should be noted that the term 'acid' used in connection with rock composition refers simply to the content of silica, and not to acidity (pH).

All silicate minerals contain some silica, and almost all except quartz some basic cations. The usual division is as follows:

Acid rocks	Over 66% silica
Intermediate rocks	52–66% silica
Basic rocks	45–52% silica
Ultrabasic rocks	Under 45% silica

Rocks are often subdivided further into those in which the alkaline cations of potassium and sodium predominate, and those more common rocks in which the calcic elements, calcium and magnesium, are the more important.

An acid rock, defined as having over 66% silica, is a rock in which there is so much silica that not all of it is combined with other elements. The surplus silica forms quartz, which consists, as seen above, of silica alone. Thus acid rocks always contain some quartz. Granites, a very important and widespread group of acid rocks, consist essentially of quartz and feldspar with some mica (white mica or biotite) and/or hornblende. The granites consisting of quartz, feldspar and white mica are relatively pale, but other granites occur containing some dark biotite or hornblende, examples of ferromagnesian minerals.

The more acid the rock, the greater the proportion of quartz and feldspars it contains. The more basic the rock, the greater the proportion of dark (mafic) ferromagnesian minerals it is likely to have. Basic and ultrabasic rocks commonly consist of olivine and pyroxene, and are dark in colour. Intermediate rocks commonly contain hornblende, biotite and plagioclase feldspar. Acid rocks contain quartz and feldspars (often mainly orthoclase) as their main minerals, sometimes with some biotite or hornblende, and are generally pale in colour.

Igneous rocks are divided into broad groups or families according to their grain size, which reflects the rate of initial cooling of the magma, and their composition, as set out in Table 6.1. Thus basic rocks are collectively termed basalts when fine grained, but dolerites when medium grained and gabbros when coarse grained. Of the acid rocks, rhyolites are very fine grained, but granites, with the same broad range of composition, are medium to coarse-grained rocks.

The weathering of igneous rocks In physical weathering, rocks are merely broken down by mechanical means to smaller and smaller particles without their chemical composition being altered, though this fragmentation may then make chemical attack easier by increasing the surface area. In chemical weathering, the rock is decomposed chemically to liberate its constituent elements and ions. These are either removed, or left to accumulate in the soil, in which case they may recombine to form new substances, as in the formation of clays.

Mechanical weathering predominates in drier climates, where changes in temperature may cause shrinking and expansion and the cracking and breaking up of rocks, and where physical forces such as winds and water break up rock particles by rolling, impact, and so on.

In most of the tropics where rainfall is moderate to heavy, the weathering of rocks is mainly chemical, the main agent being percolating soil water. Rain falling through the atmosphere dissolves very small quantities of such constituents as nitrogen oxides, sulphur dioxide, oxygen, carbon dioxide, and perhaps traces of ammonia, sodium chloride, and other compounds. The first two give nitrous, nitric and sulphurous acids, but of particular interest and importance are the oxygen and carbon dioxide which attack weathering rock by oxidation and the formation of carbonates. In addition, various organic acids and decomposition products derived from the decay of plant and animal material in the soil are added to the soil water as it seeps downwards through the soil profile.

The percolating soil water is thus a weak but complex solution. This solution attacks exposed rock fragments in the soil and penetrates into massive rocks along cracks and joints. The mechanisms of chemical weathering include the following:

1 Solution.

2 Hydration – the combining of constituents with water.

Table 6.1 The more important rock families classified according to chemical composition and grain size. Families in brackets are the alkaline families, rich in potassium and sodium; the remainder are calc-alkaline rocks, with calcium and magnesium as the dominant bases.

	Ultrabasic	Basic	Intermediate	Acid
Fine grained		Basalts	Andesites (Trachytes)	Rhyolites Felsites
Fine or medium grained		Porphyrites Dolerites	Micro-diorites	Micro-granites Granite porphyries
Medium or coarse grained	Peridotes Serpentines	Gabbros (Syenites)	Diorites	Granites Granodiorites (Alkali-granites)

3 Hydrolysis – the substitution of constituent ions by hydrogen ions.
4 Oxidation and reduction.
5 The formation of carbonates and bicarbonates.
6 Attack by acid and alkaline solutions.
7 The removal of the soluble products liberated.

Thus chemical weathering is a complex process, the details of which vary considerably according to the soils, rocks and climate involved.

In the chemical weathering of igneous rocks, one of the constituent minerals is often more easily attacked than the others. The softening and breaking down of this mineral may result in the disintegration of the rock and the separation of the remaining minerals which are then exposed to further attack. In the weathering of granites, for example, the feldspars are the first minerals to weather. The metallic bases in the feldspars may be removed and the remaining silica and alumina may then combine to form kaolin. The less easily weathered mica and the very resistant quartz may thus remain embedded in soft clay.

It is useful to put the common mineral constituents of igneous rocks in an order of weathering, as follows:

Olivine	Most easily weathered
Calcium feldspar	↓
Pyroxenes and amphiboles (hornblende)	
Sodium feldspar	
Black mica (biotite)	
Potassium feldspar	
White mica (muscovite)	
Quartz	Most resistant to weathering

Sedimentary rocks

When rocks are weathered, the sand, clay and other residual products may be deposited by wind and water to form sediments. These sediments are often size-sorted to some extent, since the load-carrying capacities of both wind and the water in streams and rivers vary according to the speed. As a result, sediments may consist mainly of clay-sized, silt-sized or of sand-sized particles, or of alternating beds of these. If weathering is rapid and the products accumulate near the place of origin, or if they are not much sorted in transit, then mixed sediments may be formed containing some rock particles not yet completely weathered. Often, however, sediments consist mainly or entirely of resistant residues that have already passed through a cycle of weathering. In highly weathered sediments, everything that can easily be broken down may already have decomposed, leaving only the most resistant constituents, particularly kaolinitic clay and quartz sand.

The sediments derived from weathering, transport and deposition form the materials from which sedimentary rocks are derived. The original deposits may be subjected to various processes which may compact and consolidate them. They may be compressed by the weight of sediments above, and any water contained (particularly by the clays) squeezed out; pore space is reduced and the rock is compacted and hardened. Chemical changes may also occur, largely through the seepage into the sediments of solutions, causing recrystallisation, cementation and other effects, including the formation within the beds of secondary minerals.

Coarse-textured sediments include conglomerates, in which gravels and pebbles are bound together by finer material, and breccias in which the stones are angular.

Medium-textured sediments consist mainly of sandstones in which resistant, well-sorted quartz sand may be cemented together by quartz or other cements, such as iron oxides. Where the sand fragments are more angular, less sorted, and contain some unweathered material, a greywacke is formed. Sandstones containing both feldspar and quartz grains are known as arkoses.

The fine-grained clay sediments form a range of clay sedimentary rocks depending on the type of clay, the content of other constituents, and the degree of alteration undergone by the original sediment. Compressed clay becomes a mudstone, but when secondary layers or laminae develop as a result of pressure, the mudstone merges into shale.

Other classes of sediments include chalk and limestone, siliceous deposits, rock salt, coal and various iron carbonate and oxide minerals, as well as the residual formations such as bauxite.

Sedimentary rocks as soil parent materials vary widely according to the two criteria discussed above for igneous rocks, namely the ease with which they weather, and the substances released on weathering. In general, they are poorer parent materials than igneous rocks. Sedimentary rocks as a class are much less likely to contain crystalline silicates which can weather to give nutrients to the soil than are igneous rocks. This is because they have often already passed through a cycle of weathering and may consist wholly or mainly of resistant minerals. These resistant residues may weather only slowly, and release little of value to the plant. Sandstones may be taken as an example: a sandstone consisting of only quartz will be very resistant and contain little of value to the plant apart perhaps from minor quantities of impurities, but a sandstone containing both quartz and feldspar (an arkose) will be a better soil-forming material because the feldspar can weather to give clay and potassium or calcium. Little-altered muds and clays will, on weathering, break down into the original clay constituents.

Metamorphic rocks

For a sediment to become a sedimentary rock, certain relatively minor changes take place. When a pre-existing rock of any type, or a sediment, is more fundamentally changed, or metamorphosed, a third class of rocks is formed, the metamorphic rocks. The forces involved are those of heat, pressure and chemical change. The changes often take place at some depth within the earth's crust, as when folding and other earth movements subject rocks to great heat and to tremendous pressures, or when existing rocks are changed by coming into contact with intrusions of molten magma.

The result of rock metamorphism is often the formation of a new structure, and a re-orientation of at least some of the components into secondary crystalline minerals, though the size of the crystals varies from very fine (discernible only with a microscope) to coarse.

Very characteristic of metamorphic rocks is an orientation of the constituents to give a banded effect. These bands or layers must not be confused with the depositional beds of sediments, but are secondary layers formed at right-angles to the main source of pressure. They may be very fine, as in phyllites, or coarse, as in schists. Where plate-like silicates, such as micas, occur, the rock may split along cleavage planes. Coarser-grained crystalline metamorphic rocks showing only a rough banding are grouped as gneisses.

Schists and gneisses are broad groups of metamorphic rocks defined according to their mode of origin and structure. They are usually further defined according to their most characteristic mineral. Thus both biotite schist and sericite schist are schists in which, secondary biotite or sericite (small white mica flakes), respectively, has formed, often mainly between the characteristic layers of the schist, which may thus show some cleavage. A quartz schist is a quartz-rich schist derived from an original deposit containing much quartz sand. Similarly, gneisses are named after a characteristic constituent, e.g. hornblende gneiss, a coarsely banded crystalline rock containing hornblende in addition to feldspar and other constituents. The bands and layers formed during the metamorphosis may be folded or contorted because of crustal movements and pressures.

From the point of view of weathering and soil formation, metamorphic rocks show a very wide range of characteristics. Some, such as quartz schist derived from quartz sand, may be nearly sterile and break down to give little except quartz. Others, in which secondary micas have formed, may be somewhat richer, while certain basic metamorphic rocks resemble basic igneous rocks and can give rise to very fertile soils.

The rates at which metamorphic rocks weather shows a similar broad range, with the poorer, quartz-rich rocks often weathering very slowly, and those containing basic minerals often being subject to a much more rapid decomposition if the climate is favourable.

Soils in relation to relief and geomorphology

The soil catena

Soils of the tropics frequently occur in a well-defined and fairly regular sequence of soils related to relief – the soil catena.

Since a catena is related to relief, with a regular sequence of different soils encountered as one goes from hilltop to valley bottom, it seems logical to assume that relief is the main soil-forming factor responsible for differentiating the soils. In practice, the influences of relief and geomorphology are closely linked, and the sequence of soils is usually due to two main factors:

1 Differences in parent material at different positions in the catena.
2 Differences in site, i.e. in slope, drainage and position in relation to other soils.

These two groups of factors will now be briefly considered in turn.

Parent material differences within the catena Even if the underlying geology is uniform, differences in the immediate parent materials of soils occur within a catena. The underlying rock is usually the parent material of only some of the soils of the catena, generally the upland ones. Other soils in the catena may have developed in transported materials which, though they may have been ultimately derived from the underlying rock material, now consist of weathered or partly weathered materials which have been transported and perhaps sorted to some extent while being moved. These transported materials are mainly either colluvial in origin, i.e. consisting of material that has moved down the slope under the action of gravity, or alluvial. Alluvial deposits may include older terrace deposits at various heights above the present floodplain in addition to more recent alluvium associated with present-day streams and rivers.

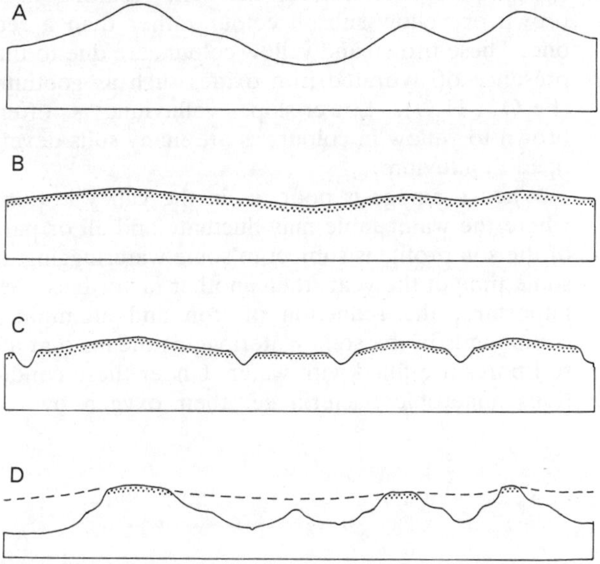

Fig. 6.2 Stages in the dissection of a landscape. In diagram A a hilly landscape is undergoing erosion which, if sufficiently prolonged, will reduce it to a near flat, peneplain stage as shown in B. Old, highly weathered soils (shaded), often rich in iron, mantle the peneplain.

In the third diagram C the rivers on the peneplain have been rejuvenated, usually by a rise in the land relative to sea level, and new valleys are rapidly being cut into the old landscape.

Diagram D shows a later stage at which the original peneplain surface (broken line) is represented only by the flattish summit areas of occasional hills. The highly weathered peneplain soils (shaded) survive on these summits, though perhaps in modified form, and constitute the oldest soil materials in the area. The valley slopes and floors are associated with younger soils, though these younger soils may incorporate some of the older material, as when former ironstone crusts break up to give ironstone gravel in soils lower down the slope.

In young, dissected landscapes a relatively simple sequence of soils is often found consisting of sedentary soils on upland sites, soils developed in local colluvium on lower slopes, and soils developed in alluvium in valley bottoms. In areas of more complicated geomorphology, the soil catenas encountered become correspondingly more varied.

In some tropical landscapes, the oldest soils are found on flat or nearly flat summit areas which represent the remains of former peneplains. A peneplain is a flattish area formed by erosion over a long period. When rejuvenated, as by unlift, erosion starts afresh, cutting into the old surface until a later stage may be reached when the former peneplain surface occurs only as flattish hill summits (Fig. 6.2). Peneplains remnants at several levels are found in West Africa and in northern South America, some of the older surfaces being associated with bauxite deposits resulting from very long periods of weathering. Other surfaces may be associated with ironstone crusts (Chapter 7), or with deep, highly weathered soils.

Catenas become further diversified when they include shallow soils on very steep slopes, or soils developed in old river terraces at various levels above the present floodplain.

The influence of site and drainage on the catena Differences between soils of a catena are related not only to differences in local parent materials, but also to differences in position and in drainage. One of the most obvious signs of drainage differences is the colour of the subsoil.

Upland, well-drained soils in many tropical areas are frequently somewhat reddish in colour – often reddish-brown or brownish-red, occasionally a very bright red or purplish-red. These red colours are due to the presence in the soil of haematite, Fe_2O_3, a non-hydrated iron oxide. The red colour thus indicates good drainage, and the presence of some haematite, though only a very small amount is sufficient to give the soil its colour. A red colour does not of itself indicate a fertile soil.

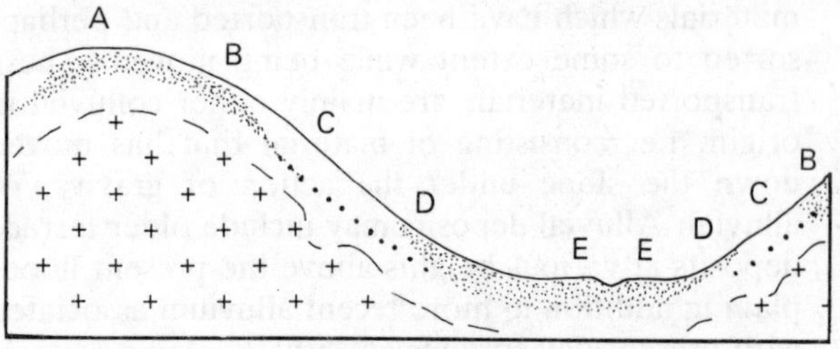

Fig. 6.3 Diagram of a simple forest catena over granite. Gravelly red and brown upland sedentary soils are found at A and B on summits and upper slopes. These grade downslope into yellow–brown sandy light clay soils developed in middle slope colluvium at C and into yellow–brown sandy loam and loamy sand colluvium at D. The soils developed in local granite-derived alluvium at E are mostly grey to white sands with subordinate areas of gritty or sandy grey clays.

In middle and lower slope positions in the catena, soil drainage is usually a little slower than on upland sites. These soils may remain moist longer than upland ones because of moisture seeping downslope. This is reflected in an increasing degree of hydration of the iron in the soil, resulting in a brown or yellow subsoil colour rather than a red one. These brown and yellow colours are due to the presence of hydrated iron oxides such as goethite ($Fe_2O_3 \cdot H_2O$). Lower-slope colluvium is often brown to yellow in colour, as are many soils developed in alluvium.

When drainage is poor, as in the valley bottom where the water table may fluctuate and all or part of the soil profile is subject to some waterlogging at some time of the year, then another factor becomes important: the reduction of iron and aluminium compounds in the soil. Waterlogging means that all soil pores are filled with water. Under these conditions anaerobic bacteria get their oxygen by re-

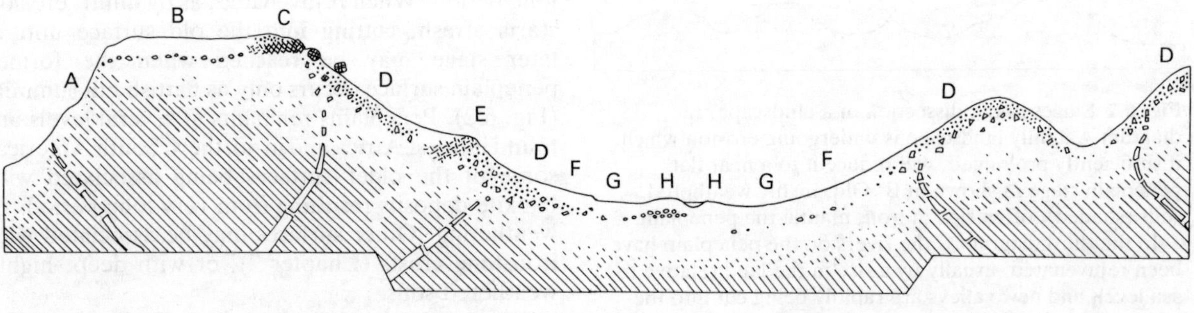

Fig. 6.4 Diagram of a tropical soil catena over schist.
A Shallow, immature soils of steep slopes.
B Upland 'drift' soils on older peneplain surfaces.
C Shallow gravelly soils over ironstone.
D Gravelly upland sedentary soils.
E 'Shoulder' where an indurated horizon comes to the surface.
F Soils developed in local light clay colluvium.
G Soils developed in clay alluvium of valley bottoms.
H Soils developed in sandy alluvium of levees.
Source: Ahn, 1970.

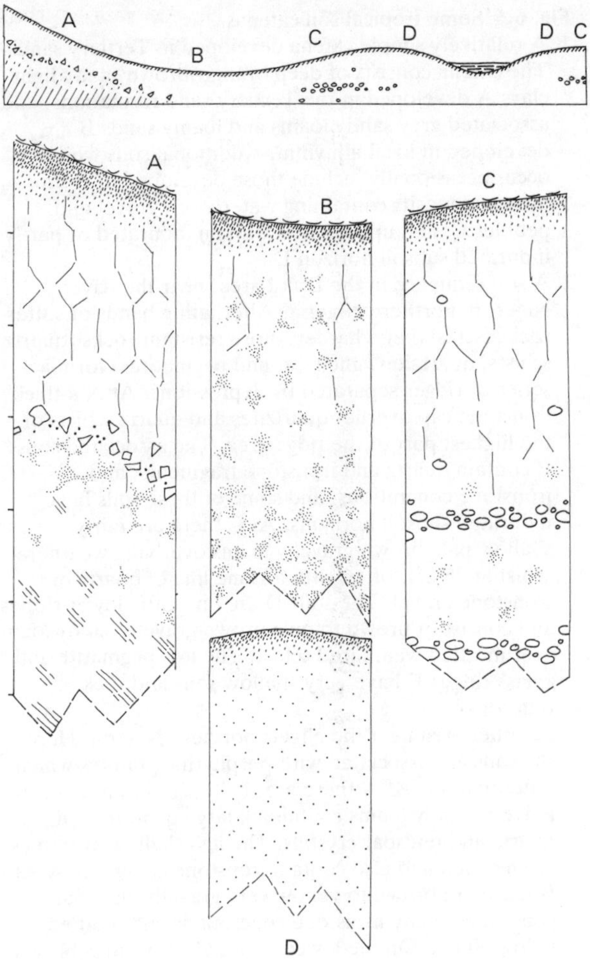

Fig. 6.5 Profile diagrams of soils developed in colluvium and alluvium. The profile on the upper left is developed in about 1 m of yellow–brown light clay colluvium which is separated by a discontinuity (often marked by a stone line, gravel, slight induration or mottling) from the weathered substratum below. The weathered substratum may contain traces or fragments of decomposing underlying rock.

The central profile is developed in clay alluvium of the valley bottom. Poor drainage is indicated by the grey colour and presence of mottles in the subsoil. The soil on the upper right is developed in somewhat deeper and better drained yellow–brown light clay alluvium containing pebble beds.

The profile at lower left is developed in yellow–brown alluvial sand of the levee.

ducing oxygen-containing compounds in the soil.

The colours typical of reducing conditions in soils are the bluish-greys, greenish-greys and neutral greys typical of poorly drained soils. These colours suggest prolonged waterlogging. Where the waterlogging is intermittent or seasonal, as in that part of the profile where the water table fluctuates, then mottled colours are likely to be produced. A mottle is a spot or small area of different colour from the surrounding material.

The process giving reduced colours and mottles to the soil is known as gleying, and poorly drained soils thus affected are known as gley soils.

Differences in drainage conditions are thus responsible for the colour catenas seen in many areas of the tropics. The subsoil colour changes from red or reddish-brown on well-drained upland sites, to warm brown, brown, yellowish-brown and brownish-yellow down the valley slope. Poorly drained valley-bottom soils are often mottled or show grey colours due to gleying.

To some extent drainage and soil colour are related to climate and rainfall. The drier the area and the more seasonal the rainfall, the greater the likely extent of reddish soils. As rainfall increases and dry seasons become shorter, so the red soils become less extensive and the brown and yellow soils more widespread.

Valley-bottom soils and soils of floodplains may receive valuable additions of material in solution and suspension when the associated stream or river floods. In low-lying areas of prolonged waterlogging, the normally rapid decomposition of plant residues is likely to be slowed down for lack of oxygen, and a thicker, more humous topsoil than normal is formed. Where water is at or above the surface for long periods, little-altered plant remains may accumulate to form organic materials such as peats.

The natural vegetation associated with poorly drained lower-slope and valley-bottom sites may differ markedly from that of associated upland soils, so that soils, relief and vegetation are closely connected. In forest areas, the normal forest of well-drained sites may be replaced in the poorly drained areas by swamp forest. In savanna areas, fringing or gallery forest may occur along watercourses. Vegetation and its influence on soils is discussed further in Chapter 7.

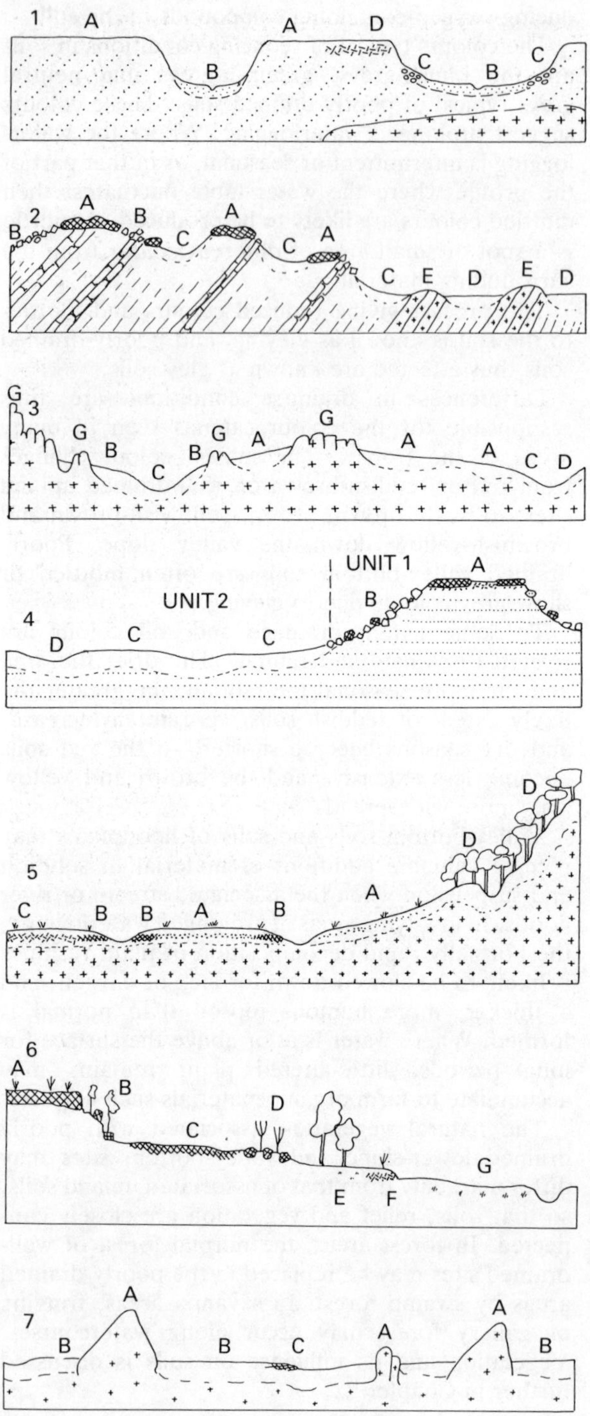

Fig. 6.6 Some tropical soil catenas.

1 A relatively simple catena developed in Tertiary sands. The catena consists of deep orange–brown sandy light clays A developed in the Tertiary sand formations, and associated grey sandy loams and loamy sands B developed in local alluvium. Additional soils which occur occasionally include those developed in river terrace deposits containing water-worn gravel and pebbles C, and upland soils with an indurated or partly indurated subsoil horizon D.

2 A soil sequence in the Duku area, near the River Niger, in northern Nigeria. Alternating bands of softer rocks (schists) and harder, more resistant rocks (quartz schists, quartzites, gneisses, and pegmatites) form a series of ridges separated by depressions. At A a thick ironstone cap overlies quartzites and quartz schists of the highest part of the ridge area. The associated soils B contain quartz and ironstone fragments, and ironstone concretions, and some of these soils have indurated subsoil horizons. At C there are fairly shallow pale brown gravelly loams overlying weathered schist at 75 cm, often with a stone line of quartz or ironstone gravel. The soils D are on gentle lower slopes and consist of brown to pale brown gravelly sandy loam over mottled weathered schist. The low pegmatite and gneiss ridges E have very shallow soils and rock outcrops.

3 Another area near the Niger, northern Nigeria. Here the soils are associated with porphyritic granites which outcrop at G. At A there are shallow red to brown very gravelly sandy loams containing large quantities of quartz and feldspar crystals. The less shallow examples are mottled and also contain ironstone concretions. At B there are brown to yellow very gravelly clay loams containing many ironstone concretions and mottled below 60 cm. On the lower slopes C are yellow–brown sandy clay loams containing much quartz and feldspar gravel. These soils grade into small areas of bottom soils D, mainly pale brown to grey–brown non-gravelly loamy sands over clay loam.

4 An area over Nupe sandstones in Niger Province, northern Nigeria, divided into two distinct soil-topographical units. Unit 1 consists of sandstone and ironstone capped flat-topped hills with steep scarps rising out of unit 2, an extensive, gently undulating sandstone peneplain with few drainage channels.

In unit 1 the flat summits at A are capped with iron impregnated hard sandstone or with ironstone outcrops over weathering sandstone. The very steep slopes B consist of a very shallow (less than 30 cm thick) layer of loamy sand over rubble of sandstone and ironstone.

In unit 2, most of the area is covered by orange–brown to red deep sandy clays C with duller, generally brown mottled sandy clay loams in the depressions D.

5 An area in Sardauna Province, north-east Nigeria, developed over acid crystalline rocks of the basement complex, mostly feldspathic granite gneisses. Forest still survives on some of the steep hills but various types of derived savanna now occupy the lower, more gently sloping areas. In these lower areas a sandy clay with a mottled, concretionary and gravelly subsoil is found at A. Where the mottled subsoil becomes exposed at the surface, at the valley edges B, it hardens to a compact ironpan. In other cases C the soil has a partly indurated subsoil. Forest survives on the generally relatively deep sandy clays and loams of the steep hill sides and summits D.

6 Ironpan formation associated with terraces of the river Milo, Guinea. The highest area shown A is 25 m above the river floodplain and is capped with an ironstone sheet 5 m thick which is breaking into blocks at the edge B. The terrace C, 7 m above the floodplain, also has an ironstone crust, partly covered by deposits from the neighbouring areas, and broken up at the edge D. The terrace below this, E, is 3 m above the present floodplain and induration is confined to its edge F, where it is due to lateral movement of iron and is only partial. The recent alluvium of the present floodplain G shows well-developed mottling and the formation of ironstone concretions, but is not indurated.

7 A diagram representing areas adjacent to the Red and White Volta rivers, Haute Volta, with granite domes and wall-like inselbergs A rising steeply from a very flat peneplain with 1 to 2% slopes. The soils of the peneplain B are generally very shallow and poorly drained, with occasional indurated horizons outcropping, as at C. Sources: see Ahn, 1970.

Fig. 6.7 A shallow lithosol, consisting of about 20 cm of soil over rock.

Fig. 6.8 A shallow soil over shale. Flattish fragments of partially weathered shale can be seen below 20–25 cm.

7 Soil formation: climate, vegetation, soil fauna and soil age

The effects of climate

Rain is the main agent of chemical weathering and is also, in the tropics, the main climatic control of vegetation and agriculture. Climate therefore has a double effect on the soil. It has direct effects on weathering, leaching and soil profile development, and it has important indirect effects on soils through its influence on the nature and mass of the vegetation which in turn affect soil characteristics.

Since most chemical reactions are speeded up by higher temperatures and wetter conditions, both the chemical weathering of the original rock and the subsequent weathering and modification of soil material in the profile are more intense and rapid in the wet tropics than in drier and cooler areas of the world, and the processes involved may not be quite the same. This results, in each case, in soils broadly related to climate. It also means that statements that may be true of high latitude soils do not necessarily apply equally well elsewhere. Since most standard soil textbooks have been written by and for workers in the temperate areas, such textbooks must be used with caution by those interested in using and understanding soils of the tropics. It is precisely in those characteristics of the soil that reflect the influence of climate, and of vegetation and fauna, that tropical soils are most likely to differ from those of other latitudes.

If the soils of the tropics are considered in a general way, they are seen to fall into broad belts related to climate and vegetation. Within these belts soils have certain common characteristics related to rainfall, to temperature, and to the nature of the plant cover, though some soils also show the influence of earlier climates. Within any broad climate–vegetation belt, soil differences occur on a more local scale owing to differences in geology and parent material, and to relief and position (Chapter 6), and to other factors that differentiate soils and their degree of development.

Weathering

Once a mineral breaks down chemically in the soil, the elements liberated may

1 accumulate in the soil with little further change,
2 combine to form new compounds in the soil, or
3 be removed out of the soil profile.

In nature, weathering of minerals and the removal or translocation of weathered products go on side by side, but it is convenient to consider the two processes separately.

The translocation of the products of weathering may involve their movement within the soil profile, as when materials move from an eluviated A horizon to accumulate in an illuvial B horizon, or their removal right out of the soil profile, as when substances are leached out in solution or in suspension and find their way to lower positions in the landscape or are removed in streams and rivers.

An analysis of stream or river water effectively indicates what is being removed at that time from the catchment area. Such studies indicate that, given sufficient rainfall, mobile elements liberated on weathering are removed in a fairly well-defined order according to their relative solubilities.

Very soluble substances and mobile ions such as chlorides and sulphates are quickly removed from weathering rock zones in the first stages of weathering. In a second stage, calcium, sodium, magnesium and potassium are progressively removed. In a third, more advanced, stage the combined silica in the silicates is attacked and removed. This includes the silica in the lattices of clay minerals, though it does not include to any extent the silica in quartz which, because of its simple, regular structure and lack of isomorphous substitution (Chapter 6), is extremely resistant to chemical weathering in acid and neutral environments. The attack and breakdown of clay minerals results in some removal of silica in relation to alumina, and thus the progressive enrichment of the remaining material in alumina.

The most resistant compounds to weathering and removal are the sesquioxides of iron and aluminium, Fe_2O_3 and Al_2O_3. Sesquioxides are 'one-and-a-half' compounds in which two of the metals combine with three of oxygen as in the two examples just given.

Young soils that are little weathered may contain in their silt and sand fractions minerals that are relatively easily decomposed and which will disappear if the soil is subjected to further weathering. These minerals associated with little-weathered soils, including soils of arid regions where lack of water inhibits chemical weathering, include gypsum and calcite, and the silicate minerals which are relatively easily decomposed such as olivine, hornblende, biotite (black mica), and calcium and sodium feldspars (Chapter 6). Young soils containing easily weathered silicate minerals in their silt and sand-sized fractions therefore possess nutrient reserves locked up in these minerals which can be expected to be released progressively as the minerals weather. The smaller, silt-sized minerals, because of their greater surface area, normally weather more easily and quickly than mineral particles that are sand-sized or larger. The presence of easily weatherable minerals in the silt fraction effectively indicates a young, little-weathered soil.

In more weathered soils, the easily weathered minerals are absent and the more resistant silicate minerals are progressively attacked. These include potassium feldspars and the considerably more resistant muscovite, and the 2:1 layer clay minerals, including illite and montmorillonite. The silica layers are progressively removed from these clay minerals, usually leaving behind the alumina (gibbsite) material which thus accumulates. Well-weathered soils do not have in their sand and silt fractions the easily weathered silicate minerals characteristic of young, little-weathered soils, and thus lack this reserve of potential nutrients.

In very advanced weathering stages the minerals still present in the soil are likely to be dominantly quartz (in sand-sized and larger particles), kaolinitic clay, gibbsite and alumina (Al_2O_3), iron oxides (haematite, Fe_2O_3) and hydrated iron oxides (goethite), and perhaps small amounts, depending on the nature of the parent rock, of very resistant silicate minerals such as zircon and anatase. A characteristic of very highly weathered tropical soils is therefore that they lack easily weatherable minerals and consist mainly of

1 kaolinitic clay,
2 quartz sand, and
3 iron and aluminium oxides.

Soils of this broad type have in the past been referred to as latosols, a term first suggested by the American soil scientist, Charles Kellogg, but the word now lacks precision, and the current approach is to classify highly weathered soils according to the absence or presence of an oxic horizon, as discussed further in Chapter 8. An oxic horizon shows features characteristic of a high degree of weathering, namely the presence of kaolinitic clays (as evidenced by a low cation exchange capacity), iron and aluminium oxides (usually associated with a stable structure and micro-aggregation and low amounts of water-dispersible clay), and the absence or near absence of weatherable minerals.

Since progressive weathering of the clay fraction of the soil results in the preferential removal of silica relative to alumina, and the relative accumulation of the latter, weathering modifies the silica/alumina ratio of the clay fraction. This is 2 for unweathered kaolin, but drops as weathering results in silica removal and alumina enrichment. The silica/alumina ratio of the clay fraction thus serves as an index of weathering, and has been used, notably by French pedologists, to classify soils.

Another effect of soil weathering is that silt-sized particles, because of their greater surface area, are attacked preferentially to sand-sized particles. Many young soils, or little-weathered soil horizons, are initially high in silt-sized particles. The C horizons of many sedentary tropical soils over phyllite, schist and other metamorphic rocks are often very high in silt-sized particles of sericite (white mica) and other easily weathered minerals, which are progressively weathered to clay. A feature of highly weathered soils is therefore that silt-sized soil particles are progressively eliminated to give a soil that consists mainly of clay and sand. This fact was made use of by Belgian pedologists working in the Congo who used the silt/clay ratio as an index of weathering. The silt/clay ratio is useful for this purpose where soils over similar parent materials are compared, but the initial texture of the freshly weathered C horizon material depends very much on the nature of the parent material which may be high or low in silt-sized particles or particles that readily weather to silt.

Translocation

Translocations within the soil profile are related to the movement of water in the profile, which are in turn related to rainfall and to factors that affect water penetration into the soil. Substances move in solution, as when they are leached (see below), but they also move in suspension.

The movement of very fine soil particles in percolating soil water constitutes one of the mechanisms of horizon formation and profile development. Small particles in suspension may be moved from a soil horizon laterally, downslope, to enrich other areas or be lost in drainage waters. Progressive removal of fine material, as when trees are uprooted, has been advanced as one of the processes responsible for the formation of very gravelly horizons in many tropical soils. Alternatively, the fine material may move downwards in the soil profile from the A horizon to the B horizon.

Material that moves in percolating soil water is most likely to be very small in diameter, and in practice translocation affects the fine clay fraction (less than 0.2 μm) more than either the rest of the clay fraction or larger size classes. The result can be the removal of fine clay particles from the upper soil horizons to a lower horizon, usually the subsoil, and the enrichment of this horizon by illuviation.

A subsoil horizon that has been enriched with clay, if it meets certain quantitative criteria, is termed an 'argillic horizon'. Argillic horizons form one of the main diagnostic horizons of current soil classification (Chapter 8).

The diagnostic feature of an argillic horizon, as defined in the US Soil Taxonomy (Chapter 8), is that it contains illuvial-layer lattice clays, i.e. clays that have moved into the horizon from an overlying eluvial horizon, and more fine clay and more total clay than the eluvial horizon. For the purposes of quantitative definition there must be a minimum increase in clay content relative to the overlying eluvial horizon, and within a distance of 30 cm. This minimum increase is defined in terms of the amount of clay in the overlying eluvial horizon. If the eluvial horizon is low in clay (less than 15%), there must be an absolute increase of 3%. If the eluvial horizon has 15–40% clay, then the illuvial, argillic horizon must have 1.2 times as much clay as in the overlying horizon. If the eluvial horizon has over 40% clay, then the absolute increase must be at least 8%. In all cases the ratio of fine clay to total clay is expected to be at least one-third higher in the argillic horizon than in the overlying eluvial horizon from which the clay is assumed to have come.

However, an increase of clay between the A and the B horizons is not of itself conclusive evidence of clay movement. Fortunately, microscope studies of thin sections of soils show that where clay has been illuviated, characteristic micromorphological features can be discerned that are due to the manner in which the illuviated clay is deposited. The fine clay particles moving down in soil water are deposited as soon as the water in which they move is absorbed by the surrounding soil. The flattish clay flakes tend to be deposited or plastered in a regular manner on the surfaces of pores and peds which absorb water. In the deposit the clay flakes are more or less aligned in the same direction, and since they are crystals with well-defined optical properties, the deposit itself has certain optical properties when the flakes are aligned in a sufficiently regular manner. Under polarised light (when the nicols of a petrographic microscope are crossed), non-oriented clay is more or less opaque, but the oriented clay of the deposits transmits polarised light and appears luminous (birefringent). In this way, clay material with a regular preferred orientation can be distinguished from other material in which the arrangement of the clay flakes is haphazard.

Clay films plastered on ped faces and the walls of soil pores and fissures are referred to as 'clay skins'. When sufficiently well developed they can be seen with the naked eye as shining films. Their identification as oriented clay films produced by illuviation is made certain by using a microscope and polarised light, as described above.

An additional diagnostic criterion for argillic horizons other than those dominated by 2:1 lattice clays is, therefore, that if peds (Chapter 2) are present in the soil, illuviated clay skins should be present, at least in the lower part of the horizon, on vertical and horizontal ped faces and also in some soil pores, or that oriented clay should occupy at least 1% of the cross-section.

The relationship between the presence of an argillic horizon and soil genesis has been the subject of discussion. An argillic horizon is evidence of clay movement downwards, and this movement is associated with translocation due to percolating soil water. Argillic horizons therefore form where conditions are wet enough for a long enough period to give rise to significant movement. What is not clear is how long the process takes. It was originally thought that clay translocation is a very slow process and that clay skins are evidence of movement over long periods of time, during which conditions remained stable. It has now been shown that at least some forms of clay skins can form rapidly in agricultural soils as a result of cultivation.

As a feature of soil morphology and a tool in soil classification, argillic horizons have been emphasised as being relatively easy to identify and as separating out soils with a certain degree of profile development from other, usually pedogenetically younger, soils lacking the argillic horizon feature. Further weathering of an argillic horizon would be expected to convert it to an oxic one.

Rainfall and leaching

When rain falls, it may

1 evaporate from the surface of the ground or from the vegetation, or
2 run off the ground surface and down the slope, to be lost as runoff, or
3 infiltrate into the soil and move down the soil profile.

The proportions of the rain which evaporate, run off or infiltrate into the soil depend on a number of factors. These include, for example, the amount and the intensity of the rain, the amount and type of the vegetation cover, the condition and the nature of the soil (whether wet or dry, compact or porous, permeable or impermeable), the slope of the ground and the temperature of the soil.

Of the rain that enters the soil, some moves towards the soil surface and is evaporated, some is absorbed by plant roots and transpired through their leaves, and some may move downwards through the soil profile. The subject of water retention and storage by soils is of the greatest practical importance to agriculture and is discussed in Chapter 4. Here, however, we are concerned with the effect of rain and soil water on soil development, and particularly whether the soil is likely to receive more water than it can hold with the result that there is an excess which percolates through and out of the profile to cause leaching.

Leaching means 'washing', and the washing or leaching caused by excess moisture moving down the soil profile occurs essentially when the soil has received more water than it can hold against gravity. For leaching to occur, a soil profile or relevant soil horizon must first receive enough water to bring it to field capacity (Chapter 4). Further water cannot, by definition, be held by the soil, and moves downwards. Where there is enough water, water moves through the soil profile and is drained away to lower areas of the landscape.

Leaching occurs therefore when there is a temporary excess of water above what the profile can store at field capacity. It may occur relatively seldom, perhaps only after particularly heavy falls of rain, or only in the wettest few weeks of the year.

Thorough-leaching involves the movement of substances in solution and perhaps in suspension right out of the soil profile. Heavy, concentrated rainfall in a short period is more likely to achieve this than the same amount of rain over a longer period.

In the study of soil–rainfall relationships it is helpful to assess the likelihood of leaching by comparing rainfall over a period with the amount likely to be lost from the soil by direct evaporation, and the amounts likely to be transpired by plants, the two processes together being referred to as 'evapotranspiration'. The potential evapotranspiration of an area refers to the amount of water that could be lost by these two processes together if it were available, and is related to the incoming energy available for water evaporation. Potential evapotranspiration is sometimes assumed to be similar to the amount of water that would be lost from an exposed water surface.

The comparison between total rainfall received in a period and the potential evapotranspiration gives a rough and ready indication of the likelihood of there being a surplus for soil leaching.

Soil characteristics that reflect rainfall are related to the movement of water in the soil profile, and in particular to whether there is a net movement of soluble substances upwards or downwards. Downward movement and leaching occur when there is an excess of soil water (see above), but where the soil profile or the upper part of it is moistened to field capacity or less, there is no movement of water downwards. Water may then rise by capillary attraction and evaporate from positions at or near the soil surface. When it evaporates, it leaves behind it all those substances dissolved in it, and these accumulate in the surface horizon, thus enriching it. In extreme cases salts may be deposited at the soil surface.

Water movement in the soil may therefore be responsible for a number of effects which include the following:

1 In areas where rainfall is less than evapotranspiration, there is probably a net movement upwards of dissolved substances which rise to the surface horizon and may be left behind there when water evaporates, thus enriching it. Net upward movement and accumulation of salts at the surface can be a feature of soils with aridic soil moisture regimes, defined (Chapter 1) as being dry in the control section for more than 180 cumulative days and moist for less than 90 consecutive days.

2 In areas of seasonal rainfall, with a dry season when rainfall is less or much less than potential evapotranspiration, and a wet season in which rainfall exceeds potential evapotranspiration, there is likely to be a movement of dissolved substances upwards in the dry season and a leaching downwards in the wet season. This leaching downwards may simply be from topsoil to subsoil, as when nitrates or sulphates are progressively moved downwards during the wet season, or it may, if the excess of rainfall is great enough, involve the movement of dissolved substances and mobile ions right out of the soil profile. In soils with an ustic soil moisture regime (Chapter 1), a marked dry season of 3–6 months may cause some net rise during the dry season which counteracts to some extent the downward movement of dissolved and mobile substances in the soil solution during the wet season. The extent of this movement will be related to the excess of rainfall over potential evapotranspiration in the wet season.

3 In wet climates where the total rainfall exceeds potential evapotranspiration, the extent of leaching downwards depends on the amount of rainfall in relation to potential evapotranspiration and its distribution. Leaching is favoured when the excess of rainfall over potential evapotranspiration is concentrated in a few very wet months during which the soil profile is first brought to field capacity and then subjected to through-leaching as successive rains fall on an already wet soil. The capillary rise of soil water to the surface and the compensating movement upwards of dissolved substances will be slight if dry seasons are short or absent. In udic soil moisture regimes (Chapter 1), the control section of the soil is dry for less than 90 consecutive days, but the extent of through-leaching varies considerably according to the amount and distribution of the wet season rainfall.

Leaching and soil reaction

Where leaching occurs, various cations and anions are moved downwards in the profile. Where thorough-leaching takes place, these ions are moved right out of the soil profile and are lost beyond recovery by plant roots. In general terms it has been found that there is a relationship in the soil leachate (the soil water leaching through the profile) between the total quantity of anions in the soil solution and the total quantity of cations. It has been assumed therefore that leaching of cations is easier when the anion supply is not limiting, and that conversely an addition of, for example, ammonium sulphate will, after plant absorption of ammonium ions, leave a net surplus of sulphate anions which will favour leaching of cations in the soil solution.

Whatever the exact mechanisms involved, it has been observed empirically that as soils become progressively more leached:

1 cations are progressively removed, but calcium followed by magnesium tend to be removed preferentially to other cations such as potassium;

2 the net result is a progressive desaturation of the exchange complex, since a lower quantity of total cations occupy the exchange positions (Chapter 2); and

3 the result of desaturation of the exchange complex is that the soil pH becomes lower, and the soil becomes progressively more acid.

Studies in the wet forest areas of many parts of the hot wet tropics have shown that it is the topsoil pH that mainly reflects the leaching effect of increasing rainfall, and that the soil saturation and pH at depth may be affected to a somewhat lesser extent.

It was observed in the forest areas of West Africa that in areas of moderate rainfall (up to 1500–1600 mm) which did not much exceed potential evapotranspiration, topsoil pH may often be in the range pH 6.0–7.0, but in heavy rainfall areas of 1750 mm and more, where total rainfall exceeded potential evapotranspiration, then the topsoil pH of most soils except those over particularly base-rich rocks became progressively more acid with increasing rainfall, and could fall to 4.0–5.0 as a result of leaching. This effect was particularly well marked where the rainfall formed a marked peak in a few months of the year. In the wettest forest areas of Ghana and the Ivory Coast there is a marked rainfall peak in May and June during which 1 m of rain may fall. Potential evaporatranspiration in the same period is unlikely to be more than 250 mm, thus giving a concentrated rainfall surplus during these two months which is thought to be responsible for the fact that the associated soils have topsoil pH values in the general range 4.0–5.0.

Percentage saturation of the subsoil

As discussed above, leaching resulting from excess rainfall appears to affect the upper soil horizons in particular and to lower both their percentage saturation and pH. For soil classification purposes, attention has been concentrated on the saturation of the subsoil rather than the topsoil, partly because the subsoil is less easily modified by cultural pratices.

The argillic horizon is a diagnostic feature of two widespread and important orders of soils in the US Soil Taxonomy (Chapter 8), the Alfisols and the Ultisols, which are separated according to whether the argillic horizon is more or less than 35% saturated at depth (usually 1.25 m below the upper boundary of the argillic horizon or 1.8 m below the soil surface). These two orders correspond closely to the Luvisols and Acrisols of the FAO/Unesco map legend, which also contain an argillic horizon, and which are also separated by their degree of saturation.

The Alfisols of the Soil Taxonomy (and the corresponding Luvisols of the FAO/Unesco map legend) are therefore those soils possessing an argillic horizon which has not been leached sufficiently to reduce the saturation of the lower part of the argillic horizon to less than 35%. Ultisols, on the other hand (and the corresponding Acrisols of the FAO/Unesco map legend), have, by definition, an argillic horizon with less than 35% saturation. The Ultisols therefore are associated with a greater degree of leaching and desaturation. Except in cases where this is the result of the action of a previous climate, Ultisols are normally associated therefore with heavier rainfalls and a greater degree of leaching and weathering than Alfisols, and with a correspondingly more acid reaction.

Soils of dry areas

In ustic and aridic moisture regimes where potential evaporatranspiration is greater than rainfall for a considerable part of the year and there is a net annual rainfall deficit, it is likely that there will be a net upward movement of substances in solution. The soil reaction can be expected to be neutral or alkaline and there may be accumulations of sodium or calcium compounds at the surface or in the subsoil.

An argillic horizon that is markedly affected by sodium is referred to in the US Soil Taxonomy, if it meets certain diagnostic criteria, as a **natric horizon**. In addition to the properties of an argillic horizon, a natric horizon normally has prisms or, more commonly, a columnar structure (Chapter 2) in some part of the horizon (usually the upper part) and must also have considerable exchangeable sodium. The minimum sodium content that must occur in some subhorizon within 40 cm of the upper natric horizon boundary is defined as that which gives at least 15% saturation of the cation exchange capacity, or a sodium absorption ratio of 13 or more. The sodium absorption ratio (SAR) is defined as half the exchangeable sodium divided by the square root of the sum of exchangeable calcium and magnesium (expressed in m.e./100 g) and is used as a measure of the extent to which sodium is dominant over other cations present.

Natric horizons are a feature of dry-area soils. Other diagnostic horizons associated with soils of dry areas are gypsic, petrogypsic, calcic, petrocalcic and salic.

A **gypsic horizon** is a horizon of enrichment with secondary sulphates, either non-cemented or weakly cemented. Diagnostic criteria in the US Soil Taxonomy (Chapter 8) require that it has at least 5% more gypsum than the C horizon and is at least 15 cm thick. A **petrogypsic horizon** is a gypsic horizon so strongly cemented that fragments do not slake in water and plant roots cannot enter. These horizons are normally associated with arid climates and parent materials rich in gypsum.

A **calcic horizon** has an accumulation of calcium carbonate or of calcium and magnesium carbonates. This accumulation can be in the topsoil, the B horizon or the C horizon. Calcic horizons are horizons of secondary carbonate enrichment, and in some forms secondary carbonates can be identified as pendants on pebbles or concretions, or occur as soft powdery forms. If a calcic horizon is so indurated or cemented that dry fragments other than pendants do not slake and fall apart in water, then the horizon is termed **petrocalcic**.

Salic horizons are horizons enriched with soluble salts. To qualify as a diagnostic salic horizon in the US Soil Taxonomy (Chapter 8), a horizon must have at least 2% of salts more soluble that gypsum and be at least 15 cm thick.

Because of the generally sparse vegetation of dry areas, arid-zone soils are typically very low in organic matter and have poorly defined A horizons. Lack of organic matter is related to poor topsoil structure. In general, weathering is slight, because of the lack of moisture, though some highly weathered soils occur in arid areas that have been weathered during an earlier period when the climate was wetter. There is a tendency for soluble salts to rise to the surface and for soluble substances to accumulate to form the types of horizon considered in the paragraphs above.

Because of the very limited possibilities of agriculture in these areas, apart from irrigated agriculture where water is available, arid-zone soils have not received the same attention as the soils of less dry areas. One noted authority on arid areas once wrote, 'In general, the more arid the climate, the less is the accuracy of the soils information'.

When irrigation is contemplated, then very careful soil assessments are usually made, particularly in regard to soil salinity and water quality, and the soil properties (especially structure, texture and consistency) that affect water movement through the profile. Soils that are sandy and extremely permeable may require too much water, but impermeable soils cannot easily be leached of any soluble salts they contain or which may build up during use. The classification of arid-area soils for irrigation purposes thus emphasises soils that are preferably initially very low in soluble salts or which can be reclaimed economically by the use of chemical amendments and leaching, and downgrades soils in which permeability and drainage are either excessively rapid or very slow.

Iron movement and accumulation: laterite

Weathering under hot, wet conditions tends to result in the accumulation of oxides of iron and aluminium in the soil. It is sometimes helpful to distinguish between

1 relative accumulation of sesquioxides, and
2 absolute accumulation.

Relative accumulation occurs when sesquioxides remain constant in amount but form an increasing proportion of the residual soil material owing to the progressive removal of other soil constituents, particularly of combined silica. Thus progressive weathering of kaolinitic clay removes silica and leaves behind alumina. Accumulation of the alumina in this instance is therefore relative.

Absolute accumulation of sesquioxides occurs when iron or aluminium oxides are mobilised and move into soil material which is thus enriched. Movements include the movement of sesquioxides downslope to lower soils, or from an upper soil horizon to a lower one in the same profile, thus giving an absolute accumulation since there are more total sesquioxides than formerly.

Under certain conditions, considered below, sesquioxides become soluble and mobile and move in percolating soil water, resulting in their accumulation and concentration in certain areas of the landscape or in certain horizons in the soil. The formation of sesquioxide-rich soil horizons which may later become hardened (indurated) owing to exposure is particularly characteristic of certain tropical soils and landscapes, particularly in parts of Brazil, West Africa, India, South-east and East Asia. Hardened horizons in turn affect relief, soil depth and agriculture.

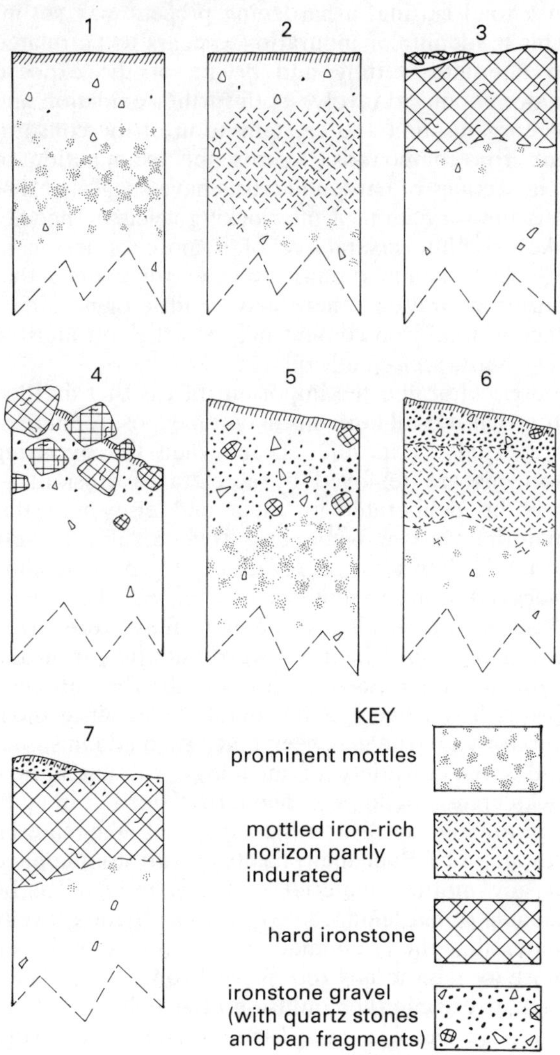

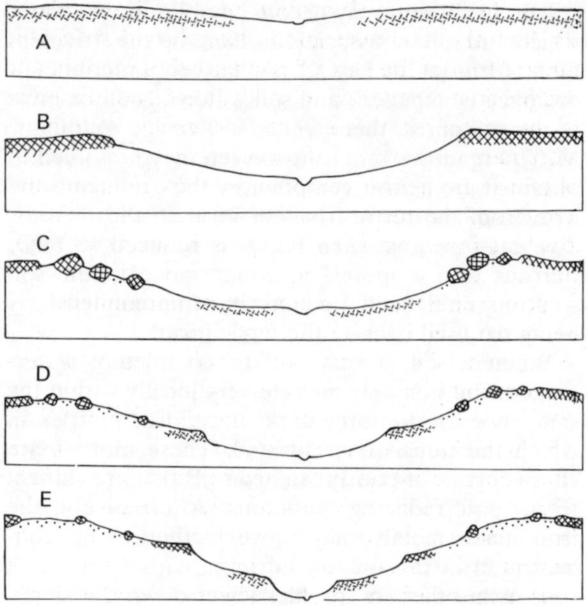

Fig. 7.1 Possible stages in the formation of ironstone crusts. The profile diagrams 1–7 illustrate possible developments in the soil profile, beginning with (1) the formation of a mottled subsoil zone which becomes better developed and partly indurated (2). Later this indurated horizon is exposed at the surface and hardens to a rock-like ironstone crust (3). The crust subsequently breaks up (4) to provide the gravel in a later profile (5) in which a very gravelly layer containing ironstone fragments and nodules overlies a mottled, iron-rich layer. Later the overlying gravel layer is progressively removed and the mottled iron-rich layer becomes indurated (6), eventually being exposed and more completely hardened to form an ironstone crust at the surface (7).

The possible landscape changes associated with the profile developments shown opposite are indicated in diagrams A–E. These show the formation of a mottled, partly indurated subsoil horizon in A, which subsequently remains as a crust capping the hills of diagram B. In C the iron derived from the break-up of this crust moves downslope in solution and contributes to the formation of a new mottled and partly indurated horizon at a lower level. Further dissection of the landscape by streams and rivers has exposed part of this lower indurated horizon as a 'shoulder' in diagram D, while diagram E shows successive 'shoulders' resulting from a repetition of these processes, the older, higher shoulders having been further indurated to become ironpan.

Iron in the soil may become mobile in two ways, which are often associated. Iron in the trivalent ferric form (as in Fe_2O_3) is relatively immobile. If drainage is impeded, and soil water fills all or most of the soil pores, there will be a shortage of soil air. Micro-organisms requiring oxygen may be forced to obtain it from iron compounds, thus reducing the iron from the ferric trivalent form to the ferrous, divalent form, as when Fe_2O_3 is reduced to FeO. Ferrous iron is mobile and may move in the soil solution until such time as it is immobilised by being oxidised back to the ferric form.

When a soil is very wet or completely waterlogged, soil iron may migrate very locally within the same horizon to form dark or reddish mottles in which the iron is concentrated. These mottles are characteristic of poorly drained soils that are subject to periodic reducing conditions. Alternatively, the iron made mobile may move further to be concentrated in certain soil horizons within the same soil, or in other soil profiles lower down the slope. In extreme cases the iron may be lost to streams and rivers.

The second cause of iron becoming mobile and capable of migrating is the mobilisation due to the action of organic compounds. These compounds include secretions from living roots and organic decomposition products.

The concentration of iron is often favoured in soils of flattish or gentle relief where there is a fluctuating water table or at least a subsoil horizon that is alternately saturated and dry. Alternate wetting and drying within the horizon appears to favour the mobilisation, local movement and concentration of iron to form well-defined iron-rich mottles. These mottles may grow and subsequently coalesce to give a matrix of reddish iron-rich material with paler areas or vesicles between. Thick mottled horizons are characteristic of many tropical soils, particularly in wet (udic) and seasonally wet (ustic) environments. They may occur in the subsoil and continue downwards to a C horizon containing weathering rock. In the latter case they may reflect the structure of the weathered rock.

The iron-rich mottles and mottled horizons probably also contain aluminium oxides, but these are colourless and their presence or movement is less easily noticed.

The hardening of iron-rich layers If the sesquioxide-rich mottled horizon in the subsoil remains protected by an upper soil layer, it may remain soft, or only slightly hardened, so that it can still be penetrated by plant roots and does not form a barrier. Such matrial can be dug out quite easily, though it may be firm or slightly hard when undisturbed, and prove brittle and break into angular fragments when dropped.

If the mottled, sesquioxide-rich horizon becomes exposed at the surface, as when the soil is truncated and the upper horizons removed, or when exposed in a road cutting, a hardening process may set in. This hardening or ınduration appears to be related to alternate wetting and drying of the exposed material, and to involve both further oxidation and a reorganisation and crystallisation of the constituent iron compounds. Microscope examination of thin sections of hardened crusts have suggested that it is the formation of interlocking acicular (needle-like) goethite crystals or of a more or less continuous crystalline phase of the iron oxides that causes induration. There may be little or no difference in total iron content between the soft and the very hard, rock-like forms.

Agriculturally, the important fact is that mottled iron-rich subsoil horizons may, on exposure, harden irreversibly. Exposure occurs when the overlying horizons are removed by accelerated erosion, or may be deliberate. In India and elsewhere this material has long been cut out while still relatively soft and then allowed to harden to form building blocks. A visitor to southern India in 1807, Buchanan, described these blocks and how they were made and used, and coined the word 'laterite', from the Latin word *later*, a brick, to describe the hardened iron-rich soil material he found there. Since then, the word 'laterite' has been much used and misused, and given a variety of meanings, so that it now lacks precise definition. The term 'laterite', in addition to being applied to the hard, rock-like material described by Buchanan, has been variously applied to any mottled material that was thought to be capable of hardening, to concretionary and gravelly material applied to unmade road surfaces, and even more loosely, to any red tropical soil.

To overcome the confusion attached to the term 'laterite', the US Soil Taxonomy uses the word 'plinthite', from the Greek word *plinthos*, a brick, to describe mottled iron-rich material that can become indurated, but plinthite remains difficult to define exactly. Since plinthite can be either soft or hard, these adjectives need to be used to describe it. Very hard, rock-like horizons, sheets or boulders may be described as 'ironstone'. Iron-rich pea gravel in the soil may be described as 'ironstone nodules' or 'concretions', the latter term implying that the nodules have been formed by accretion. Well-weathered, iron-rich and often red soils of the tropics have in the past been grouped as Latosols, but this term has now been superseded by more precisely defined ones, as considered further in Chapter 8.

Fig. 7.2 A soil with an indurated horizon. The upper layers of the soil consist of gravelly loam to clay loam, but at 55 cm there is a sharp transition to indurated mottled plinthite which forms a barrier to root penetration.

Fig. 7.3 A close-up of hard plinthite showing mottles and small holes (vesicles).

The importance and extent of indurated sheets and horizons The hardening of an iron-rich mottled soil layer to something resembling a sheet of rock represents a dramatic change which has no precise counterpart in the temperate regions of the world. The use of 'laterite' and the presence of sheets and boulders of hard ironstone in tropical landscapes has aroused enormous interest among travellers, soil scientists, agronomists and others. An exaggerated impression has been engendered as to the importance and extent of laterite in the tropics, to the point where some writers have given the impression that all or most tropical soils are so high in iron that careless use or exposure threatens to turn them to useless rock or 'worthless brick pavement'. Fortunately the truth is less dramatic.

Although ironstone crusts are occasionally a conspicuous feature of some tropical landscapes and occur in some soil catenas, only a very small proportion of tropical soils have horizons sufficiently iron-rich that they are likely to harden irreversibly on exposure. Although precise estimates are lacking, soils containing plinthite layers probably cover less than 5% of the tropics. The plinthite layers do not usually harden and interfere with agriculture unless exposed by removal of the overlying horizons.

It appears that horizons of soft plinthite form most readily under hot and very wet environments, often under a forest vegetation. The final hardening into rock-like ironstone sheets and layers may occur much later, possibly under a changed, more seasonal climate. In West Africa, soft plinthite is found at depth in many forest-zone soils, but really hard sheets and crusts are much more common in the savanna areas of gentler relief and lower, more seasonal rainfall to the north. It has been suggested that the original formation of the mottled iron-rich layers occurred in the past under a wetter climate, but the final hardening was the product of seasonal wetting and drying. The present distribution of ironstone is thus related to past as well as present climate. In Australia and elsewhere, fossil ironstone crusts (hard plinthite) are found in areas that were once hot and wet but which are now dry.

Ironstone crusts, once formed, are in time broken up by erosional forces as the landscape as a whole is dissected and lowered. Investigators have shown that in some landscapes the disintegration of crusts and the reduction and mobilisation of the iron they contain leads to the formation of further iron-rich layers and crusts at positions down the slope. Some tropical soil catenas are thus marked by a series of crusts at different levels which outcrop or form distinctive breaks of slope.

In many tropical landscapes, and particularly in West Africa and parts of India, the break-up of crusts appears to be responsible for the presence in many subsoils of large quantities of what is sometimes loosely referred to as 'pea gravel'. This pea gravel consists mainly of hard, ironstone nodules, typically 5–15 mm in diameter, often with varying proportions of quartz gravel. In some cases the nodules are concretions that have grown up in the soil, and have a concentric structure, but in most cases they have been formed by the break-up and progressive attrition of fragments of old crusts. When broken open they may show the original plinthite horizon structure.

In the US Soil Taxonomy, soils containing plinthite are not separated at the order or suborder level, but are separated at the great group and lower levels. At the great group level (see Table 8.1), Alfisols with plinthite are classified as Plinthustalfs or Plinthaqualfs according to the soil moisture regime. Ultisols with plinthite include Plinthudults, Plinthustults, Plinthohumults and Plinthaquults. In other cases plinthite-containing subgroups of great groups are designated by the adjective 'plinthic'. A contact between soil material and hard ironstone is described as a 'petroferric contact'.

Fig. 7.4 A sheet of ironstone capping a hill.

Fig. 7.5 A boulder of pisolithic ironstone.

In the FAO/Unesco world soil map legend, a few of the 26 major world classes are subdivided into units containing plinthite which are designated by using the adjective 'plinthic', e.g. Plinthic Luvisols.

The effects of vegetation

The most obvious and important division of the vegetation of the tropics relating to soil formation and properties is between the closed forest of areas of generally higher and better distributed rainfall, and the various savanna formations, more or less modified by human activity (Chapter 1). The latter are associated with generally lower rainfall, or with rainfall with a more marked dry season or seasons. Various subdivisions of both these major formations occur, and are useful in studying the relationship between vegetation and soil formation and properties.

In broad terms the nature of the vegetation in any place is determined by both the climate and the soil, since soil characteristics modify, among other things, the effective rainfall. The interactions between climate, vegetation and soil are therefore complex. In practice it is often useful to think in terms of climate–vegetation–soil belts, in which the properties of mature soils are related to climate and vegetation.

Vegetation as a soil-forming factor

Unconsolidated material does not develop as a soil until it is colonised by plants. Where bare or cleared ground is involved, a succession of plant communities may established themselves and succeed each other, as when regrowth vegetation takes over abandoned farmland (Chapter 17), until, given stable conditions for long enough, a climax vegetation is established. The development of the vegetation is accompanied by parallel modifications of the soil.

The mechanisms by which plants affect the formation and properties of soils include the following:

1 The supplying of humus to the soil from the decomposition of plant parts, including leaves and roots, usually with a concentration of organic matter in the topsoil, or A horizon, whose properties change as a result.

2 The removal of nutrient ions from the whole root zone and their concentration, via the leaf litter layer, in the upper soil horizons, and mainly in the topsoil.

Plants also influence the soil in many other ways. They have a role in the weathering of rocks and minerals, in anchoring the soil, in shading it and modifying microclimates, and in protecting the soil from the direct force of the rain. Plants also influence the nature and activities of soil fauna, but the two main mechanisms listed above are the most important from the point of view of soil formation.

Vegetation and humus amount and distribution

The important differences as regards soil formation between the two broad classes of tropical vegetation – forest and savanna – are related to

1 the much greater weight of vegetation and greater quantity of residues returned to the soil in the forest as compared with the sparser vegetation of savanna areas, and
2 differences in the rooting systems of trees and grasses.

In tropical forest areas the total weight of vegetation (wood, leaves, roots, litter and associated lianes) varies with the nature of the forest and the species concerned, the climate and the soil. Total dry-matter weights of between 50 and over 150 t/ha have been reported in various studies of tropical forest vegetation, with mature secondary or original forest often being near the upper part of this range.

A study of mature secondary evergreen forest at Kade, Ghana showed a total weight of about 150 t/ha, of which a little over half was wood and bark, and a further third was stumps and dead wood. Roots accounted for about 11 t/ha, and litter about 1 t/ha. A comparable study of little-disturbed original moist semi-deciduous forest in the Congo showed a similar total weight of vegetation of about 150 t/ha, mostly wood and bark. Secondary forest formations have lower total dry-matter weights, depending on age, structure and composition. An 18-year-old secondary forest at Yangambi, Zaire, was found to have a total weight of about 65 t/ha, and a 5-year regrowth forest in the same area about 40 t/ha. A 20-year-old secondary forest at Kumasi, Ghana was estimated to weigh about 55 t/ha, and a

6-year-old *Acioa barteri* fallow in Benin, southern Nigeria, about 25 t/ha.

These figures indicate that the total dry-matter weight of tropical forest is very considerable, and that little-disturbed original forest and mature secondary forest can be expected to produce 100–150 t/ha, with secondary formations, depending on age, often being in the range 25–75 t/ha. These tonnages are much higher than those of most savanna formations, and represent a very considerable nutrient reserve locked up in the vegetation, as discussed further in Chapter 17.

Savanna formations in the tropics, as outlined in Chapter 1, show a very considerable range from sparse formations of low rainfall areas to the savanna woodland of the wetter areas. Savanna vegetation is characterised botanically by a stratum of grasses and sedges, but with a shrub, tree and palm component that is very variable in amount. Since most savannas are regarded as being products of human activities, profoundly modified by annual burning in the dry season, the extent of the human influence has added to the variation to be expected from climate and soil factors. It is not surprising therefore to find that the total dry-matter weight of savanna vegetation

1 varies within very wide limits, but
2 is generally very much less than that of forest formations.

A *miombo* woodland in Zambia was found to produce a dry weight of woody matter of about 45 t/ha, but a natural grassland in Marandellas, Zimbabwe was found to have a total dry weight of only about 2 t/ha. An investigation of an *Imperata cylindrica* grassland at Ejura, Ghana, showed total dry weight to be about 11 t/ha, but of this total over two-thirds was represented by roots and stolons. A high-grass *Andropogon* savanna in the same area, undisturbed for over 20 years and with a good growth of savanna trees, had a herb layer weighing about 4 t/ha, of which roots accounted for nearly half, and a tree weight of about 24 t/ha.

The amount and the distribution of organic matter in forest and savanna soils is related both to the total quantity of organic residues available for decomposition, and the nature and source of those residues. In forest areas, the humus is derived mainly from the surface litter layer which consists mainly of leaves, twigs and small branches with some flowers and fruits. Under hot wet conditions, decomposition and turnover is rapid, so that despite the large quantities of organic material deposited on the soil surface, the leaf litter layer is typically thin. The leaf litter layer is decomposed with the help of worms, bacteria, fungi, and a variety of soil fauna. Most of these, with the exception of nest-building termites, live, leave their droppings and die within the upper layers of the soil. In practice the organic matter in tropical forest soils is typically concentrated in the upper 5–15 cm of the soil to form an A horizon which is dark and humus-stained, masking the colour of the mineral soil, in contrast to a subsoil in which the organic-matter content is much lower. The division between topsoil and subsoil in many forest soils is quite sharp, in contrast to savanna soils in which it is typically much more diffuse.

In tropical savanna areas the amount of plant material that is available to form humus is much less than in forest soils because

1 the total weight of savanna vegetation is usually only a fraction of the weight of forest vegetation,
2 the above-ground plant parts are grazed by animals, and
3 dead plant material remaining in the dry season is normally set on fire in the annual burn and thus converted to ash rather than being available for humus formation.

A further important difference between forest and savanna vegetation concerns the nature of the root system. The root system of forest trees is usually, apart from some tap roots, shallow and concentrated in the top 25 cm of the soil. The root system of grasses is generally much more evenly distributed down the profile, and is dense and fibrous; moreover, grass roots are renewed more frequently than tree roots.

In practice a greater proportion of the humus in savanna soils is derived from roots than it is in the forest areas, and though total amounts of organic matter are usually less or very much less, the distribution is different. In savanna soils the humus is spread more gradually down the upper 100 cm of the soil, in contrast to the very marked topsoil concentration of organic matter in forest soils where the main source is the surface litter layer.

Figure 3.2 shows typical organic-matter distributions in West African forest and savanna soils. In the forest examples given, organic matter reaches 6% in the upper 0–5 cm layer in two secondary forest profiles, and is 6% and 12% in two high-forest soils. Below 5 cm, however, organic-matter levels drop very sharply in all four examples, falling to 1–3% at 15 cm and to 1–2% at 30 cm. Three of the four profiles have organic-matter contents of about 1% between 30 and 100 cm.

In the four examples of savanna soils given in Fig. 3.2, the amounts of organic matter in the surface horizons are much less than in the forest soils, but the organic matter is spread out more evenly down the profile. As seen in the field, this difference results in a much less well-defined topsoil as regards humus content and colour, and a more diffuse transition to the subsoil.

In most of the tropics the organic-matter content of savanna soils is lower than that of forest soils unless the latter have been much degraded. On well-drained sites, the highest organic-matter contents are found in little-disturbed high closed forest, though higher organic-matter accumulations may be found in some poorly drained soils in which organic matter accumulates.

In the temperate areas of the world the position is generally reversed, with higher organic-matter contents being found under temperate grasslands such as the prairies of North America and the steppes of Russia, with their dark, humous, calcium-rich topsoils, than under forest soils. In these temperate soils organic residues accumulate during the growing season but decomposition slows down or ceases during the winter.

Differences between the soils of temperate and tropical areas as regards organic-matter contents in relation to vegetation mean that some soil textbooks written primarily by and for workers in the temperate zone might be misleading in this respect if taken to apply to tropical soils.

Nutrient cycling by vegetation and topsoil enrichment

Although most plant feeding usually occurs in the upper soil horizons where feeding roots are most numerous and nutrients are released by organic-matter decomposition, plants remove nutrients from the whole of the root zone, including the lower layers tapped by the deepest roots. In many of the more highly weathered tropical soils a relatively fertile topsoil, due to its organic-matter content, overlies a subsoil that is leached and weathered and relatively less fertile. In the C horizon, however, chemical fertility may increase if there are weatherable minerals which decompose to release nutrients. In addition, in areas of heavy rainfall, nutrients may be leached down the profile to the lower horizons. Deep plant roots may therefore be able to tap C horizons where nutrients are released from chemical weathering or may have been washed down in percolating soil water.

The effects of vegetation on soil formation therefore include a cycling action from the lower soil horizons tapped by roots, via the plant and leaf fall, to the topsoil where nutrients are released again when plant residues decompose.

The generally upward movement of nutrients and their concentration in the topsoil due to plant action contrasts with the generally downward movement of substances in solution and suspension provoked by soil leaching, as considered in the first section of this chapter.

A knowledge of the root systems of natural vegetation and of crops is of interest in assessing from what depths nutrients can be brought up by plants. It is known that some plants such as grasses with a fibrous root system, and grass crops such as sorghum, are more effective than others at trapping nutrients such as calcium and magnesium moving downwards in the soil solution during periods of heavy rain and soil leaching, but shrubs and trees are normally considered more efficient at bringing up nutrient ions from the lower soil horizons. One advantage claimed for cropping systems that associate annual crops with deep-rooted perennials is that the tree component of the association reaches soil depths beyond the reach of the annuals.

The role of soil fauna

Soil fauna are usually considered together with vegetation as one of the five major soil-forming factors (see p. 59).

Soil fauna range in size from large burrowing animals, through worms, termites and other soil insects that can be seen with the naked eye, to microscopic soil fauna including bacteria and fungi.

Most agricultural soils that are well supplied with organic matter are so teeming with life that the soil has been described as a 'living thing'. It has been estimated that a single teaspoonful of humous top-soil may contain as many bacteria as the world does people, though it should be emphasised that the bacteria population can rise or fall dramatically in a short space of time as conditions, including a water supply and a food substrate, change.

Tropical soils when examined under a microscope often show abundant signs of faunal activity. In temperate regions of the world considerable attention has been given to the role of earthworms since Darwin first drew attention over a hundred years ago to their effects on soil formation and leaf litter decomposition in his classic book *The Formation of Vegetable Mould through the Action of Worms*, published in 1881. In tropical soils earthworms, including giant varieties, are important, but in many soils their activities are subordinate to those of ants and, particularly, termites, which are rare or absent in most temperate zones. In the tropics termites often dominate the soil fauna. They appear to have influenced soil formation and properties in the tropics over long periods of time. They have given rise in many cases to distinctive soil morphologies, due to their ability to separate and transport upwards the fine-earth fraction of the soil, leaving behind the gravel and stones. They have important effects, on the cycling of organic matter and its distribution in the soil because of their ability to digest structural plant polysaccharides and thus to decompose plant tissue.

Soil organisms are grouped according to size into the following:

1 **Macrofauna** – vertebrate animals, mainly of the burrowing type, that live wholly or partly underground.
2 **Mesofauna** – small invertebrate animals including arthropods (mites, springtails, beetles, ants, termites, millipedes, centipedes and various insect larvae), annelids (earthworms), molluscs (snails and slugs) and nematodes.
3 **Micro-organisms** (microfauna and microflora) comprising bacteria, fungi, actinomycetes, algae and protozoa.

The weight of the organisms is referred to as the biomass. For the macrofauna and mesofauna the biomass is usually given in terms of liveweight in kg/ha to a stated soil depth. The micro-organisms are intimately mixed with the soil organic matter and are difficult to count or weigh, but their biomass may be estimated indirectly by measuring the carbon dioxide they produce or the oxygen they consume, and is expressed as weight of organic carbon in kg/ha or as a percentage of the total organic matter. The microbial biomass may be 2–3% of the organic matter, and in productive soils well supplied with humus this may correspond to 0.5–2.0 t/ha to a depth of 25 cm.

The mesofauna

Arthropods other than termites and ants The arthropods include the mites, springtails, beetles, various insect larvae, ants, termites, millipedes and centipedes.

Mites and springtails are very numerous, feeding on dead organic material in the lower part of the litter layer and also occurring in the larger soil pores (0.3–5.0 mm in diameter).

Many beetles and insect larvae live in the soil. Some, the saprophytes, live on dead organic material, but others feed on living tissues and can cause serious damage to agricultural crops. The dung beetles (coprophagous beetles) of Africa and other parts of the tropics contribute to the rate at which the dung of large herbivores is comminuted and physically incorporated into the soil.

Centipedes are carnivorous, while millipedes feed on vegetation, including living roots, bulbs and tubers.

Earthworms In temperate regions of the world earthworms, because of their biomass and activity, are generally more important in the consumption of leaf litter than all the other soil invertebrates together, though their biomass varies considerably from only a few ka/ha to a maximum in soils under pasture of up to 1500 kg/ha.

Earthworms feed only on dead organic matter, the majority of species migrating between the leaf litter layer and the soil. In the course of their feeding they take in large quantities of clay and silt-sized soil particles, which may be as much as 90 t/ha annually for the large earthworm populations under grass. Leaves from deciduous tree species and of grass are more palatable to earthworms than are pine needles. The soil and organic matter are intimately mixed when these materials are deposited in the worm faeces, which may appear as casts on the soil surface in the case of some species and in burrows in the case of others. In this way organic matter from the soil surface becomes distributed and incorporated in the upper part of the soil profile. The earthworm burrows also help aerate the soil and improve rainfall infiltration.

Normally the size of the earthworm population is limited by the amount of organic matter available; hence the high populations under temperate grasslands, although earthworms are also influenced by moisture, temperature and soil reaction, earthworms rarely being found in soils more acid than pH 4.5. In the tropics worms are more important in areas with udic or seasonal moisture regimes than in dry environments, and in soils with a good humus supply. They avoid waterlogged soils. Some species, including giant worms up to 1 m in length, form conspicuous casts at the soil surface, but in general their occurrence is often patchy. In some areas where worms were absent, their introduction has increased soil productivity. One estimate of wormcast production in an area of Nigeria with a six-month season and vegetation providing abundant leaf fall showed that worms of the species *Hippopera nigeriae* produced 50 t/ha of wormcasts in the wet season, consisting of soil material finer than 0.5 mm and mostly finer than 0.2 mm in diameter.

Limited data indicate that earthworm biomass is lower in many tropical habitats than in many temperate ones. In one investigation in a Nigerian forest area, worm biomass was estimated at 10 g/m^2, and in another on various habitats in Uganda worm biomass was found to range between 0.06 and 4.55 g/m^2. In contrast to this, termites are thought to exceed 50 g/m^2 in some tropical areas.

Termites The importance of termites in the tropics has recently received increasing attention from soil scientists and biologists. The termite, unimportant or absent in most temperate regions, has been called the tropical analogue of the earthworm because of its comparable role in comminuting all forms of litter in both the forest and savanna regions of the tropics. An estimate of grass-harvesting termite biomass in Zaire showed that this may exceed 10 g/m^2, which is similar to the highest recorded biomass of grazing animals.

Fig. 7.6 *Macrotermes* mound, Niger State, Nigeria

Termites are social insects that live in nests, known as termitaria, which they construct. The termitaria serve to house the colony, store food and maintain an optimum environment. The functioning of the society is based on the division of labour among different castes within the colony, including the reproductive caste which swarm out of the nest at intervals, and the sterile castes of workers, the most numerous, and soldiers. The food collected by the workers, both for their own use and for the dependent castes, is the basic energy resource of the colony and consists of plant material, either living or dead, and if dead including partially or almost completely decomposed material. The source and decomposition of this food affects the cycling of organic matter and nutrients, and also its distribution, since food and waste products are concentrated in the termitaria.

The food includes (a) wood, (b) grass, herbs and plant litter (c) soil humus, and (d) fungi. Wood is only rarely consumed as living wood, most species eating only sound dead wood or wood that is rotten or decaying (and thus probably rich in fungi). The true foraging or harvester termites are those species that forage on the surface and cut down standing grass and herbs, storing the harvested material, sometimes in large quantities, in their nests. The stored fragments of grass, for example, are often 4–10 mm long. Other species, the scavengers, collect leaves, seeds, twigs and other plant litter lying on the soil surface, while some species act both as harvesters and scavengers. Included among the scavengers are those species eating well-decomposed plant material and dung. The humus-eating termites (humivores) are those that ingest appreciable quantities of humus-rich mineral soil. Many termites eat fungi in rotting wood and other decomposing plant debris, but in the Macrotermitinae the association between termites and fungi has reached a high degree of specialisation. These termites construct special structures known as 'fungus combs' which support a growth of various fungi belonging to the genus *Termitamyces*. The fungus combs are continually eaten by the termites and replaced, and supply them with nitrogenous materials and vitamins. In one study in West Africa it was found that fungus-eating termites were rare or absent in dry areas, their preferred habitats being the wetter savanna areas and the forest.

Termite mounds range in size from small domed, conical or mushroom-shaped structures only a few centimetres in height to the colossal mounds of some species of African Macrotermitinae which are up to 9 m in height and may reach 20–30 m in diameter at the base. In one investigation in Zaire the termite mounds were found to occupy 30% of the soil surface and to weight 2.4 million kg/ha, equivalent to a soil layer 20 cm thick over the whole surface.

Soil particles are often the dominant constituent of the termite nest. Many species show a preference in selecting the finer soil particles for nest construction, with the result that mounds may have up to 20% more clay in them than any of the soil horizons, though mound composition varies with the species as well as with the soil. Termites usually avoid very sandy soils, and also black cracking clays (Vertisols). Clay soils are preferred by many species. A study in Zaire showed that the numbers of large *Macrotermes* mounds per hectare increased progressively with the clay content of the soil from 2.7 mounds/ha on soils with 30–35% clay to 4.9 mounds/ha on soils with 60% clay.

Certain species of termites construct their nests of soil from well-defined soil horizons. The Macrotermitinae of Africa, noted for their large and spectacular mounds, construct them from subsoil material, usually containing a high proportion of clay-sized particles. Soil particles are ingested and then regurgitated. Saliva may be used as a cementing material, particularly by the Macrotermitinae which plaster their workings with rolls of regurgitated soil mixed with saliva. Microscope examination of thin sections of newly constructed *Macrotermes*

Fig. 7.7 Small conical mounds of *Amitermes vitiosus* on shallow soils in Australia. Source: Lee and Wood, 1971.

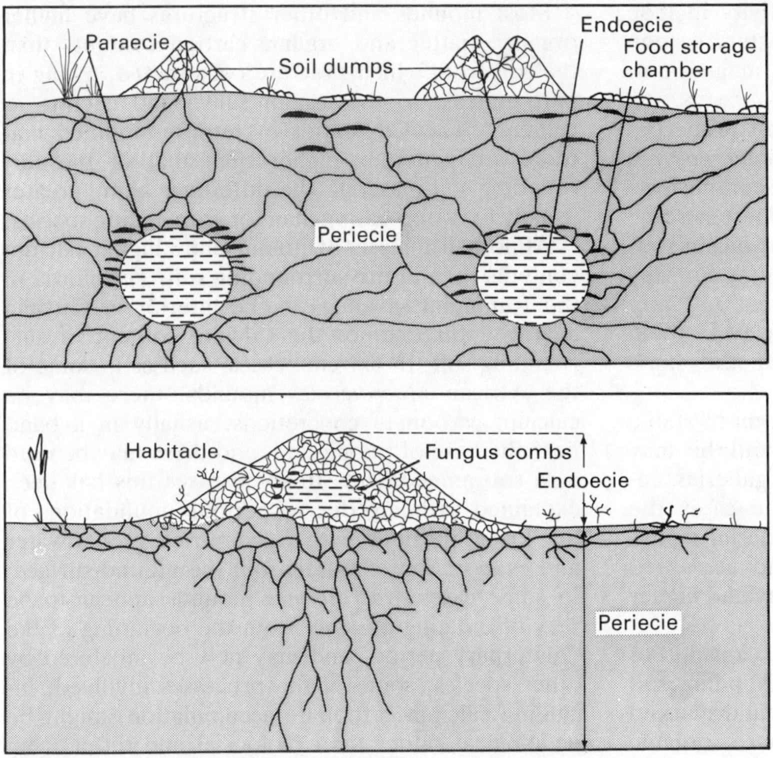

Fig. 7.8 Diagrams of termite nest systems. Top: *Hodotermes mossambicus*. Below: *Macrotermes subhyalinus*. Source: Lee and Wood, 1971.

material shows an open pellet structure with pores remaining, but later the open structure is filled in and the fabric of the completed capping has a dense, structureless appearance. Other groups of termites use excrement as a cement.

Particularly important is the fact that the size of particles is limited by what a worker can carry, and normally gravel-sized particles (over 2 mm in diameter) are rare or absent from termite constructions. From the point of view of soil formation this means that termites have a sorting action, moving large quantities of fine earth upwards and leaving gravel and stones behind. Some soil scientists and geomorphologists have considered that this sorting action over long periods of time has resulted in the formation of gravel-free soil layers overlying a horizon of stones or gravel (sometimes referred to as a stone line). If this is so, then termites have contributed to the morphology of some tropical soils in a spectacular way.

The effects of termites on soil properties The effects of termites on the physical and chemical characteristics of soils are of direct interest to soil scientists and agronomists. In some cases it has been observed that crops grow better on old termite mounds than on the surrounding soil, and in others that they grow less well. These differences can be explained partly in terms of the species of termite involved, partly in terms of the soil concerned, and partly in terms of the relative importance of the different physical and chemical modifications brought about by termite activity.

The physical effects of termite activity include effects on soil particle size, structure, bulk density, water-holding capacity, porosity, water infiltration, aeration and the penetration of plant roots.

The chemical effects of termite activity include modifications of soil pH, organic matter, carbon and nitrogen, and content of calcium, magnesium, potassium and phosphorus.

(a) Effects of termites on physical soil properties Particle-size analyses of mounds of *Macrotermes* species in Africa have shown that the composition is close to that of the subsoil but that there is some selection in favour of the finer-size fractions in soils that are not high in clay, and the reverse for clay subsoils. In Tanzania it was noted that tall thin mounds had a greater proportion of clay in them than larger, dome-shaped mounds of the same species.

Soil material in mounds often has a more stable aggregation than the surrounding soil, and this may be particularly marked in nests and galleries cemented with termite excreta. In some cases the mound is cemented into a massive material that is very resistant to erosion. Packed and cemented mound material may have lower porosity and higher bulk density than the surrounding soil.

Comparisons of the water-holding capacity of mound material with that of the surrounding soil have shown no increase in some cases and marked increases in others, particularly where organic matter was increased.

The gallery systems associated with mounds, nests and foraged areas themselves locally increase pore space and aeration, water infiltration and plant root penetration, but some termites bring fine material to the surface, cement it and thus form a layer not easily infiltrated by water. In one study on coffee roots in Kenya which penetrated infilled termite galleries at depths of 60–120 cm, it was found that galleries were filled with darker, richer soil derived from upper horizons in which fine roots proliferated. In an investigation in West Africa, grey soil from upper horizons appeared to have been washed down into termite galleries at depths up to 8.5 m and roots were found to follow this less compact material to penetrate the deeper soil layers.

(b) Effects of termites on chemical soil properties Many investigators have noted an increase in the pH of the termite soil relative to the surrounding soil. Increased pH, especially in mounds of *Macrotermes* species, is often associated with accumulations of calcium carbonate.

Most mounds and other structures have higher organic matter and organic carbon contents than the soil from which they are constructed, owing in part at least to the use of saliva and excreta as cements. The C/N ratios of termite-modified material are generally higher than that of the surrounding soil material, the difference being greater usually for wood-eating than for grass-eating species.

Almost all analyses of mounds show higher calcium contents than in the surrounding soil. Termitaria in the Rupununi savannas in the interior of Guyana averaged three times the calcium content of surrounding soil. In extreme cases, such as in some of the African *Macrotermes* mounds, there may be calcium carbonate concretions, usually in a band near the base of the mound, and these may be used as a soil amendment. In some cases this has been explained as being due to periodic inundations of the lower portions with calcium-rich groundwater and evaporation of water from the mound surface. In some cases *Macrotermes* mounds appear to be very old, dating possibly from the beginning of the Quaternary period, and may now be inhabited by other species; some of the processes involved, including calcium carbonate accumulation, might be pedological rather than biological and differ from those in surrounding soils.

Magnesium and potassium increases, usually small in relation to those of calcium, have been noted in many mounds, and appear to be due mainly to organic-matter incorporation during mound construction. Phosphorus levels in termite mounds often do not differ significantly from those in the surrounding soil, though higher levels have been found in some Indian termitaria. In a study of East African mounds formed by species of *Macrotermes*, it was found that phosphorus was generally uniformly distributed in amounts similar to those of the surrounding soil, but in some cases near areas of impeded drainage, phosphorus increased down the mound, reaching a maximum at about ground level, and then decreased with depth below the mound.

The use of termite mounds for cropping The practice of growing crops on termite mounds or on levelled mound patches is widespread in parts of Africa and Asia. Reports on this use of termite mounds for cropping give conflicting results. This may be related to the varied properties of the subsoil material generally used in constructing mounds as compared with the properties of the surrounding topsoil.

It has been reported that sisal grows better on large mounds in Java, and also on large *Macrotermes* mounds in East Africa, than on the flat land between them. In East Africa the better growth was attributed to the higher calcium content and better drainage and soil structure in the mounds. In Thailand various crops are grown on termite mounds rising above flat rice paddy fields, and in shifting cultivation (*chena*) areas of Ceylon many crops have been reported to grow better on *Ondontotermes* mounds than on surrounding land. Annuals such as maize, on the other hand, were reported to grow relatively poorly on levelled *Macrotermes* mounds in a Nigerian forest area, but in many West African forest areas maize is greener and taller on mounds, suggesting a better nitrogen supply.

Where large mounds are particularly abundant they may present a serious obstacle to cultivation. In an area of central Zaire, large mounds were found to occupy 30% of the ground surface and to be constructed of relatively infertile subsoil material, so that levelling without topsoil redistribution resulted in sterile patches of ground.

Effects of termites on decomposition of organic matter and on nutrient cycling The ability to digest cellulose and sometimes lignin sets termites apart as a particularly important group of soil animals. Where termites are the dominant soil animals, as in much of the tropical regions of the world, the organic cycle, which is elsewhere dependent mainly on soil microflora, is drastically modified. Large quantities of plant tissue are taken from fallen litter and sometimes from living plants, often moved for storage to a nest or mound, and so intensely degraded that the residue may contain little of use to other organisms. In the digestive process termites are largely dependent for cellulose digestion on symbiotic microfauna, which are present in their digestive systems, mostly in enlargements of the hind-gut. The symbiotic microfauna are mainly flagellate protozoa and bacteria, and include species that are able to attack sound wood. The symbiotic processes which aid plant digestion in termites are in some ways similar to processes of digestion in ruminants. The principal component of wood decomposed by termites during digestion is cellulose, but some lignin is also digested.

In soils with large populations of mound-building termites the distribution and quantity of organic matter, and the cycling of plant nutrients it contains, can be considerably modified. This is related to the fact that termites forage for food over a wide radius but then return it to the nest. The use of excreted end-products of digestion for mound-building effectively locks up this material for long periods in the mound. The recycling of nutrients in termite bodies requires further research. Some recycling occurs when termites swarm from the nests in mass flights. Assuming the termite body to contain 10% nitrogen, it has been calculated that nitrogen leaving the colony annually in this form amounts to 4.0 mg/m^2 for *Cubitermes* termites in a savanna area in Zaire, and 7.0 mg/m^2 for *Trinervitermes* in a savanna area of Nigeria.

Termites and vegetation In some parts of Africa, including inland Natal and the Accra plains of Ghana, clumps of thicket or of trees growing on old termite mounds separated by savanna grassland form a distinctive vegetation pattern. The mounds are raised a little above the surrounding flat grassland and have deeper, darker and more fertile soils than adjacent soils under grass. The microclimate in the thicket or tree mounds is cooler and more shaded than in the exposed savanna which surrounds them.

In Africa the principal *Macrotermes* species building large mounds are considered to favour open savanna woodland conditions, and some observers consider that these species do not occur in primary rainforest, and that where they are found under forest vegetation this indicates that the vegetation is secondary. Large mounds in southern Nigeria and the Congo basin under forest are considered to have been built during a drier period in the past.

In Thailand large trees were observed on termite mounds which did not grow well, if at all, on surrounding areas of paddy and forest. On the Loita plains of Kenya most of the species characteristic of termite mounds are exclusive to them in that area. In many cases the plants characteristic of mounds are restricted to the special habitat of the mounds because they are outside their climatic areas of optimum growth, but in some cases the mounds have specialised vegetation because they are protected from the annual burning of the surrounding grasslands. Some investigators have traced the succession of plant growth on mounds in Oubangui–Chari and on the Sese Islands in Lake Victoria from grasses to shrubs to trees, and suggested that where conditions are favourable these islands of woodland expand and merge to produce a forest, so that termite mounds may serve as the starting-points for recolonisation of areas denuded by clearing, burning and overgrazing.

In contrast to the large mounds of Africa and Asia which support a distinctive vegetation are the smaller ones that support no vegetation at all, either while still occupied, possibly due to the steep and hard outer surfaces, or even when abandoned and eroded. Toxic substances may be involved in some cases of eroded mounds and bare areas around the base of mounds, but the hardness and poor physical properties of the eroded material and of the hard, compacted zone around the mound are probably the most common causes of a lack of plant cover.

The micro-organisms

Bacteria Bacteria live in water films around soil particles in all but the finest pores. They can be as small as 1 μm in length and 0.2 μm in breadth, and round or rod-shaped. Under favourable conditions they can reproduce very rapidly in the soil, as when two daughter organisms are produced from a single parent by fission in 8–24 hours. They can increase to very large numbers in the soil provided there is a suitable food supply and that other conditions are favourable.

Bacteria can be divided on the basis of their feeding habits into heterotrophs, which are the more numerous and which attack complex substances and break them down to simple end products which they can absorb, and autotrophs, which synthesise simple inorganic molecules using either the energy of sunlight, in photosynthesis, or chemical energy obtained by oxidising inorganic compounds.

Bacteria can also be divided on the basis of their need for air oxygen. Aerobes required air oxygen for respiration, while anaerobes obtain oxygen by reducing oxygen-containing compounds such as nitrate. The latter can be subdivided into facultative anaerobes, which normally use air oxygen but which can become anaerobic when necessary, and obligate anaerobes which grow only in the absence of air oxygen.

Many bacteria are of the greatest importance to the breakdown of organic matter in the soil (Chapter 3) and to the various nutrient cycles in the soil which help to maintain soil fertility. These include the various nitrogen-fixing and nitrifying bacteria discussed in Chapter 9.

Bacteria are adapted to a very wide range of soil conditions, but modifications in those conditions (such as a change in aeration, or in soil pH) bring about corresponding changes in the relative numbers and importance of the species involved. In general terms bacteria are highest in moist soils that are well supplied with organic matter and moderately acid to slightly alkaline in pH. They are less varied and numerous in acid soils, in dry areas and in soils that are low in organic matter.

Fungi (including mycorrhiza), actinomycetes, algae and protozoa A germinating fungal spore develops a long colourless thread, the hypha. This may grow straight, or branch to form a network of hyphae called the mycelium. Fruiting bodies on the mycelium produce further spores. The mycelium of fungi are often easily seen in leaf litter and decaying wood. Some fungi live in the soil and feed on dead organic matter. The spores of others germinate and grow only when living tissue of a suitable host plant occurs nearby, and some of these are pathogenic, growing as parasites on the host. Fungi are thought to be particularly important in breaking down woody tissue that is low in nitrogen and not readily attacked by other micro-organisms.

The **mycorrhizal fungi** are beneficial since they live symbiotically on the host roots, deriving carbon compounds from the host but in return supplying it with mineral nutrients, especially phosphorus, and thus effectively enlarging the root system. There are two forms of mycorrhiza, the ectotrophic and the endotrophic. The ectotrophic forms occur on some trees, the mycelium growing externally and forming a sheath around the root cylinder; infected roots live longer than non-infected ones and continue to absorb phosphorus long after the non-infected roots have ceased uptake. The endotrophic form of

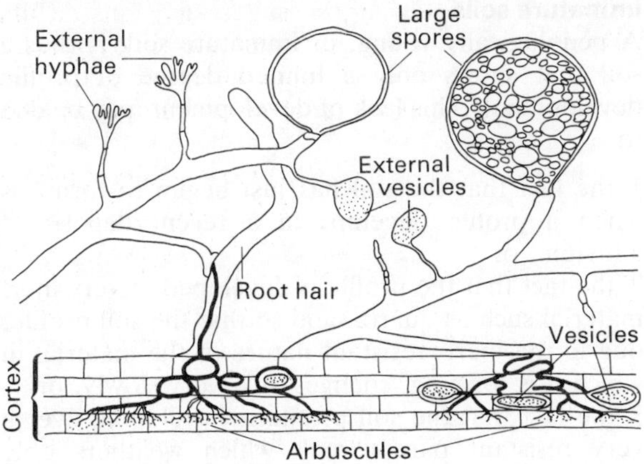

Fig. 7.9 Diagram of vesicular–arbuscular mycorrhiza. Source: Nicholson, 1967, as given in White, 1979.

mycorrhiza, commonly called vesicular–arbuscular, has hyphae that ramify widely outside the host tissue once the young mycelium from the germinating spore has penetrated a host root and established itself. The vesicles are storage bodies which occur both within the root cortex and on the external hyphae, while the arbuscules are branched structures growing between the cells of the root cortex of the host plant which probably release solutes. The external hyphae add up to 50% to the length of absorbing roots and provide a pathway for the transport of phosphate across the depletion zone around the root, thus conferring an advantage in phosphorus uptake. The incidence of infection is sometimes particularly high in impoverished soils, but declines with cultivated crops as phosphatic fertilisers are applied. Of tropical crops, cassava is thought to benefit widely from mycorrhizal infection, and mycorrhizal inoculation has been used to establish pines and other species in poor soils.

Actinomycetes form a group of organisms that are intermediate in some respects between fungi and bacteria, forming colonies similar to fungi with abundant spore formation and typically being most abundant in the pores of moist surface soil horizons. They are aerobic, and less tolerant of soil acidity than the fungi, but are well able to decompose lignin and other complex organic compounds synthesised in the soil.

Fig. 7.10 Azolla in a rice paddy at the International Rice Research Institute's central farm, the Philippines. Azolla incorporated into the soil forms a useful source of nitrogen.

Soil algae include the blue-green algae which are important for their ability to fix atmospheric nitrogen. They have been studied mainly in connection with nitrogen fixation in rice paddies, where they may form a symbiotic association with the small floating *Azolla* fern, much used in rice paddies of Vietnam and southern China, and are sometimes found to fix more nitrogen than is removed by the growing crop. Blue-greens tolerate a wide range of moisture conditions and can live in moist and even dry soils, so that they may be found to have a wide application in tropical agriculture, and deserve further research. They are photosynthetic and mainly confined to the surface soil. Their main limitation appears to be that they do not like acid soils. The blue-greens are remarkably self-sufficient in that they can obtain all their needs from air, water and mineral salts, and thus can colonise areas which organisms needing organic residues cannot.

Protozoa are the smallest of the soil animals, being single-celled, and ranging from 5 to 40 μm in length. They include amoebae, flagellates and ciliates. They live in water films in the soil, moving by means of cytoplasmic streaming, flagellae or cilia, and cannot move or be active in dry soil, though many withstand dry periods by forming cysts or by going into a state of suspended animation. Their numbers are probably greatest in the moist topsoils of forest soils. Some protozoa (such as the flagellates *Euglena*) contain chlorophyll and can live autotrophically, and some feed on dead organic material, but most of the protozoa can feed only by preying upon other small organisms such as some species of bacteria, algae and nematodes, and play a role in regulating the bacterial population. Protozoa were once regarded as being entirely harmful to bacterial life in the soil and as reducing bacterial activity in maintaining soil fertility, but more recent research has shown that this is an oversimplification and that some forms of bacterial activity (including nitrogen fixation by *Azotobacter* and bacterial oxidation of sugars) might be stimulated by the presence of some protozoa.

Soil age: mature and immature soils

Due to processes of weathering and soil formation, soil profiles undergo certain changes with time. Since these changes serve to produce differences between soils even where the other factors of soil formation are similar, then time or soil age is itself a soil-forming factor that differentiates soils.

When talking about time as a soil-forming factor, or the 'age' of a soil, the soil scientist is not thinking usually in terms of absolute age, but rather in terms of the stage of development a profile has reached. Thus an immature soil is a 'young' soil in the sense that its profile shows very little development. It may consist simply of a humous A horizon overlying parent material, or it may consist only of horizons that are thought to form relatively quickly. If a soil contains horizons thought to form slowly, such as argillic horizons or oxic horizons, then the soil is not pedologically young. The emphasis is on the degree of development of the profile and not usually on absolute age, which is generally not known, nor is the absolute age usually of much practical concern to the pedologist or agriculturalist.

Immature soils

A pedologically young, or immature soil, is thus a soil that shows only a limited degree of profile development. This lack of development may be due to

1 the fact that the soil has just begun to form, as when a profile develops in a recent deposit of alluvium, or
2 the fact that the profile is developed in very inert material such as quartz sand so that the soil profile, due to the inert, resistant nature of the material in which it is forming, changes only very slowly, or
3 the fact that the soil profile is developing over a very resistant parent rock which weathers only slowly giving a shallow profile, a situation that is commonly also associated with
4 the fact that the soil is developing on a steep slope so that soil material is removed almost as fast as it is formed and the resultant profile is shallow, perhaps consisting of a thin topsoil over rock.

Of these four possible cases resulting in profiles with little development, only the first is likely to be a soil that is young in absolute terms. The other examples are immature because development is held back by the resistant nature of the parent material or rock, or by the steep slope, or by a combination of these two factors.

Young soils in recent deposits A soil profile developing in recent river alluvium newly colonised by vegetation will at first consist of a humous A horizon, which becomes better defined with time, overlying undifferentiated depositional material below it, as yet little affected by soil-forming processes. Any horizons present in the subsoil at this stage are depositional horizons related to the layers formed when the material was laid down, and are not pedological horizons resulting from soil-forming processes and changes and translocations within the profile. The profile will thus technically be an A C profile in which an A horizon overlies little-altered C material, in this case alluvium, volcanic ash, or other recently deposited material newly colonised by plants. The parent material may consist mainly of quartz sand, in which case profile development will be slow and the soil remain poor, but it may consist instead, as is often the case with young alluvium, of material containing weatherable minerals. The presence of easily weathered minerals such as olivines and feldspars (Chapter 6) indicates a young soil, and agriculturally a soil with nutrient reserves that will become available as these minerals progressively weather.

Soils developed in recent alluvium containing weatherable minerals are often among the more productive soils in the tropics. In some cases additional nutrients are supplied when associated streams or rivers flood, and the low topographic position often ensures a better water supply than that of upland soils.

Immature soils developed in recent river alluvium are classified in the US Soil Taxonomy (Chapter 8) in the Fluvent suborder of the Entisols, or as Fluvaquents (in the Aquent suborder of the Entisols) if they have an aquic moisture regime. In the FAO/Unesco soil map legend they are classified as Fluvisols, defined as soils developed in recent alluvium with no diagnostic horizon or with no diagnostic horizon other than an ochric horizon, an O horizon, a hydromorphic horizon or a sulphuric horizon, all of which are thought to develop rapidly.

Immature soils developed in quartz sand (regosols, Psamments, Arenosols) Soil profiles developed in quartz sand change little because the quartz sand is inert and very resistant to chemical weathering. If there is little or no clay in the profile and an absence of other minerals that can weather to give clay, the profile remains very sandy and there can be little or no translocation to form a B horizon of illuviation. Such a profile may consist of a humous topsoil overlying a sandy subsoil with little or no profile differentiation. In extreme cases the profile consists of white sand and little else. The associated natural vegetation is likely to reflect the poverty of the soil.

Agriculturally these soils are unproductive because of their lack of colloids, low water and nutrient-holding capacity, and lack of weatherable minerals capable of releasing nutrients. Such apparently poor soils, however, can be used for undemanding crops such as cashew nuts, and, depending on how poor and sandy they are, for coconuts, groundnuts and various climatically suited timber species.

Immature sandy soils of this type have long been referred to as regosols, a term that is still widely used and understood.

In the US Soil Taxonomy (Chapter 8), immature soils are placed in the Entisol order, and the very sandy soils considered here are in the Psamments suborder. At Great Group level very quartz sand-rich examples are grouped as Quartzipsamments.

In the FAO/Unesco world soil map legend, sandy soils are mapped as Arenosols, which are defined as soils developed in over 50 cm of loose sand other than recent alluvium.

Immature soils that are shallow and rocky (lithosols) Where slopes are very steep, removal of soil material as it is formed may be rapid enough to keep the soil shallow. Such soils may consist of a topsoil more or less overlying hard rock, or a topsoil over a shallow weathered substratum containing rock fragments.

Shallow soils over rock or rock fragments may also occur in areas of gentle or flat relief where the parent rock is particularly resistant to chemical weathering or where the climate is dry and weathering is slow. In practice, resistant parent rocks are often associated with hilly relief since the resistant rocks remain as residual hills, often with steep slopes. A combination thus frequently occurs of steep slopes and a parent rock resistant to chemical weathering, and it is on these sites that shallow, rocky soils are likely to occur.

Shallow, stony or rocky soils have long been known as lithosols (from *lithos*, a stone), a term that is still widely used and understood.

The term Lithosol has been retained in the FAO/Unesco world soil map legend, though narrowly defined there as including soils with hard rock within 10 cm of the surface. Shallow soils on steep slopes over siliceous parent materials with a dark A horizon less than 50 cm thick are described as Rankers, a term first applied to shallow rocky soils in the mountainous areas of Switzerland and Austria, while shallow soils with a dark humous horizon over hard limestone at less than 50 cm are classed in the FAO/Unesco legend as Rendzinas.

In the US Soil Taxonomy, soils that were known as lithosols are not separated at the order, suborder or Great Group level, but form various lithic subgroups, e.g. Lithic Ustorthents are Ustorthents (tropical Entisols with an ustic moisture regime) having a lithic contact (i.e. hard rock) at less than 50 cm.

Fig. 7.11 Cultivation patterns in the Peruvian Andes, with small narrow fields on shallow soils of the steep mountain slopes contrasting with more extensive cultivation in the flat alluvial soils of the valley floor.

Agriculturally, shallow soils over rock have the obvious disadvantage of very limited root room and soil volume, resulting in correspondingly reduced water-holding capacity as compared with deeper soils. The fact that they are often on sloping sites increases the need to maintain a vegetative cover to protect the soil against accelerated erosion, and makes all agricultural operations (from hoeing to rubber tapping) more difficult. However, in high rainfall areas of the tropics where many of the deeper soils are leached and low in nutrients, lithosols that are over rocks containing weatherable minerals may be more chemically fertile than more deeply weathered and leached soils. Local farmers are often aware of their fertility and are prepared to cultivate very sloping sites, in steep areas of the Andes in South America, for example, despite the difficulties and risks involved. In Ghana forest areas, excellent cocoa (which needs chemically fertile soils) is sometimes found on lithosols that would have been condemned in most conventional capability classifications (Chapter 8) as too steep for cultivation.

Mature soils

The term 'mature soils' lacks precise definition. In general terms mature soils are all those soils that show a greater degree of profile development than the immature soils considered above. Mature soils are considerably more varied than immature soils since profile development is related to a wide range of parent material, climate, vegetation and relief factors and can take a wide range of forms. In general, the formation of horizons other than an A horizon and certain little altered or relatively quickly formed subsoil horizons is taken as an indication that the soil is no longer an immature one. Horizons characteristic of mature soils include argillic horizons, formed by movement of clay-sized particles from the A to the B horizon, and oxic horizons in which intense weathering has produced a soil high in kaolinitic clay and sesquioxides and very low in weatherable minerals.

Since the features of mature soils include those related to climate and vegetation, mature soils associated with particular climate and vegetation belts were regarded in earlier classification systems as zonal soils. A zonal soil is a mature soil that shows features related to its climate–vegetation zone, as distinct from azonal soils (soils that are too young to show zonal features) and intrazonal soils (soils, such as those with hydromorphic properties, that occur in more than one zone). The concept of zonal, azonal and intrazonal soils, though no longer central to soil classification, usefully draws attention to three broad groups of soils with properties related to genesis:

1 Immature (azonal) soils that lack profile development and do not belong to any particular climate–vegetation zone.
2 Mature (zonal) soils that bear the stamp of the local climate and vegetation.
3 Soils (intrazonal soils) that are formed under special conditions of poor drainage and thus have certain dominant hydromorphic characteristics wherever they occur.

Soils with hydromorphic properties are considered further in the earlier part of this chapter dealing with relief and drainage as a soil-forming factor.

Senile soils

By analogy with youth, maturity and old age in humans, some authors have described certain soils as 'senile'. Even more than in the case of mature soils, the term lacks definition. A senile soil is one that in some respects can be considered over-mature, worn-out and past its best. A very highly weathered soil might qualify, but here care is needed, because although the most highly weathered soils in the US Soil Taxonomy are considered those with an oxic horizon (Chapter 8), and these soils are chemically relatively inactive and usually leached and low in nutrients, they often have very good physical properties due to stable micro-aggregation of the clay fraction (Chapter 2). A senile soil might, with somewhat more justification, be taken as a soil that is over-deep. If a soil is highly weathered to 10–15 m, then the weathering rock zone below this deep weathered subsoil, if present, is too deep to be tapped by plant roots. Truncation of this soil profile, as by accelerated erosion, might therefore make the soil more productive if it brings within rooting depth a previously inaccessible horizon in which the weathering of silicate minerals releases plant nutrients. Such cases are likely to occur mainly in hot, very wet environments or on old landscapes that have been subjected to very long periods of weathering.

8 Soil surveys and soil classification

Soil surveys

Soil surveys have been carried out in great detail in the more developed parts of the world, particularly the United States and Europe, and are recognised as being of great value to agriculture, as well as to other activities connected with the planned use of soils such as forestry, town planning and road building. In the tropics systematic detailed soil surveys are less widespread and common, though especially since the 1950s they have become an increasingly important feature, and are a normal requirement for most major agricultural developments involving international aid and finance. With them have come corresponding increases in our knowledge of tropical soils, their properties and classification.

A soil survey is not an end in itself, but a tool for agricultural development, providing a basic inventory of one of the most vital of all natural resources.

The fundamental aim of the soil surveyor is to examine, describe, classify and map the soils occurring in the area selected for survey. The result is normally

1 a map showing the distribution of the soils or groups of soils mapped, at a scale related to the density of observations on the ground, and
2 a soil survey report which describes the soils mapped and which, in accordance with the aims of the survey, usually but not necessarily interprets the soils for agricultural and other purposes.

Types of soil survey

Although the nomenclature and definitions of the types of survey vary between countries and organisations, we can distinguish between the following types of survey on the basis of the mapping scale:

1 Small-scale surveys include exploratory surveys designed to give a preliminary indication of the main soil features of a relatively large area. Scales are usually in the range of 1:1 000 000 down to 1:500 000. Only broad groups of soils are mapped, sometimes in relation to broad divisions of relief, geology and vegetation.
2 Medium-scale surveys include reconnaissance surveys with scales usually in the range 1:250 000 to 1:100 000. The mapping unit is commonly a group of soils. In many tropical areas the mapping unit may be a soil catena, a sequence of soils related to relief (Chapter 6).
3 Large and very large-scale surveys include detailed and intensive soil surveys at scales between 1:25 000 and 1:1000. These maps usually show individual soils (soil series) or subdivisions of them. In general they are used for feasibility studies in connection with the development of a project, and to provide information relevant to detailed soil use and management. The largest-scale, most intensive surveys are usually for irrigation studies and are intended to show relatively slight variations in relief and soil characteristics (including salinity) that might affect management.

Soil survey methods

Of primary importance is the size of the area surveyed, the purpose for which it is surveyed, and the scale at which it is mapped. The scale in particular dictates the survey procedures best used.

The survey of very large areas, such as whole countries or provinces to be mapped on a small scale relies heavily nowadays on air-photo interpretation (API) and the interpretation of existing information on climate, geology, relief, vegetation and land use. The accompanying fieldwork is usually limited, and field observations confined to points selected on the basis of the photographs and maps.

At the other extreme, a very detailed soil survey of a small area to be mapped in detail at a large scale will rely much less on secondary information and will be based very largely on the direct examination and description of the soil at frequent intervals, often on a fixed grid or pattern of observations.

An 'observation' refers to the examination of the soil in the field for the purpose of description and mapping. Various types of observation are possible,

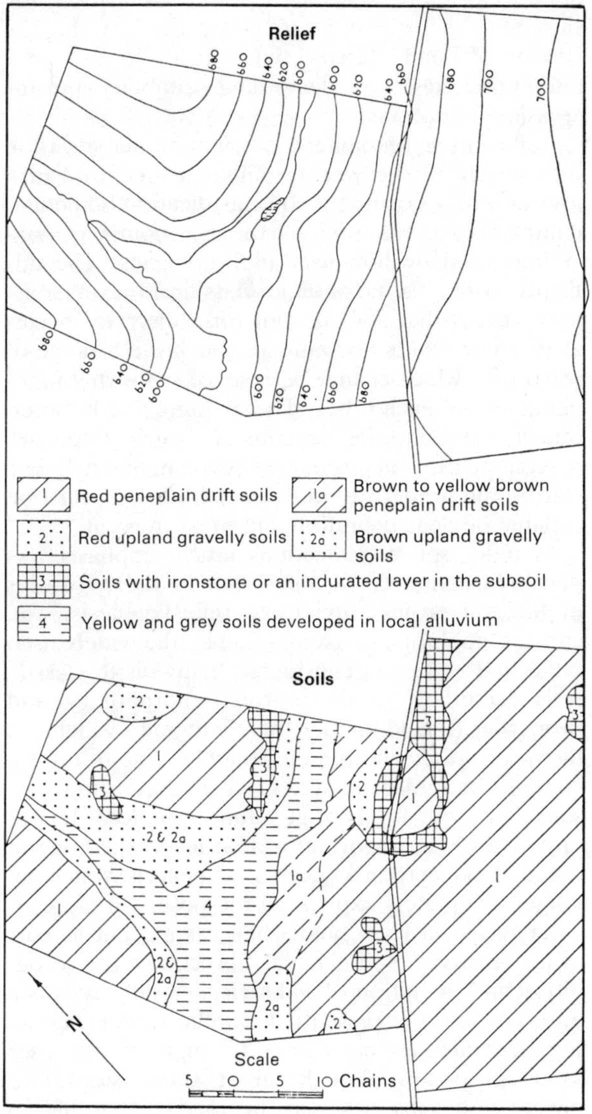

Fig. 8.1 A detailed soil map: Goaso Cocoa Station, Ghana. The scale as reproduced here is about 1:15 000, which is large enough to show individual soil series. Source Ahn, 1970.

including the examination of soils in ditches and road cuts, but most observations carried out by professional soil surveyors involve either auger borings or the digging of soil profile pits for more careful description and sampling.

The density of observations is related to the scale of the soil map. A general working rule is that there should be, on average, about one soil observation for every cm^2 of the final map. This means that for a very detailed soil map published at a final scale of 1:1000, field observations are correspondingly close, at every 10 m. For a reconnaissance survey at a scale of 1:100 000 (1 cm being equivalent to 1 km) the average distance between observations should be something like one every 1 km^2, and for exploratory surveys published at a scale of 1:1 000 000, observations will be about 10 km apart, or about one to every 100 km^2. This general rule has in practice to be modified according to the complexity of the area surveyed, more observations being needed for areas with a complicated soil pattern that changes over short distances, and somewhat less for more uniform areas with large expanses of similar soils.

A soil surveyor usually begins by studying all existing available information on the soils of the area and on those aspects of the environment that influence soil formation, particularly geology and geomorphology, relief, climate, and vegetation. He also studies the agriculture, land use and general geography of the area to be surveyed. The collection and study of this information is often referred to as the 'pre-fieldwork' since it precedes the examination of the soils in the field, and greatly helps in planning the exact fieldwork and location of observations to be aimed at.

Nowadays almost all soil survey work makes use of air photographs on a suitably large scale, sometimes backed up by satellite imagery. Air photographs as normally used consist of black-and-white or panchromatic vertical photographs, 23 cm^2, flown in parallel strips with considerable overlap to allow stereoscopic interpretation of adjacent pairs. Air photographs are used

1 as base maps, for planning routes and recording the location of field observations and soil boundaries, and

2 for interpretation, providing primary information on spatial distributions which supplements the field observations.

Information provided includes that on land use at the time the photographs were taken, and the proportions of arable land, pasture, forest, plantations, and the patterns of cultivation and settlement. In areas of sparse vegetation, rocks, dunes and other features are clearly seen and can be mapped. Differences in reflectance can indicate some soil surface characteristics and the presence of eroded or depositional areas.

For small-scale maps, soils observations and sampling are likely to be along the more accessible access routes, but for medium-scale maps use will be made of an increasingly close network of traverses. A traverse is a route, such as a road, footpath, compass-bearing or cut line, along which observations are made. Large-scale surveys usually involve auger and pit observations sited on a regular dense grid. Free survey involving only limited fieldwork and making extensive use of air-photo interpretation and of existing secondary information requires the greatest ability and experience on the part of the soil surveyor, whereas a survey involving sampling on a relatively close grid requires the least.

Soil classification

The classification of tropical soils has undergone very considerable changes during the last 30 years as increasing numbers of detailed soil surveys, descriptions and analyses have added to our knowledge of tropical soils. This new knowledge has shown that tropical soils are at least as varied as temperate soils, and that many of the older generalisations about tropical soils were misleading and inaccurate.

Soil classification is a technique whereby soils are grouped together in various ways and according to various criteria to form categories or classes. These classes are defined in such a way as to be useful in understanding the formation, properties and uses of soils. The type of classification employed depends fundamentally on the purpose of the classification.

Some soil classifications are deliberately simple and practical, as when soils are classified according to major properties affecting their agricultural value, such as depth, drainage, slope, texture, or salinity. A simple but useful example of soil classification emphasising limitations to use is provided by the land capability classification of the US Soil Conservation Service considered later in this chapter. Somewhat more specific is the classification for suitability for irrigated agriculture used by the US Bureau of Land Classification which also puts land into broad classes of decreasing suitability and increasing limitations.

For a more fundamental, scientific classification into which all the world's soils can be fitted in a logical way, a taxonomic soil classification is needed, comparable in its way to the taxonomic classifications used by botanists and zoologists. The difficulty with the taxonomic classification of soils, however, is that soils do not form discrete species as do most plants and animals, but have a range of properties which change with space somewhat independently of each other. The dividing line between one soil and another depends on which properties or combination of properties are emphasised. For clarity, the properties used to separate soils must be capable of clear definition and measurement.

In older soil classifications much emphasis was placed on assumed soil genesis and on those features of the soil profile thought to reflect genesis. The current emphasis, as exemplified by the widely used US Soil Taxonomy, considered below, is to classify soils according to their profile morphology and properties (including the climatic regime) which can be measured by accepted techniques. The result is a complex but objective system which groups together soils at various levels of generalisation according to properties held important at that level, shedding in the process certain concepts of genesis which no longer adequately explain all the observed facts.

At the local level, and when carried out as part of a soil survey, soil classification entails the recognition and definition of **soil series**. A soil series is a group of soils with essentially similar profile characteristics which are considered for practical purposes as a single soil, though minor subdivisions into phases and subseries may be made according to such features as variations in the texture of the A horizon, slope, stoniness, and other factors. In theory soils of a single series with similar profile characteristics are similar because they have been formed in a similar way. This implies (Chapters 6 and 7) that all the factors of soil formation are similar. Soils of a single series are thus developed from similar parent materials, with similar relief, under similar climate and vegetation, and showing a similar degree of profile development (age). A soil survey report should not only describe the soil series mapped but should include a description of the range of characteristics permitted for each series, including the range of variation of depth, texture, colour and other morphological characteristics included in the series.

Classification at a higher level involves the grouping of soil series into broader groups of soils or classes based on common criteria selected as important. In the US Soil Taxonomy, for example, over 10 000 recognised soil series are grouped first into some 4500 families. The families form about 1000 subgroups, which in turn form part of about 185 great groups. These great groups are classified finally into 47 suborders and, at the highest level of generalisation, into 10 orders.

The criteria used for differentiating first the orders, and then the suborders, great groups and subgroups, are of the highest possible interest and importance to soil scientists and agronomists, particularly when these criteria reflect properties of agricultural value and succeed in separating groups of soils with common properties as regards soil management.

The United States Soil Taxonomy

The *US Soil Taxonomy* was published in 1970 after about 20 years of development and modification, and further changes are envisaged to reflect new knowledge and experience. Soils are separated into orders, suborders, great groups, subgroups and families. Temperate and tropical soils of similar morphology are normally separated at the subgroup or family level by use of temperature regimes (Chapter 1).

There are several important features of the Soil Taxonomy:

1 The system is based on soil properties that can be measured quantitatively.

2 The classification makes extensive use of diagnostic horizons.

3 The nomenclature used in designating orders, suborders and lower classes follows a regular system using various roots which indicate the concepts used in classification.

Quantitative measurement of criteria A difficulty with many other soil classifications has been that some criteria used in separating soils have not been capable of exact definition and measurement. In the Soil Taxonomy the soil properties used to describe and separate soils are all defined by accepted techniques. For soil colour, the standard Munsell soil colour chart is used. For soil texture, organic matter content, cation exchange capacity, percentage base saturation, pH, water dispersible clay, extent of soil cracking, and presence of clay skins and other properties, standard laboratory procedures and other measurements are used and classes defined accordingly. In addition to profile morphology, considerable use is made of soil moisture and temperature regimes. These cannot be obtained by examining and analysing the soil but have to be calculated from climatic data.

Diagnostic horizons A major feature of the Soil Taxonomy which has in itself much advanced the study of soil formation and classification is the use of carefully defined diagnostic horizons. A diagnostic horizon is a soil horizon that has certain carefully defined properties. The absence or presence of these horizons is used in classifying soils. The presence of an oxic horizon, for example, is a requirement of all soils in the Oxisol order. The use of diagnostic horizons is a convenient way of drawing attention to certain soil characteristics used in classification.

Diagnostic horizons are of two types, the surface horizons, known as epipedons (from the Greek, *epi*, over, and *pedon*, soil) and the subsurface horizons. The more important diagnostic horizons are the following:

Surface diagnostic horizons (epipedons)

1 **Mollic** A thick, dark-coloured humous surface horizon with a high base saturation and a good crumb structure (from *mollic*, soft). Typical of the black earths (Chernozems) of humid temperate grasslands.

2 **Umbric** Thick dark humous surface horizon similar to mollic but with lower base saturation (from *umbric*, dark).
3 **Ochric** Surface horizon with lower organic matter and lighter colour than mollic or umbric, sometimes with weaker structure, or massive (from *ochric*, light coloured).
4 **Histic** Surface horizon that is very high in organic matter and which is wet during part of the year.

Subsurface diagnostic horizons
1 **Argillic** Subsurface horizon having a significantly higher clay content that the horizon above, suggesting movement of clay from the A to the B horizon as shown by the presence of clay skins (coatings of oriented clay deposited on ped faces and on walls of soil pores and channels).
2 **Natric** An argillic horizon modified by high sodium; has columnar or prismatic structure.
3 **Spodic** The dark horizon of accumulation of a podsol; accumulation of organic matter, iron and aluminium oxides usually below an overlying eluviated bleached horizon known as albic.
4 **Cambic** A changed or altered horizon; a horizon other than argillic, natric or spodic in which parent material has been subjected to chemical or physical changes as shown by changes in colour, texture or structure.
5 **Oxic** Highly weathered horizon high in iron and aluminium oxides. Has dominantly 1:1 type clay minerals, little or no water-dispersible clay, little or no weatherable minerals, and low cation exchange capacity. Horizon boundaries are usually diffuse.

Orders, suborders, and lower classes of the Soil Taxonomy The 10 orders of the Soil Taxonomy separate soils according to the degree and type of development shown by the soil profile.

1 **Entisols** Entisols are very recent soils, so young that they have not yet developed any genetic horizons other than an ochric surface horizon. The dominant feature is a lack of profile development. Included are soils that have been deposited too recently for there to be time for profile development, and also soils which have for some reason been unable to develop genetic horizons in the subsoil. The latter include shallow rocky soils, often associated with steep slopes and resistant parent material, previously known as lithosols, and soil profiles developed in very inert quartz sand previously known as regosols but now termed Psamments in the Soil Taxonomy.

2 **Vertisols** The name (from 'invert') suggests the inverting or self-mixing or self-mulching that is typical of these soils, caused by topsoil fragments falling down or being washed down the wide cracks and thus being mixed with other material below. Vertisols are heavy cracking clays with more than 35% clay, of which over 50% of the clay minerals are of the expanding 2:1 lattice type. They are defined by the presence of cracks at least 1 cm wide, present at some time in most years to at least 50 cm. They shrink and swell considerably with changes in moisture content, and have gilgai microrelief. The peds are commonly marked by striated pressure faces termed slickensides and usually include wedge-shaped nature structural aggregates that have their long axes tilted 10–60° from the horizontal.

3 **Inceptisols** Inceptisols (from 'inception', beginning) are young soils with more profile development than the Entisols but with less than most of the remaining orders. The only diagnostic subsoil horizon permitted is a cambic horizon.

4 **Aridisols** Aridisols are soils of dry regions with little profile development. They have an aridic moisture regime (Chapter 1); i.e. the moisture control section of the profile is dry throughout more than half the time, and is never moist in any part for as long as 90 consecutive days. They normally have a light-coloured surface horizon that is low in organic matter (ochric epipedon) and may have a calcic, gypsic or saline horizon, or contain appreciable amounts of soluble salts.

5 **Mollisols** Mollisols are characterised by a mollic epipedon, i.e. a surface horizon that is high in organic matter, soft when dry and with over 50% base saturation. The surface horizon is usually dark and has a strong crumb or granular structure. Mollisols are associated mainly with temperate grasslands.

6 **Spodosols** Spodosols are defined as having a spodic horizon, a dark, sometimes indurated subsurface horizon that is high in organic matter, aluminium and usually iron oxides which have been leached downwards by organic acids and then preci-

pitated to form the spodic horizon. The spodic horizon is often overlain by a bleached, eluviated horizon, the albic. Spodosols (formerly known as podsols or podzols and still referred to as such in other classification systems) are widespread in temperate zones, and are particularly associated with pine forests whose litter is low in bases and therefore acid. In the tropics, podsols and giant (very deep) podsols are not areally important but occur in a number of areas (including West Africa, Suriname and Sarawak), and are found mainly developed in areas of high rainfall over very sandy materials.

7 **Alfisols** Alfisols are a varied group of moderately to well-weathered soils generally associated with moderate to heavy rainfalls, being more weathered and showing more profile development than Inceptisols but less weathered than Ultisols and Oxisols. They are very widespread and important in the tropics. They are defined as having an argillic horizon with more than 35% base saturation, or a natric horizon. They lack mollic, spodic or oxic horizons. Morphologically they are similar to Ultisols, which also have an argillic horizon, but because of their higher saturation and pH they are often more fertile than the Ultisols which are generally associated with higher rainfalls and a greater degree of leaching.

8 **Ultisols** Ultisols (from the Latin, *ultimus*, the last) are highly weathered soils, but they are not, as the name might suggest, the last in the weathering sequence – a place reserved for the Oxisols. Morphologically they are comparable to the Alfisols since they also have an argillic horizon, but are separated from Alfisols by the lower base saturation of the argillic horizon which must be less than 35%. They are typically associated with warm to hot, wet climates and are important in the wetter tropics. They often contain enough iron oxides to give them red, brown or yellow colours depending on the degree of hydration (Chapter 6). The clay mineral is generally kaolinitic. The cation exchange capacity is therefore low but varies with the soil reaction. The physical properties of these soils are often good.

9 **Oxisols** Oxisols are defined as soils containing an oxic horizon. Oxic horizons have a low cation exchange capacity (defined as less than 16 m.e./100 g) but since the clay fraction is dominated by kaolinitic minerals and iron and aluminium oxides with pH dependent charges, the low cation exchange capacity varies with pH and is even lower in acid oxisols. Oxisols are considered the most highly weathered soils, the end product of thorough weathering in hot wet tropical environments. They contain little or no weatherable minerals and only traces of water-dispersible clay. The definition of an Oxisol is such that only the most highly weathered tropical soils qualify, and even in the hot wet tropics these soils may be subordinate in extent to Ultisols, often being confined to old land surfaces where weathering has been particularly long and severe. Chemically these soils are not very active, and contain little or no reserves of unweathered minerals, but physically they are very stable and often have an excellent microstructure, and good porosity and permeability. Their good physical properties compensate for their usually poor to moderate fertility status after the original vegetation has been lost.

10 **Histosols** Histosols are organic soils containing a histic epipedon (over 20% organic matter). They are often associated with poor drainage and depression sites, and are subdivided according to the proportion of fibre contained in the organic matter.

The important orders in the tropics, in order of relative extent, are thought to be approximately as follows:

Oxisols about 22% of the tropics, mainly in the Amazon bason and the *campo cerrado* areas of South America, in Central Africa and on old, highly weathered surfaces in West Africa. No large Oxisol areas have been reported in Asia.

Aridisols about 18% of the tropics, mainly in the deserts of Africa with scatterd areas in tropical America and Asia.

Alfisols about 16% of the tropics, including many areas of West Africa, India and Sri Lanka previously mapped as 'latosols' or 'laterites'.

Ultisols about 11% of the tropics. They are widespread and important in many of the wetter areas of Africa and tropical America, and are often the dominant upland soils of South-east Asia.

Inceptisols and Entisols each cover about 8% of the tropics. The Entisols include large areas of sandy soils, Psamments, in the drier parts of West Africa and elsewhere.

Vertisols cover about 2% of the tropics. Large areas are found in Sudan, Ethiopia, India, West Africa and Java.

The 10 orders of the Soil Taxonomy are divided into 47 suborders and about 185 great groups. The **suborders** often reflect a climatic or drainage influence. Their names are constructed by taking the roots of the orders and adding a prefix to indicate salient properties. Thus the root *Ox* in Oxisols is used with the prefixes *Orth*, *Ust*, *Torr*, *Hum* and *Aqu* to give:

Orthox Oxisols with udic soil moisture regimes
Ustox Oxisols with ustic soil moisture regimes
Torrox Oxisols with aridic soil moisture regimes
Humox Oxisols with high organic matter
Aquox Oxisols with aquic moisture regimes.

The **great group** names are formed by putting a prefix before the suborder name. For example, the Aquox suborder may be divided into great groups by adding the prefixes *Ochr*, *Umbr*, *Plinth*, and *Gibbsi*, to give:

Ochraquox Aquoxes with an ochric A horizon
Umbraquox Aquoxes with an umbric A horizon
Plinthaquox Aquoxes with plinthite
Gibbsiaquox Aquoxes with gibbsite.

Simple definitions of some of the more important suborders and great groups of the Soil Taxonomy found in the tropics are listed in Table 8.1.

Table 8.1 Some suborders and great groups of the Soil Taxonomy of importance in the tropics, with very brief indications of salient features.(For more details and more precise definitions, readers are referred to the *Soil Taxonomy*, Soil Survey Staff, (1975) and to *Keys to Soil Taxonomy*, Soil Management Support Services (1987)).

Orders	Suborders	Great groups
1 Entisols	*Aquents*	Entisols with aquic moisture regimes
		Tropaquents – Aquents of the tropics
		Fluvaquents – alluvial Aquents
		Hydraquents – high moisture content, thixotropic
		Sulfaquents – acid sulphate soils or cat clays
	Fluvents	Entisols of recent alluvial origin
		Tropofluvents – tropical Fluvents
		Ustifluvents – Fluvents with ustic moisture regimes
		Torrifluvents – Fluvents with aridic moisture regimes
	Psamments	Very sandy Entisols
		Tropopsamments – simple tropical Psamments
		Ustipsamments – with ustic soil moisture regimes
		Torripsamments – with aridic soil moisture regimes
		Quartzipsamments – with mainly quartz sand
	Orthents	other tropical Entisols
		Troporthents – typical
		Ustorthents – with ustic soil moisture regimes
		Torriorthents – with aridic soil moisture regimes
2 Vertisols	*Usterts*	Vertisols with ustic soil moisture regimes
		Chromusterts – with a chromic A horizon
		Pellusterts – dark coloured
		Uderts – Vertisols with udic soil moisture regimes
		Torrerts – Vertisols with aridic soil moisture regimes
3 Inceptisols	*Andepts*	Inceptisols derived from volcanic materials
		Cryandepts – cold Andepts
		Vitrandepts – high in volcanic glass
		Dystrandepts – low base saturation
		Eutrandepts – high base saturation
		Hydrandepts – well drained, with high water content
		Durandepts – with duripan

Table 8.1 (cont)

Orders	Suborders	Great groups
	Aquepts	Inceptisols with aquic soil moisture regimes
		Tropaquepts – tropical aquepts
		Andaquepts – affected by volcanic ash
		Halaquepts – saline Aquepts
		Sulfaquepts – acid sulphate soils or cat clays
		Plinthaquepts – with plinthite
	Tropepts	other tropical Inceptisols
		Dystropepts – low base status
		Eutropepts – high base status
		Ustropepts – with ustic soil moisture regimes
		Humitropepts – high in organic matter (high altitudes)
4 Aridisols	*Argids*	Aridisols with an argillic horizon
		Haplargids – typical
		Paleargids – with a deep argillic horizon
		Natrargids – with a natric horizon
		Durargids – with duripan
	Orthids	Aridisols without an argillic horizon
		Salorthids – with a salic horizon
		Paleorthids – with an argillic horizon at depth
		Calciorthids – with a calcareous horizon
		Gypsiorthids – with a gypsic horizon
		Camborthids – with a cambic horizon
		Durorthids – with duripan
5 Mollisols	*Rendolls*	Mollisols over limestone (formerly called rendzinas)
	Ustolls	Mollisols with ustic moisture regimes
		Haplustolls – typical
		Argiustolls – with an argillic horizon
		Paleustolls – with a deep argillic horizon
		Calciustolls – with a calcic horizon
		Durustolls – with duripan
		Natrustolls – with a natric horizon
	Aquolls	Mollisols with aquic moisture regimes
		Haplaquolls – typical
		Argiaquolls – with an argillic horizon
		Calciaquolls – with a calcic horizon
		Duraquolls – with duripan
		Natraquolls – with a natric horizon
6 Spodosols	*Aquods*	Spodosols with an aquic moisture regime
		Tropaquods – tropical Aquods
		Duraquods – with hard pan
	Humods	Spodosols high in organic matter
		Tropohumods – tropical Humods
	Orthods	other Spodosols
		Troporthods – tropical Orthods
7 Alfisols	*Udalfs*	Alfisols with udic soil moisture regimes

Table 8.1 (cont)

Orders	Suborders	Great groups
	Ustalfs	Alfisols with ustic soil moisture regimes
		Haplustalfs – typical
		Paleustalfs – with a deep argillic horizon
		Rhodustalfs – dusky red colour, high in oxides
		Plinthustalfs – with plinthite
		Natrustalfs – with a natric horizon
		Durustalfs – with duripan
	Aqualfs	Alfisols with aquic soil moisture regimes
		Plinthaqualfs – with plinthite
		Natraqualfs – with a natric horizon
		Duraqualfs – with duripan
8 Ultisols	*Udults*	Ultisols with udic soil moisture regimes
		Paleudults – with a very deep argillic horizon
		Rhodudults – dusky red in colour, high in oxides
		Plinthudults – with plinthite
		Fragiudults – with fragipan
	Ustults	Ultisols with ustic soil moisture regimes
		Haplustults – typical
		Paleustults – with a deep argillic horizon
		Rhodustults – dusky red colour, high in oxides
		Plinthustults – with plinthite
	Humults	Ultisols high in organic matter (usually at high altitudes)
		Palehumults – with a deep argillic horizon
		Plinthohumults – with plinthite
	Aquults	Ultisols with aquic soil moisture regimes
		Paleaquults – with a deep argillic horizon
		Plinthaquults – with plinthite
		Fragaquults – with fragipan
		Albaquults – light coloured A horizon
9 Oxisols	*Orthox*	Oxisols with udic soil moisture regimes
		Haplorthox – typical
		Eutrorthox – high base saturation
		Acrorthox – very low base saturation, acid
		Gibbsiorthox – with gibbsite
	Ustox	Oxisols with ustic soil moisture regimes
		Haplustox – typical
		Eutrustox – high base saturation
		Acrustox – very low base saturation, acid
	Torrox	Oxisols with aridic soil moisture regimes
	Humox	Oxisols with high organic matter (usually at high altitudes)
		Haplohumox – typical
		Acrohumox – very low base saturation, acid
		Gibbsihumox – with gibbsite
	Aquox	Oxisols with aquic soil moisture regimes
		Ochroquox – with an ochric A horizon
		Umbraquox – with an umbric A horizon
		Plinthaquox – with plinthite
		Gibbsiaquox – with gibbsite

Table 8.1 (cont)

Orders	Suborders	Great groups
10 Histosols		Fibrists – with over two-thirds fibre Folists – not saturated with water Hemists – between one-third and two-thirds fibre Saprists – less than half fibre.

The FAO/Unesco World Soil Map Legend

The FAO/Unesco World Soil Map Legend, sometimes more loosely referred to as the FAO classification system, is not a system of soil classification as such but, as its correct name suggests, is the legend to a soil map of the whole world published at a scale of 1:5 000 000 (1 cm on the map corresponds to 50 km on the ground).

Aims of the FAO/Unesco 1:5 000 000 soil map The FAO/Unesco soil map of the world was started in 1961 as a result of a recommendation of the International Society of Soil Science. It was, as stated in the publication, 'a first attempt to prepare, on the basis of international cooperation, a soil map covering all continents of the world in a uniform legend, thus enabling the comparison of soils on a global scale'.

It was recognised that a major obstacle to the comparative study of soil resources in the past had been the fact that similar soils had been given a wide variety of names in different areas. The diversity in the nomenclature itself reflected different approaches to soil classification and the selection of different criteria as being important.

The new map was not intended to replace local classification systems but to serve as a common denominator which would lead to better understanding and, hopefully, to an internationally accepted system of soil classification which would 'considerably strengthen the status and impact of soil science in the world.

The objectives of the soil map of the world included the following:

1 To make a first appraisal of the world's soil resources.
2 To supply a scientific basis for the transfer of experience between areas with similar environments and to strengthen international contacts in the field of soil science.
3 To promote the establishment of a generally accepted soil classification and nomenclature.

The FAO/Unesco soil map and its legend The complete publication consists of a booklet and a map sheet explaining the legend itself, and nine sets of maps, each set with an accompanying explanatory text. The nine sets of maps total 18 map sheets and, as indicated above, are all on the same scale of 1:5 000 000. The publication of the set was completed in 1974. A revised legend was published in 1988.

The project aimed at mapping the world's soils, and not at setting up a hierarchical system of soil classification. The list of soil units used is monocategorical, and does not aim to be a taxonomic system subdivided into categories at different levels of generalisation. However, for the sake of a logical presentation, the 153 mapping units used in the revised legend have been grouped into 28 groups of units (World Classes) on the basis of generally accepted principles of soil formation.

In contrast to the US Soil Taxonomy, which coined completely new names for orders, suborders and lower categories of soils, the FAO/Unesco soil map used the existing soil terminology as far as was practicable. Thus traditional soil terms such as Chernozems, Kastanozems, Podzols, Planosols, Solonetz, Solonchak, Rendzinas, Regosols, Vertisols, Andosols, Gleysols, and Lithosols were retained, though these soils were defined in such a way as to narrow down the meanings sometimes given to them. At the same time, a number of well-established and widely used terms had to be dropped because they could not be retained without perpetuating the confusion created by their dissimilar use in different countries. Among the terms that had to be dropped for this reason were Lateritic and Alluvial soils.

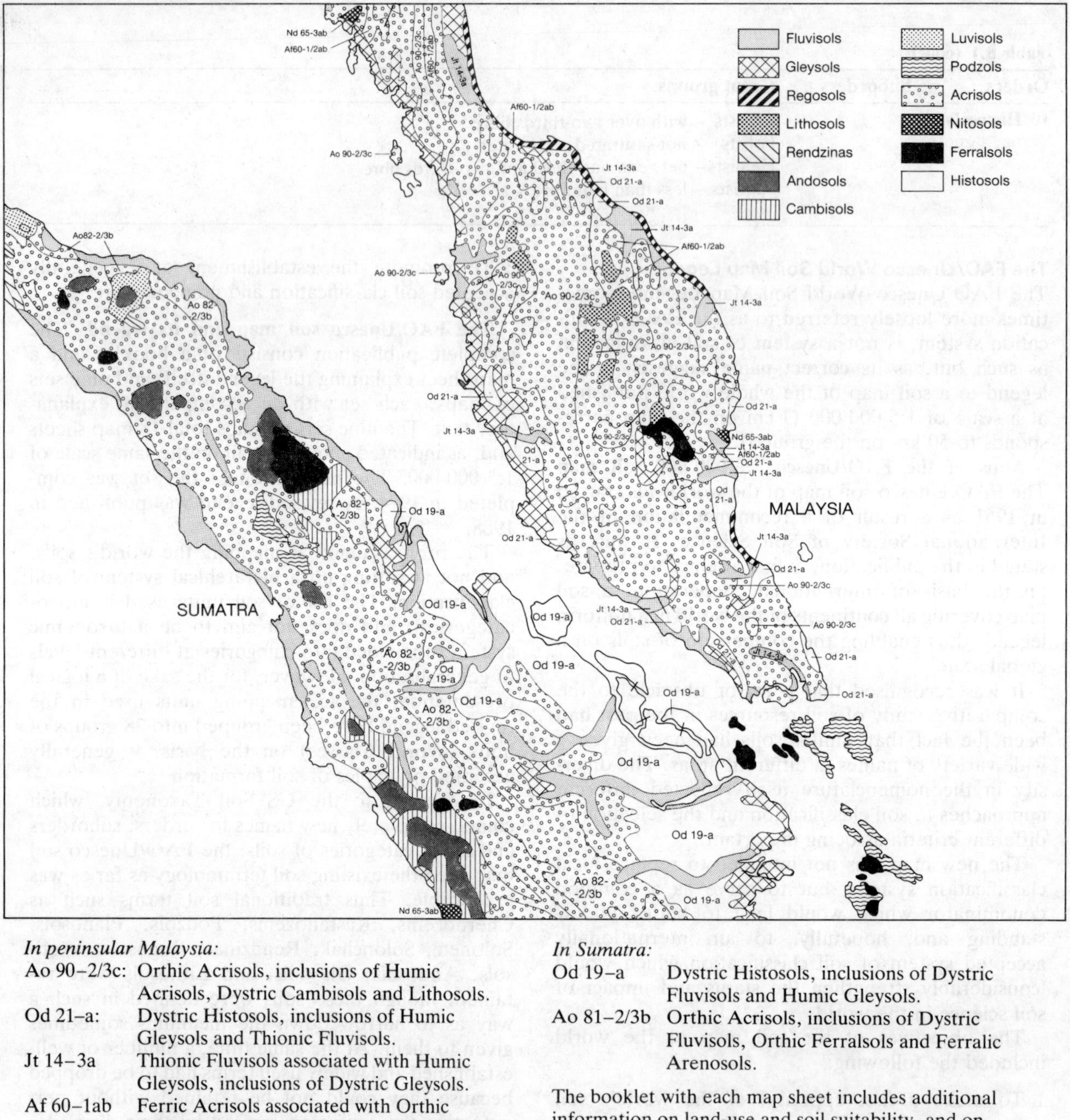

In peninsular Malaysia:

Ao 90–2/3c:	Orthic Acrisols, inclusions of Humic Acrisols, Dystric Cambisols and Lithosols.
Od 21–a:	Dystric Histosols, inclusions of Humic Gleysols and Thionic Fluvisols.
Jt 14–3a	Thionic Fluvisols associated with Humic Gleysols, inclusions of Dystric Gleysols.
Af 60–1ab	Ferric Acrisols associated with Orthic Acrisols and Ferralic Cambisols, inclusions of Gleyic Acrisols and Ferralic Arenosols.
Nd 65–3ab	Dystric Nitosols associated with Dystric Cambisols, inclusions of Chromic Vertisols, Rendzinas and Lithosols.

In Sumatra:

Od 19–a	Dystric Histosols, inclusions of Dystric Fluvisols and Humic Gleysols.
Ao 81–2/3b	Orthic Acrisols, inclusions of Dystric Fluvisols, Orthic Ferralsols and Ferralic Arenosols.

The booklet with each map sheet includes additional information on land-use and soil suitability, and on environmental conditions. Tables give the climate and area of each mapping unit shown on the map.

Fig. 8.2 Part of the South Asia map sheet (sheet IX) of the FAO/Unesco 1:5 000 000 soil map published in 1979, simplified and redrawn in black and white.

The map is an example of a small-scale soil map, with each centimetre on the map representating 50 km on the ground. At such a small scale only the dominant soil associations can be mapped.

The original map uses different colours for each soil unit. Each mapping unit is designated with a symbol indicating the dominant soil unit, together with associated soils, if present, and inclusions. The symbol may also indicate the dominant textural class (1 coarse, 2 medium, 3 fine-textured) and slope (a: level to undulating, b: rolling to hilly, c: steeply dissected to mountainous). Where Lithosols are dominant, a stripped pattern of shading is used, and various overlays indicate soil phases such as stony, lithic, petrocalcic, saline or sodic.

Each mapping unit symbol used is explained at the back of the map sheet. The examples listed below the figure illustrate the information given in addition to the textural and slope symbols.

For a limited number of soils it was found imperative to create new names. The new soils named included Luvisols, Acrisols, Cambisols, Phaeozems, Yermosols, Xerosols, Ferralsols, Nitosols, Fluvisols, Histosols and Arenosols (Table 8.2).

Table 8.2 The 26 original major categories of the FAO/Unesco map legend. These are subdivided into 106 mapping units by using the adjectives listed in Table 8.3. In 1988 a revised legend (without new maps) was published which increased the number of units to 153, and the number of major soil groupings to 28.

Class	Brief description
A Acrisols	(Latin *acris*, very acid) Soils with an argillic horizon which is less than 50% saturated with bases (compare Luvisols, where the argillic horizon is over 50% saturated). About equivalent to the Ultisols of the US Soil Taxonomy.
B Cambisols	(Latin *cambiare*, to change) Soils with a cambic B horizon, and with no other diagnostic subsurface horizon except a calcic or gypsic one. A cambic horizon suggests relatively slight changes in colour, structure and consistency due to weathering, as shown by the absence of diagnostic subsurface horizons other than those listed.

Table 8.2 (cont)

Class	Brief description
C Chernozems	(Russian *chern*, black) Soils rich in organic matter and black in colour, and usually developed under natural temperate grasslands with cold winters. Similar to Kastanozems except that the mollic A horizon is darker (moist chroma of 1.5 or less on the Munsell colour chart).
D Podzoluvisols	Soils with an argillic horizon which has a very irregular upper boundary with tongues of the overlying pale eluviated horizon penetrating into it (intermediate in development between Luvisols and Podzols).
E Rendzinas	(Polish *rzedzic*, noise, suggesting the noise of a plough scraping through a stony limestone soil) Soil with a mollic A horizon overlying calcareous material within 50 cm.
F Ferralsols	(Latin *ferrum*, iron, and *aluminium*) Soils with an oxic B horizon, i.e. highly weathered soils high in sesquioxides. These include the Oxisols of the US Soil Taxonomy, and the French *Sols ferralitiques*.
G Gleysols	(Russian *gley*, a mucky soil mass) Soils affected by water, with hydromorphic properties. Must have a gleyic horizon with its upper boundary within 50 cm of the surface (may also have a cambic, calcic, gypsic, or plinthic subsurface horizon).
H Phaeozems	(Greek *phaios*, dusky) Soils rich in organic matter with a mollic, dark A horizon. However, there are no calcic, gypsic or calcereous layers as are found in Kastanozems and in Chernozems.
I Lithosols	(Greek *lithos*, a stone) Soils with hard rock within 10 cm of the surface.
J Fluvisols	(Latin *fluvius*, a river; French *sol*, a soil) Soils developed in recent alluvium with no diagnostic horizon, or with none other than an ochric horizon, an O horizon, a hydromorphic horizon or a sulphuric horizon. Connotative of floodplains and alluvial deposits relatively little weathered.

Table 8.2 (cont)

Class	Brief description
K Kastanozems	(Latin *castaneo*, chestnut; Russian *zemlja*, earth) Soils rich in organic matter and brown or chestnut in colour. Must have a mollic A horizon at least 15 cm thick with either a calcic or gypsic horizon or with soft powdery lime within 125 cm of the surface.
L Luvisols	(Latin *luvi*, from *luo* wash) Soils with an argillic horizon within 125 cm of the surface which is more than 50% saturated with bases (compare with Acrisols which are similar except that the argillic horizon is less than 50% saturated). About equivalent to the Alfisols of the US Soil Taxonomy.
M Greyzems	(English *grey*, the colour) Soils rich in organic matter, grey in colour with mollic A horizon, bleached coatings on structural ped faces giving a greyish colour. Usually with an argillic horizon.
N Nitosols	(Latin *nitidus*, shining) Soils with a deep argillic horizon in which the clay content decreases by less than 20% from its maximum within 125 cm of the surface. Usually with bright, lustrous coatings on the ped faces, particularly in the lower part of the argillic horizon. No plinthite within 125 cm of the surface.
O Histosols	(Greek *histos*, tissue) Soils rich in fresh or partially decomposed organic matter. Must have an H horizon at least 40 cm thick.
P Podzols	(Russian *pod*, under, and *zola*, ash) Soils with a spodic B horizon. Normally a bleached eluvial horizon overlies a dark horizon of accumulation, the spodic B.
Q Arenosols	(Latin *arena*, sand) Soils developed in over 50 cm of loose sand (other than recent alluvium).
R Regosols	(Greek *rhegos*, blanket) Young soils developed in unconsolidated materials other than recent alluvium; no gleyic horizon within 50 cm of the surface and only an ochric surface horizon. Implies a loose mantle overlying hard rock.

Table 8.2 (cont)

Class	Brief description
S Solonetz	Salt-affected soils; must have a natric B horizon.
T Andosols	(Japanese *An*, dark, and *Do*, soil) Soils developed in volcanic ash, usually with a dark surface horizon, characterised by a particularly low bulk density (less than 0.85) or by over 60% vitric (glassy) volcanic ash in the silt, sand and gravel fractions.
U Rankers	(Austrian *rank*, steep) Shallow soils of steep slopes on siliceous parent material; must have an umbric A horizon less than 25 cm thick and no other diagnostic horizon, or an umbric A horizon 25–50 cm thick and a cambic B horizon.
V Vertisols	(Latin *verto*, turn, connotative of a churning surface soil due to swelling and shrinking with moisture changes) Soils with over 30% clay to at least 50 cm and with cracks at least 1 cm wide and 50 cm deep forming at some period in most years.
W Planosols	(Latin *planus* flat) Poorly drained soils of flat or depressed topography. Must have an albic E horizon overlying a slowly permeable horizon (such as an argillic or natric horizon, or a clay pan, but not a spodic horizon) within 125 cm of the surface.
X Xerosols	(Greek *xerox*, dry) Similar to Yermosols, but with a semi-arid rather than arid climate, and a weak rather than very weak ochric A horizon.
Y Yermosols	(Latin *eremos*, desolate; Spanish *yermo*, a desert) Soils of dry areas with a very weak ochric A horizon and one or more of the following: a cambic or argillic B, a calcic horizon, a gypsic horizon.
Z Solonchaks	(Russian *sol*, salt) Soils with high salinity and no diagnostic horizon other than an A horizon, an H horizon, a cambic B or a calcic or gypsic horizon.

Diagnostic horizons The FAO/Unesco soil map legend makes considerable use of diagnostic horizons in designating soil properties and defining mapping units. These diagnostic horizons include the mollic, umbric, histic and ochric surface horizons. The subsurface diagnostic horizons used include the argillic, natric, cambic, spodic, oxic, calcic, gypsic and salic subsurface horizons. All these surface and subsurface horizons are essentially similar to those used in the US Soil Taxonomy and are defined in a fairly similar way, though definitions differ in a few details.

In addition to the surface and subsurface horizons used in common with the US Soil Taxonomy, the FAO/Unesco soil map legend makes use of four additional diagnostic horizons:

H horizon	organic horizon with over 30% organic matter if texture is clay, and over 20% if it is lighter than clay. Saturated with water for long periods.
O horizon	organic horizon with over 35% organic matter that is not usually saturated with water.
E horizon	eluvial horizon, high in silt and sand, usually light in colour.
Sulphuric horizon	horizon resulting from artificial drainage which is high in sulphides and has a pH of less than 3.5 in water.

The 106 mapping units The 106 original mapping units of the published map were grouped into 26 World Classes, conveniently designated by the 26 letters of the alphabet, as set out in Table 8.2. Within each group, units are designated by the use of the adjectives listed in Table 8.3. Thus Fluvisols, for example, are pedologically young soils developed in recent alluvium or in alluvium enriched by fresh sediments. They have no diagnostic horizon, or no other diagnostic horizon than an ochric horizon, an O horizon, a hydromorphic horizon or a sulphuric horizon. Fluvisols are then subdivided into four mapping units using the adjectives Eutric (good or fertile), Calcaric (with an accumulation of calcium carbonate), Dystric (poor or infertile) and Thionic (containing sulphur) to give the four units Eutric Fluvisols, Calcaric Fluvisols, Dystric Fluvisols and Thionic Fluvisols.

Table 8.3 Adjectives used in the FAO/Unesco soil map legend to qualify soil units

Adjective	Meaning
Acric	very acid (Latin *acris*, very acid)
Albic	bleached (Latin *albus*, white)
Calcaric	with $CaCO_3$ accumulation (Latin *calcium*)
Chromic	high colour (chroma) (Greek *chromos*, colour)
Dystric	infertile (Greek *dys*, ill)
Eutric	good, fertile (Greek *eu*, good)
Ferralic	high sesquioxide content (Latin *ferrum*, iron, and *aluminium*)
Ferric	high in iron (Latin *ferrum*, iron)
Gelic	permafrost layer (Latin, *gelu*, frost)
Gleyic	gleyed (Russian *gley*, mucky soil mass)
Glossic	tonguing of A horizon into the B horizon (Greek *glossa*, tongue)
Gypsic	containing gypsum (Latin *gypsum*)
Haplic	normal, simple type (Greek *haplos*, simple)
Humic	rich in organic matter (Latin *humus*, earth)
Leptic	weak development (Greek *leptos*, shallow)
Luvic	illuvial clay layer (Latin *luvi*, from *luo* wash)
Mollic	good surface structure (Latin *mollis*, soft)
Orthic	the common type (Greek *orthos*, true, right)
Placic	with a thin iron pan (Greek *plax*, flat stone)
Plinthic	containing mottled iron-rich material that hardens (Greek *plinthos*, brick)
Rhodic	reddish (Greek *rhodon*, rose)
Solodic	affected by salts (Russian *sol*, salt)
Takyric	soils of a barren plain in Russia (Uzbek *takyr*, barren plain)
Thionic	presence of sulphur (Greek *theion*, sulphur)
Vertic	turning over of surface soil (Latin *verto*, turn)
Vitric	rich in volcanic glass (Latin *vitrum*, glass)
Xanthic	pale yellow (Greek *xanthos*, yellow)

Common adjectives used in designating mapping units within groups are Eutric and Dystric to denote relatively fertile or infertile units respectively, Acric to denote acid soils, Gleyic to denote gleying and hydromorphic properties, Luvic to denote an illuvial clay horizon, and Plinthic to denote the presence of plinthite (mottled, iron-rich soil material capable of hardening). The term 'Orthic' refers to the common type or central concept, to distinguish it from special cases. The complete list of adjectives employed in designating the 106 original mapping units, and their meanings, is given in Table 8.3.

Differences between the FAO/Unesco map legend and the United States Soil Taxonomy

The foregoing outlines of the US Soil Taxonomy and the FAO/Unesco world soil map legend indicate that the major differences between the two include the following:

1 The US Soil Taxonomy is a hierarchical taxonomic classification. It includes 10 orders, 47 suborders, 185 great groups, about 1000 subgroups and 4500 families, comprising over 10 000 series, and as new soils are described and classified, these are incorporated. The FAO/Unesco map legend, on the other hand, is a legend to a published map at the 1: 5000 000 scale with only 106 mapping units grouped into 26 World Classes. There are at present no internationally recognised subdivisions of these mapping units, though national soil survey organisation are free to make such subdivisions for local use as they find convenient. The question of officially recognised lower categories is under discussion and has been agreed as desirable.

2 Both systems make use of diagnostic surface and subsurface horizons, but climatic criteria, including soil moisture regimes, much used in the US Soil Taxonomy, are generally absent from the FAO/Unesco map legend definitions.

3 Whereas the FAO/Unesco map legend made extensive use of existing soil classification terms, the US Soil Taxonomy considered that their continued use would be confusing and coined entirely new terms for orders, suborders and lower categories. The new terminology, though logical and making use of Latin, Greek and other roots, appears complex and difficult to non-specialists.

4 Classification in the US Soil Taxonomy is based on profile morphology and on soil profile characteristics (including climatic parameters) which can be quantified and measured. The FAO/Unesco map legend incorporates older classification concepts which, though defined, are often less rigorously quantifiable than those used in the Soil Taxonomy.

Land evaluation and land capability classification

Land evaluation

Land evaluation is the process of estimating the potential of land for different types of land use. In this context the word 'land' includes all elements of the physical environment that influence land use. It includes not only the soil, but relevant features of geology, landforms, climate, hydrology, vegetation and fauna, and such basic features as position in relation to other areas. Land mapping units are mapped areas of land with specified characteristics, possessing sufficient uniformity of physical features.

The demand for land evaluation stems from the realisation by users that the mapping of soils and other natural resources such as vegetation and climate does not of itself provide sufficient guidance on how the land can be used, and the profitability and other consequences of the various alternatives possible. The need was felt for a further stage – the evaluation of soils, vegetation, climate, hydrology and other factors in relation to the requirements of different types of land use such as rainfed arable farming, irrigation agriculture, grazing, forestry, recreation land, and other uses. In essence, land evaluation involves the comparison between the requirements of these different types of land use and the qualities of the land evaluated. It also involves a comparison of benefits obtained with the inputs needed, for land is only suitable for a given use if the expenditure on inputs is justified by the returns obtained. Evaluation also involves comparison between alternatives. These comparisons must be in terms relevant to the conditions of the country or region concerned, including the prevailing economic conditions. Thus a land evaluation is not of universal applicability, but geared to local physical, social and economic conditions. The physical data and assessments in the evaluation are likely to remain valid for some time, but the economic assessments may quickly become outdated. For this reason it is practical to separate the two, the physical data representing a permanent record.

In practice there is considerable overlap between a comprehensive general-purpose soil survey and a land evaluation, for in a particular climatic or agro-ecological zone, the local pattern of variation between land of different qualities and potential uses is set mainly by landforms and soils with their associated hydrology and vegetation.

Land capability classification

Land capability classifications attempt to provide users with a relatively simple and untechnical interpretation of soil surveys for farm planning and related activities. The argument is that it is unrealistic to expect anybody other than a soil surveyor to produce interpretations of a soil map and soil report which use technical terminology beyond the grasp of the layman. There is a great need for simple land capability classifications that can be understood by non-technical users, and these have proved of immense benefit in drawing attention to the limitations of land and the need, for example, for adequate erosion control. It should nevertheless be remembered that land capability classifications represent a simplification, and an inevitable loss of some information relevant to wise and profitable utilisation. What they do normally is draw attention mainly to the limitations of the land, and based on the degree of limitation, suggest an appropriate category of land use capability. The intelligent farmer, however, will often want to know more. He will want to know not only that his land is put in Class II or Class III, but why. Sometimes land with different qualities and limitations is lumped together in the same class, but for different reasons. In that case, management of two areas in the same class may be different.

In practice land capability classifications are essentially usually a grading of land primarily for arable agriculture, emphasising limiting factors that restrict the range of activities which can be carried out. The assumption is usually that fertility deficiencies will be corrected by appropriate fertiliser use, so that fertility is not of itself a limiting factor.

The best-known land capability classification is that of the US Soil Conservation Service which puts land into eight broad classes. Of these, classes I-IV are suitable for arable agriculture, classes V-VII are suitable for grazing or forestry which provide a permanent vegetative cover to protect the land, and Class VIII is rough or steep land which should be reserved for wildlife or recreational purposes.

These classifications have been applied with various degrees of success to the tropics. They are most useful for areas where the primary need is to separate arable land from that which should be kept under a permanent vegetation cover of grass or tree crops, and in drawing attention to erosion hazards. However, the land capability assessment can become irrelevant or misleading. In the forest areas of West Africa, flat areas of poor or mediocre fertility with no other obvious limitations to mechanised agriculture might qualify as Class I land, whereas steep land fertile enough to support cocoa, and potentially much more valuable, might be relegated to Classes V, VI or even VII. Restrictions might prevent use of machinery but represent no problems to a peasant farmer using a hoe or cutlass. Many farmers happily cultivate slopes that would be considered uncultivable in temperate areas, and manage to control erosion by mixed cropping and measures such as partial initial clearing.

These examples suggest that not only have land capability classifications to be adapted to local needs in the tropics but also that although they are simple, they must be well understood by the user. For example, a farmer should realise that classification does not usually emphasise positive features such as soil fertility, but concentrates on negative limitations such as slope, which may not necessarily be relevant. The farmer should also realise that with intelligent use more money can sometimes be made from Class IV or Class VI land than from Class I land. The dangers from land capability classification thus spring from the tendency to simplify still further on the part of the user, to assume that Class I land is 'first class' and necessarily the best, and Class V or VI is very inferior, whereas the more intelligent and scientific approach is not to think of good, poor or bad soils but of soils with different combinations of properties, each of which has certain advantages which the farmer should understand and make use of.

9 General fertility concepts: nitrogen and sulphur

General fertility concepts

Soil fertility refers to the ability of the soil to supply the nutrients needed by plants. The study of soil fertility involves examining the forms in which plant nutrients occur in the soil, how these become available to the plant, and factors that influence their uptake. This in turn leads to a study of the measures that can be taken to improve soil fertility and crop yields by supplying nutrients to the soil–plant system, usually by adding fertilisers, manures and amendments to the soil but sometimes by supplying nutrients directly to the plant by means of leaf sprays.

Elements needed by plants

Plants need water, oxygen and carbon dioxide for growth and for carbohydrate synthesis. Other elements are also essential for growth. If plants are analysed they are found to contain up to 90 or more elements. Of these, 13 in addition to carbon, hydrogen and oxygen are at present known to be essential to plants in the sense that they cannot grow without them. Of these 13, 6 are needed in large or moderate amounts, and the other 7, the so-called trace elements, in only very minor amounts or traces.

Apart from carbon, hydrogen and oxygen, which are obtained from air and water, the three nutrients needed in relatively large quantities are nitrogen (N), phosphorus (P) and potassium (K). These three elements are often referred to as primary or major nutrients. They are the nutrients most studied in fertiliser trials and applied in the largest quantities in fertilisers.

A second group of three nutrients, calcium (Ca), magnesium (Mg) and sulphur (S), are sometimes referred to as secondary nutrients. Of these, sulphur has received increasing attention in recent years and in the tropics may be as important as potassium as a limiting nutrient. It is often taken up in quantities about equal to phosphorus, though plants and crops differ considerably in their nutrient demands.

Calcium and magnesium are needed in moderate amounts but their presence is also connected with the reaction of the soil and they may be supplied to make soils less acid.

The remaining elements, known as trace elements or micronutrients, are iron (Fe), copper (Cu), zinc (Zn), manganese (Mn), boron (B) molybdenum (Mo) and chlorine (Cl). These elements are required in very small quantities indeed and large amounts in the soil may be harmful.

Cobalt (Co) is not an essential element to plants but is needed by animals grazing them. Vanadium (V) is needed by *Rhizobium* bacteria in the fixation of nitrogen.

Other elements which are not classed as essential are taken up by plants, including some that are taken up in considerable amounts and some that have produced yield increases when given to crops. Sodium (Na) is taken up by plants and may to some extent substitute for potassium. Silicon (Si) is taken up in large amounts by grasses, including rice and sugarcane, and yield increases have been obtained from applications to these crops.

Plant nutrient uptake

Plants feed mainly by taking in essential elements through their roots, but nutrients can also be absorbed by the leaves and other plant parts, particularly through leaf stomata.

Nutrient uptake by the roots is mainly through root hairs, which are minute extensions of the outside cells of the plant root. The root hair obtains most of its needs from the soil solution, i.e. the moisture in the soil containing small quantities of nutrients. Nutrient ions (and occasionally larger molecules) are absorbed as water is taken in by the plant, but most nutrient ions are taken in by a process that involves the root or root hair exchanging ions across its outer surface, as when a hydrogen ion from the plant is exchanged for a nutrient cation, or by the action of organic carriers which move with the nutrient ion across the outer membrane of the cell.

Energy is required for nutrient absorption. Limiting the oxygen supply to the root reduces its supply of metabolic energy and thus the rate of nutrient absorption. The oxygen supply can be reduced by soil compaction (Chapter 4), thus reducing nutrient uptake, though not all nutrient ions are equally affected. Soil compaction and a lowered oxygen supply to the root seems to have most effect on potassium uptake, an intermediate effect on phosphorus and nitrogen, and least effect on calcium.

Mass flow and diffusion

Nutrients move to positions very close to the plant root by two processes: mass flow, and diffusion.

In mass flow, nutrients are carried along by soil moisture moving towards the root. Mass flow is brought about by plant transpiration and the need to take in water to replace that lost, and the amounts of nutrients taken up as a result of mass flow are related to the amount of water taken in by the plant.

In diffusion, the ion moves of itself through the soil solution to a position where it is absorbed. This is a result of nutrient ions being taken up by the root faster than they are replaced, thus causing a concentration gradient which results in ions moving from an area of higher concentration away from the root to one of lower concentration close to it.

Individual nutrient ion species differ considerably in their rates of diffusion and thus the extent to which this process results in their moving to the plant root to replace the ions taken up by it. Phosphate moves relatively slowly, and in practice is taken up only when it is very close to the root or root hair, so that for this nutrient the extent of the root system is particularly important.

Root development is less important for nutrient ions that diffuse more rapidly, such as nitrate, since their greater ability to diffuse means that they move more easily even to a limited root system, in order to replace what has been absorbed. Potassium has a diffusion rate intermediate between that of phosphate and of nitrate. Expressed in another way this means that of the three ions mentioned, a stunted or deficient root system has the least effect on nitrate uptake, an intermediate effect on potassium uptake, and the greatest effect on phosphate absorption.

Mass flow usually plays a dominant role in the movement of calcium and sulphur to the root surface, while diffusion is particularly important for phosphorus. Both mass flow and diffusion are important for potassium, magnesium and nitrogen.

Total nutrients and available nutrients

An important distinction must be made between the total amount of nutrient present in a soil and the amount available to the plant. For many nutrients relatively large quantities may be present in the soil but in a form unavailable to the plant. Phosphorus in many soils may be present in considerable amounts but in forms such as iron phosphate and aluminium phosphate which are insoluble or only very sparingly soluble and thus not available. In such cases an anlysis of the total amount of the nutrient element in a soil would give little or no idea of its availability to the crop or of the likelihood of a fertiliser response were that element supplied.

The forms of plant nutrients that can be taken up by the plant are listed in Table 9.1. Of these, potassium, calcium, magnesium and some trace elements are absorbed as cations. Phosphorus and the remaining trace elements are taken up as anions. Nitrogen is of particular interest since it is taken up with equal ease by most plants as either the ammonium cation or the nitrate anion.

To some extent cations and anions behave differently in the soil. Cations are retained in the soil by the cation exchange positions on the soil colloids, as discussed in Chapter 2. Exchangeable cations held in this way are available to the plant, though ease of uptake is influenced to some extent by the amounts present in relation both to other cations and to the cation exchange capacity of the soil. Cations are held to some extent against leaching. Although soils also have an anion exchange capacity, this is generally lower than the cation exchange capacity, and in general the anions in the soil such as nitrate and sulphate are held more loosely. They can be translocated by moving soil water and are more subject to leaching losses. The phosphate anion, in contrast, is relatively immobile in the soil.

Table 9.1 Essential nutrients required by plants and the main forms in which they are absorbed

	Element	Forms in which absorbed by plant
C	Carbon	Carbon dioxide (CO_2) – mostly through leaves, carbonate (CO_3^{2-}) and bicarbonate (HCO_3^-) anions
H	Hydrogen	Water (HOH), hydrogen ions (H^+)
O	Oxygen	Hydroxyl ion (OH^-), carbonate anion (CO_3^{2-}), sulphate anion (SO_4^{2-}), carbon dioxide (CO_2) – mainly through leaves
P	Phosphorus	Phosphate anion (mostly the dihydrogen orthophosphate ion, $H_2PO_4^-$, but also the monohydrogen orthophosphate ion, HPO_4^{2-})
K	Potassium	Potassium cation (K^+)
N	Nitrogen	Nitrate anion (NO_3^-) and ammonium cation (NH_4^+), also more complex forms such as urea (NH_2CONH_2) and soluble amino acids
S	Sulphur	Mainly as the sulphate anion (SO_4^{2-}); sulphur dioxide (SO_2) through leaves
Ca	Calcium	Calcium cation (Ca^{2+})
Mg	Magnesium	Magnesium cation (Mg^{2+})
Fe	Iron	Iron II (Fe^{2+}) or iron III (Fe^{3+}) cations, also as complex organic salts (chelates)
Mn	Manganese	Manganese II cation (Mn^{2+}), also as organic salts
Cu	Copper	Copper II cation (Cu^{2+}), also as organic salts
Zn	Zinc	Zinc cation (Zn^{2+}), also as organic salts
B	Boron	Borate anions ($B_4O_7^{2-}$, $H_2BO_3^-$, HBO_3^{2-} or BO_3^{3-})
Mo	Molybdenum	Molybdate anion (MoO_4^{2-})
Cl	Chlorine	Chloride anion (Cl^-)

Nutrient reserves in the soil

The plant nutrient reserves in the soil can be divided into those that occur in inorganic form and those that occur in organic combination in soil organic matter.

The nutrient reserves in inorganic form include various inorganic compounds in the soil such as the forms of phosphate mentioned above (considered in more detail in Chapter 10). They also occur in partly weathered or unweathered minerals in soil originally derived from the parent rock, such as mica flakes or feldspar crystals, which may form a source of potassium that is progressively released as the mineral weathers. This explains why relatively young soils which still contain weatherable minerals are more fertile than old, highly weathered soils which do not. The nature of the parent rock is also important, since basic rocks are richer in plant nutrients than acid rocks, and some parent rocks such as sandstone consisting of little except silica provide little or nothing for the plant (Chapter 6).

Soil humus contains a complete range of essential elements, some elements in considerable quantity. The nutrients occurring in organic combinations in the soil humus are not generally available until the humus is mineralised, i.e. until it decomposes to liberate its constituent elements and ions. The amount of soil organic matter varies considerably between soils according to the climate, the vegetation, the texture and fertility of the soil, and its history of cultivation. In some very poor sandy soils, and in organic soils, most of the nutrient reserves may be in the organic form.

Soil fertility is much affected by the amount of soil organic matter present, its composition, and the rate at which it mineralises to liberate its constituent nutrients to plants. The rate of mineralisation itself varies between fairly wide limits according to environmental factors such as climate and vegetation which influence soil moisture and temperature and thus the activity of soil micro-organisms responsible for organic matter decomposition. In practice, humus mineralisation rates of 1–3% per annum are those most frequently encountered, but they can be higher or lower than this. Because humus mineralisation rates are influenced by temperature, they are slower in high-altitude areas such as the Andes and the highlands of East Africa than in hot tropical lowlands. In these latter areas mineralisation rates might be relatively moderate under a high forest but speed up when clearing of the forest removes shade and surface soil temperatures increase.

Factors influencing nutrient availability

Nutrient availability is influenced by a range of climatic and soil factors and in practice uptake is also related to the ability of the plant to extract its needs. Some plants are better able to extract particular nutrients than others, sometimes because of the extent and nature of their root system. In the case of phosphorus, uptake may be increased when mycorrhiza are present on the roots. The nitrogen nutrition of legumes and some non-legumes is often helped by the presence of symbiotic bacteria which fix this element from the atmosphere.

Climate The local climate affects nutrient availability since soil moisture is necessary for nutrient uptake. Most nutrient ions taken up by the roots come from the soil solution, and nutrient uptake from a dry soil horizon virtually ceases. The moisture content of the soil influences plant water absorption and thus the rate of movement of nutrient ions by mass flow.

Within the soil profile, nutrients might be redistributed by moving soil water. When water rises to the soil surface it brings nutrients with it and these may be left at or near the soil surface when that water evaporates. When rainfall enters the soil and moves down within it, movement of nutrient ions occurs in the opposite direction. Nitrate anions are particularly mobile. Nitrate anions in the upper few centimetres of the soil at the end of the dry season may be moved progressively down into the subsoil with the rains. Sulphate in the soil is commonly greater in the subsoil, where the anion exchange capacity may be higher, than in the topsoil, and plants that show sulphur deficiency symptoms when young may outgrow them as the roots get deeper.

If the water received by the soil is enough to bring the soil profile to field capacity (Chapter 4), then additional water will move through the profile carrying nutrient ions with it which are thus lost by leaching. The cations most likely to be lost in this way are calcium and magnesium.

Soil reaction The acidity or alkalinity of the soil has a considerable influence on nutrient availability. Most major nutrients are considered to be most easily available to the plant within the pH range 5.5–7.5, and liming has long been used to make very acid soils less so for this reason. All the trace elements except molybdenum are more available in acid soils than in alkaline soils, and overliming or even the change of soil pH when ash is burnt on clearing may make these trace elements less available. In acid soils with a pH below 5.5, free aluminium is usually present and this may severely limit phosphate availability by forming very sparingly soluble aluminium phosphates. Plants that prefer alkaline soils usually have a high calcium requirement. Those that prefer very acid soils, such as tea, may have a high aluminium need.

Balance between the elements The relative amounts of elements in the soil affects their uptake by plants to a considerable degree. To some extent cations compete with other cations for entry into the plant, and anions with anions, and there may be antagonisms between them. Particular attention is often given to the ratio between exchangeable magnesium and exchangeable potassium in the soil – the Mg/K ratio, since too high levels of either element can suppress uptake of the other. A ratio of 2:4 is considered generally satisfactory, and higher ratios might reflect low potassium availability and lower ones a shortage of available magnesium. Sometimes imbalances are caused by fertiliser applications which provoke an induced deficiency, as when too high applications of calcium or of potassium induce a deficiency of magnesium. The balance between trace elements also appears critical in some instances, as when very high levels of copper in some Florida soils were associated with a deficiency of iron.

The relative proportions of cations in the leaf and in some other plant parts are related to the proportions of available cations in the soil. Adding one cation to the soil, such as potassium or calcium, will increase its concentration in the leaf and decrease that of other cations, including the concentration of any cation that is low enough to be approaching the minimum for healthy growth. This is another way in which a deficiency may be induced and growth limited. Adding one ion has reduced the plant content of another, and in practice both might need to be supplied.

Nutrient interactions When more than one nutrient is applied to a crop, there may be an interaction. The phrase 'a balanced nutrient supply' refers to the need for a plant to be able to take up nutrients in amounts according to its needs, without any one nutrient being limiting. Many hypotheses have been put forward on the general relationship between the supply of plant nutrients and the growth and yield of the plant. One of the earliest was that of Liebig, the German soil chemist (1803–73), who expressed it as the 'law of the minimum'. He thought that plant growth is limited by the one factor present in minimum amount, and increases or decreases as that factor changes. The relationship between growth and the supply of a limiting nutrient can be represented by a curve which rises at the beginning as the supply increases, then flattens out as the supply becomes adequate and growth is independent of the nutrient in question, until finally the curve drops again as the nutrient is present in such amounts as to be toxic or to interfere with others. Many fertiliser response curves are similar in that they rise with increasing increments of a nutrient and then flatten out as subsequent increments fail to bring about a further yield increase.

The law implies that if two factors are equally low, adding one will get no result without the other, since both are needed. Conversely, adding two together may cause a very big growth and yield increase. If the increase achieved by two together is very much greater than the sum of the increases achieved by each limiting nutrient separately, there has been an interaction between the two, and the two factors are said to have a positive interaction. On the other hand, if the response to adding both was equal to the sum of adding each separately, then there is no interaction between them and they are considered to have worked independently of each other.

Liebig's law has only a limited applicability and in some cases may represent a misleading oversimplication of nutrient supply–yield relationships. If several factors are low, for example, but none excessively low, then the yield will go up if any of them is supplied, independently of the others.

In practice, a better nutrient supply might lead to other modifications being desirable, such as a higher planting density, or the substitution of an undemanding and unresponsive crop variety by a more responsive one.

Nitrogen

Nitrogen in the plant

Nitrogen is required by plants in amounts that are relatively large compared with their needs of other nutrients. The growth of crops is more often limited by nitrogen deficiency than by deficiency of any other nutrient. Nitrogen-containing fertilisers are used in larger quantities than are any other class of fertiliser.

Nitrogen is a constituent of protein, and proteins are present in the growing parts of all plants, and in all parts of some plants such as the legumes. Nitrogen encourages vegetative growth of the plant and gives the plant a good green colour. Apart from its need for protein formation it performs numerous regulating functions and influences the utilisation of other nutrients.

Nitrogen deficiency leads to stunted growth, with yellowish leaves and poorly developed root systems.

The plant absorbs nitrogen almost entirely as the nitrate anion, NO_3^-, and as the ammonium cation, NH_4^+. The nitrate occurs mainly as ions in the soil solution. The ammonium cation occurs mainly as an exchangeable cation held by the soil colloids, with only a minor proportion in the soil solution at any one time.

The natural supply of nitrogen in the soil

Almost all the nitrogen used by plants is thought to have come originally from the air, of which about four-fifths is dinitrogen (nitrogen gas, N_2). Nitrogen is not available in this form and has first to be combined to form ammonium or nitrates before it can be taken up by the plant, and its nitrogen nutrition therefore depends on the availability of these ions in the soil.

The natural supply of nitrogen in the soil comes from three sources:

1 nitrates from the atmosphere washed down in the rain;

2 nitrogen fixed by biological processes; and

3 the decomposition of plant and animal residues and the mineralisation of soil humus.

Soil
Soil water

Table 9.3 Estimated annual nitrogen additions in five tropical ecosystems (kg N/ha)

Source	Rainforest	Tall-grass savanna	Short-grass savanna	Sugarcane	Rice paddy
Rain and dust	4–8	4–8	4–8	4–8	4–8
Non-symbiotic fixation					
– on leaves	12–40	0–12	0–4	0–2	0–4
– by blue-green algae	0	0–10	0–10	0–10	14–70
– in rhizosphere	0–6	0–13	0–6	0–9	0–10
– in litter	0–25	0–10	0–6	12–50	0–10
Symbiotic fixation	34–68	0	0–10	0	0
TOTAL	46–147	4–53	4–44	16–89	18–102

Source: **Kass, D. L. and Drosdoff, M.** (1970) 'Sources of nitrogen in tropical environments', Agronomy mimeo 70–90. Cornell University, Ithaca: New York.

Most of the total nitrogen in the soil occurs in the soil organic matter and is not available to plants until the organic matter is mineralised to the NH_4 or NO_3 forms. Amounts of organic matter in the soil are related to vegetation, climate, soil texture and drainage, and the past cropping history.

Nitrates from the atmosphere Small but important amounts of nitrates are obtained from rainwater because lightning flashes during storms cause some atmospheric nitrogen to combine with oxygen to form nitrates. It is thought that additions of nitrate from this source are probably in the range of 2–10 kg/ha per year except in very dry areas, and probably average 4–8 kg/ha. The highest additions are recorded in tropical areas, probably because of the level of electrical activity during thunderstorms.

Nitrogen fixed by biological processes The second natural source of nitrogen is that fixed from atmospheric nitrogen through the activity of micro-organisms. The micro-organisms involved are mainly symbiotic, i.e. they live in symbiosis with a host plant. Nitrogen is also fixed to a lesser extent non-symbiotically by free-living soil organisms, both aerobic (requiring air oxygen) and anaerobic (not needing air oxygen). The main nitrogen-fixing bacteria are the following:

1 Symbiotic: *Rhizobium*
2 Non-symbiotic: aerobic: *Azotobacter* *Beijerinckia*
anaerobic: *Clostridium*
3 Blue-green algae also fix nitrogen and are of widespread importance in the tropics and subtropics.

Symbiotic fixation is carried out by bacteria of the genus *Rhizobium*, which includes many species. Some of these species are associated with particular groups or species of plants. The plants generally associated with symbiotic bacteria are the legumes, but not all legumes have symbiotic bacteria, while nitrogen-fixing bacteria are also associated with a number of plants outside the legume family. Symbiosis depends on the mutual acceptability of the host plant and the particular strain of microbe. Some plants accept only one strain, others several. If a suitable strain is not present in the soil, or not brought in artificially by inoculation, no nodulation takes place. Lack of suitable bacteria can therefore explain disappointing results with plants that might nodulate and add nitrogen to the soil if the right bacteria were present. Even if suitable bacteria are present, nodulation can be restricted by shortages of molybdenum, calcium or boron.

As is the case in temperate regions, symbiotic nitrogen fixation is the main mechanism by which nitrogen is added to the soil–plant system in the tropics.

Estimates of the nitrogen fixed by tropical forests vary between 12 and 60 kg/ha (Table 9.2, p. 120), and these substantial amounts are probably due mainly to the leguminous trees present in tropical forests, though they also include non-symbiotically fixed nitrogen as discussed in the next section. *Rhizobium* are also associated with non-legumes, such as the Casuarina family, trees that superficially resemble pines. Stands of introduced Casuarina in Senegal have been reported as fixing 60 kg N/ha. There is

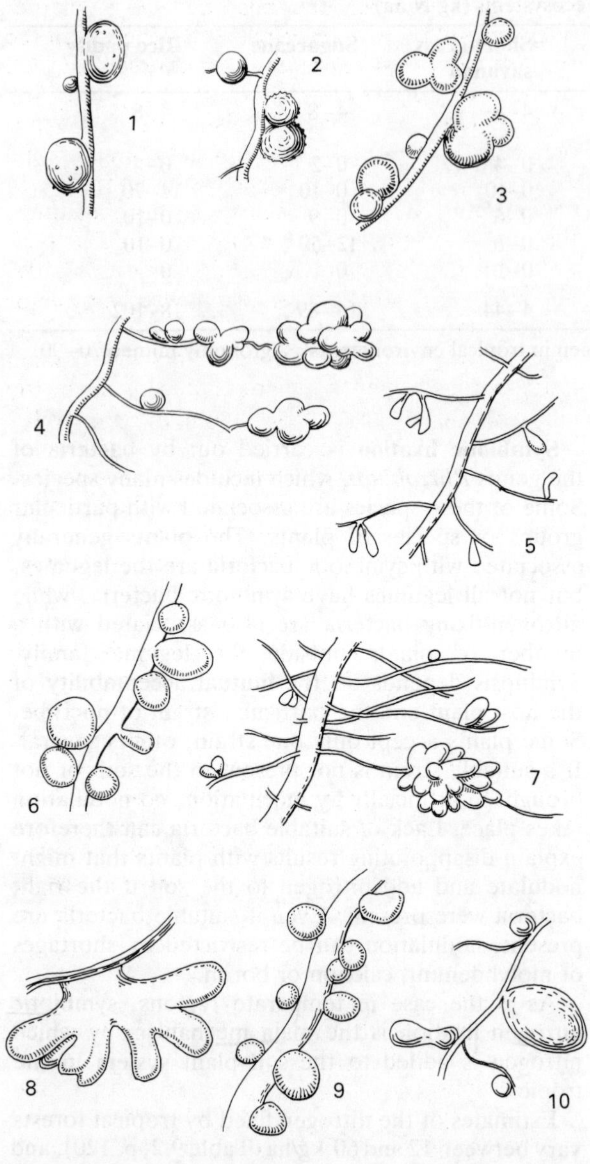

Fig. 9.1 Root nodules of some tropical legumes (not all to the same scale) showing the variety of shapes found. 1 *Arachis hypogaea*, 2 *Glycine soja*, 3 *Dolichos lab-lab*, 4 *Lupinus termis*, 5 *Medicago sativa*, 6 *Vigna unguiculata*, 7 *Crotolaria juncea*, 8 *Pisum sativum*, 9 *Phaseolus mungo* and 10 *Cajanus indicus*.
Re-drawn from **Cobley, L. S.** (1956) *An Introduction to the Botany of Tropical Crops*. Longman: London.

also increasing evidence of symbiotic fixation in the rhizosphere of tropical grasses, possibly involving other bacteria than *Rhizobium*.

In tropical savanna areas, estimated nitrogen fixation rates are much lower than in forests (Table 9.2) and these lower rates are associated with the sparser tree and shrub vegetation. A leguminous tree that has attracted much attention in Africa is *Acacia albida* which is traditionally protected by cultivators because crops are observed to do better beneath it than away from it, although in some areas it has suffered from overcutting for feeding to livestock in the dry season. Analyses of the soil beneath these trees have shown higher nitrogen and organic matter contents than in surrounding areas, and sorghum trials indicated that sorghum below the *Acacia albida* was getting 15–20 kg/ha more nitrogen than that grown away from the trees.

Table 9.2 Estimates of annual nitrogen increments in tropical surface soils

Location	Annual soil nitrogen increments (kg/ha)
Tropical lowland forests	
Ghana	22–55
India	24–60
Zaïre	23–58
Colombia	12–30
Tropical highland forests	
Colombia	23–57
Indonesia	18–45
Madagascar	15–38
Tropical savannas	
Ghana (with 1250 mm rain)	5–15
Ghana (with 850 mm rain)	2–4

Source: **Greenland, D. J. and Nye, P.** (1959) 'Increases in carbon and nitrogen contents of tropical soils under natural fallows', *Journal of Soil Science* **9**, 284–89.

Non-symbiotic fixation is known to occur not only in the soil and the rhizosphere, but also on the leaves of plants and in the litter layer. The amounts fixed appear to be greater in tropical forest environments than in savanna. *Azotobacter* is widespread in soils, but it ceases activity when the soil pH falls below 6.0. *Beijerinckia* has a wider pH range, from 3.0 to 9.0, and is more important in acid tropical soils than is *Azotobacter*. Its tolerance for acid conditions may be connected with the fact that it needs no calcium and only a little magnesium, which are lower in the more acid soils, but molybdenum is necessary for nitrogen fixation. In neutral or alkaline soils, *Azotobacter* is usually dominant, and may also appear in normally acid soils when the pH has been raised by the ash produced by the burning of dead vegetation after clearing. Both *Azotobacter* and *Beijerinckia* may occur in large populations on the leaves of many tropical trees and shrubs.

Blue-green algae sometimes fix more nitrogen than the bacteria. They have been studied particularly in connection with nitrogen fixation in rice paddies, where they are thought to fix 14–70 kg/ha under favourable conditions. They also occur widely but more erratically in well-drained soils, and are thought to contribute up to 10 kg N/ha under some conditions (Table 9.3, p.119). The blue-greens tolerate a very wide range of soil moisture conditions and can live in moist and even dry soils. Their main limitation seems to be that they do not like soils that are acid. They are remarkably self-sufficient in that they can obtain all their needs from air, water and mineral salts and can thus colonise areas where organisms needing organic residues cannot grow.

The decomposition of residues and the mineralisation of soil humus The third, and major source of natural nitrogen in the soil is provided by the decomposition of plant and animal residues added to the soil, and by the mineralisation of soil humus. **The decomposition of plant and animal residues** added to the soil is influenced by

1 the C/N ratio of the organic material, and
2 the amount of nitrogen it contains.

The carbon of the added material acts as a source of energy to the soil micro-organisms and stimulates their activity and growth. Much of the initial decomposition that takes place after organic residues are added to the soil, occurs within the first 2–4 weeks. After this initial surge of decomposition dies down, further decay continues at a slower rate.

During the initial flush of microbial activity following the addition of organic residues to the soil, nitrogen is either released or immobilised by being tied up as part of microbial tissue, to be released later when the microbes die. Whether nitrogen will be tied up or released depends on the ratio of carbon to nitrogen in the residue, and the proportion of nitrogen.

If the added residues contain less nitrogen than is needed by micro-organisms, they get the additional nitrogen they need by taking up ammonium (NH_4^+) and nitrate (NO_3^-) from the soil. This constitutes the 'tie-up' or immbolisation of nitrogen. This tied-up nitrogen may then be released again only slowly, and much of it probably remains tied up for one or two seasons.

If, on the other hand, the added residues contain more nitrogen than the micro-organisms need, the excess nitrogen is mineralised and appears as NH_4^+ and NO_3^- in the soil. This constitutes a release of soil nitrogen. The release is greatest during the initial decomposition surge but continues more slowly thereafter for two or three seasons.

The critical values of C/N ratio and percentage nitrogen that affect whether nitrogen will be tied up or released are not constant, but appear to depend to some extent on the nature of the material added. The critical C/N ratio is probably in the range 15 to 33. The 'critical limit' in this case means that residues with a C/N ratio above the limit result in tie-up, because they are relatively high in carbon and low in nitrogen, but those below the limit result in nitrogen release. For the critical nitrogen percentage, a figure of 2% and a range of 1.2–2.6% has been suggested. Residues containing more nitrogen than the critical limit, such as groundnut stover, will release nitrogen on decomposition.

These considerations regarding nitrogen release and tie-up explain the well-known fact that raw organic matter additions to the soil sometimes result in a temporary nitrogen shortage. These shortages occur when the added material has a high C/N ratio or low nitrogen content. The added material stimulates bacterial activity but fails to supply as much nitrogen as the micro-organisms need. The micro-organisms then compete with the crop for the nitrate and ammonium ions in the soil and temporarily deplete the supply, though in the long run nitrogen is added to the humus and will be released later as that humus mineralises. Farmers have found that to offset the temporary nitrogen shortage that follows the incorporation of low nitrogen residues such as rice straw and maize stover into the soil, they need to add a dressing of nitrogenous fertiliser.

The mineralisation of soil humus normally provides most of the natural nitrogen taken up by a crop. The rate at which this organic matter mineralises is related to the amount of microbial activity, which in turn depends on soil temperature and soil moisture and the presence of a food supply. Rainfall distribution as well as amount is important. The initial wetting of a dry soil after a dry period results in a flush of organic matter decomposition (the so-called Birch effect, named after one of the scientists who first studied this phenomenon). Humus mineralisation and nitrogen supply is therefore related to the seasons. Available nitrogen may be low but rising in the dry season, and be at a maximum at the beginning of the wet season. During the rains it may go down again owing to plant removal and leaching. However, minor peaks may continue to occur if the soil is alternately wetted and dried.

The microbial processes that transform nitrogen in soil humus to ammonium are termed 'ammonification'. The subsequent oxidation of ammonium to nitrite then to nitrate is termed 'nitrification'.

The mineralisation of nitrogen in soil organic matter follows these steps:

$$\underset{\text{nitrogen}}{\text{organic}} \rightarrow \mathbf{1} \rightarrow \underset{\text{(ammonium)}}{NH_4} \rightarrow \mathbf{2} \rightarrow \underset{\text{(nitrite)}}{NO_2} \rightarrow \mathbf{3} \rightarrow \underset{\text{(nitrate)}}{NO_3}$$

Step 1, the ammonification, is carried out by various non-specialised organisms, and is relatively slow. The rate of ammonification, being the slowest of the three steps, controls the overall rate of nitrogen production from organic matter. Under anaerobic conditions, as in a flooded rice paddy, the process stops here, with the formation of ammonium. Under normal aerobic conditions, steps 2 and 3 follow.

Step 2, the conversion of ammonium to nitrite, is carried out by specialised organisms, especially by *Nitrosomonas*, and the rate of conversion is relatively fast. Because of this, ammonium released from step 1, if not taken up by plants and soil organisms, is quickly converted to nitrite. The population of *Nitrosomonas* in the soil changes rapidly according to conditions and to the supply of ions, but is normally adequate to ensure a rapid conversion.

Step 3, the conversion of nitrite to nitrate, is also carried out by aerobic micro-organisms, especially *Nitrobacter*. The rate of conversion is normally relatively very fast, this being the most rapid of the three steps, so that nitrite, which is thought to be toxic, does not accumulate in soils.

The amount of nitrogen released from soil organic matter can be estimated relatively simply if we know, or can estimate:

1 the organic matter content of the soil (often the plough layer);
2 its C/N ratio; and
3 the annual or seasonal rate of humus mineralisation.

If the plough layer of a soil contains 5% of organic matter, with a C/N ratio of 10:1, then it contains about 60 000 kg/ha of carbon and 6000 kg/ha of nitrogen. These figures are obtained by assuming that soil organic matter contains about 60% carbon, and that the weight of the plough layer can be conveniently taken as 2 million kg/ha.

If the organic matter mineralises at 5% per annum, then 5% of the nitrogen it contains will be released each year. Nitrogen release in the case of the example given will therefore be 5% of 6000 kg, or 300 kg. This figure is relatively high and sufficient to give good crop yields, and should be compared with the figures for nitrogen uptake by crops given in a later section.

The figures for nitrogen release given in Table 9.4 explain why there is often a good rate of nitrogen release, and no response to fertilisers, when original forest vegetation or a good mature forest fallow is cut down. The soil has a moderate to good organic matter content, and removal of the protective shade

Table 9.4 Nitrogen released from the plough layer (kg/ha per year) at different organic matter contents and rates of mineralisation

% organic matter	Annual rate of mineralisation				
in plough layer	1%	2%	3%	4%	5%
1	12	24	36	48	60
2	24	48	72	96	120
3	36	72	108	144	180
4	48	96	144	192	240
5	60	120	180	240	300

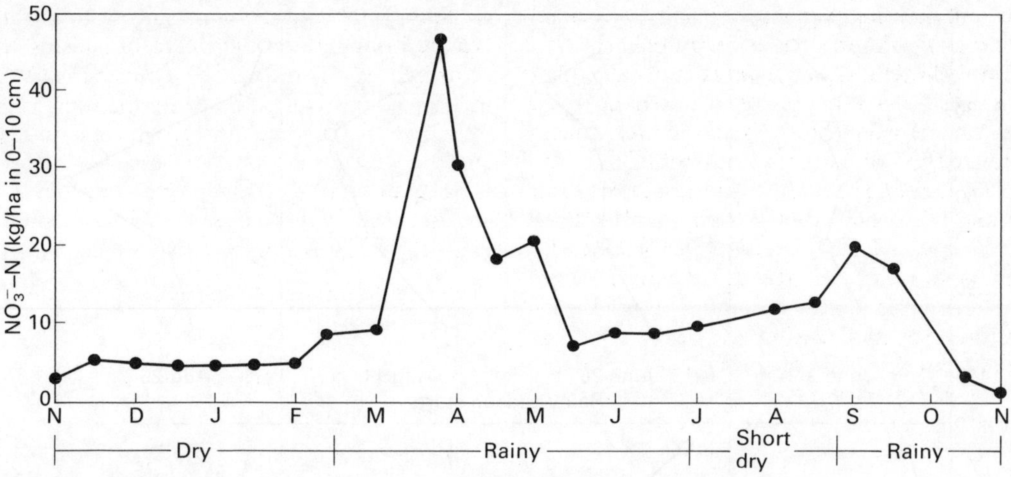

Fig. 9.2 Seasonal pattern of nitrate–nitrogen fluctuation in the top 10 cm of a cultivated Alfisol in Ghana. Source: **Greenland, D. J.** (1958) 'Nitrate fluctuations in tropical soils', *Journal of Agricultural Science* **50**, 82–91.

and exposure of the soil to the sun, with consequent rise in soil temperatures, speeds up the rate of mineralisation. In savanna soils, in contrast, organic matter contents are generally much lower than in soils newly cleared from forest, and nitrogen responses are more frequent.

Since there is a flush of microbial activity and organic matter mineralisation when a dry soil is wetted, the annual rate of mineralisation might be higher in a climate with frequent dry spells followed by rain than in one where the soil is fairly continuously moist throughout the year. More research is needed on the practical implications of the 'Birch effect'.

The movement of soil nitrogen in relation to rainfall

It appears that plants normally recover no more than about 50–60% of available soil nitrogen, both native and that added as fertiliser. What happens to the remaining 40–50% which appears to escape from the soil–plant system? Except in dry areas, much appears to be leached away by soil water. Some nitrogen is also lost by denitrification, i.e. by conversion back to nitrogen gas. A part remains unaccounted for.

The amount and the movement of soil water appear to have an overriding influence on nitrogen uptake and nitrogen losses. An understanding of the processes involved is helpful in arriving at management practices designed to reduce losses and to increase nitrogen recovery rates.

Nitrate nitrogen absorption by the plant appears closely connected with water movement and absorption. Very little nitrate if any is taken up from a dry soil horizon. The nitrate anion moves with the soil water and is thus moved to the root surface. Part of the soil water, held with tensions greater than 15 atmospheres, is not available to the plant (Chapter 4) but nevertheless contains nitrate anions. In the field, rain entering the soil and moving down through it may displace the older soil water, causing it to move to lower horizons. If the crop is taking all or almost all of its water from the upper soil horizons, then nitrate nitrogen in the lower parts of the profile may not be available, or may be tapped to only a small extent.

Investigations on the nitrate content of an Alfisol in northern Nigeria illustrate nitrate movements caused by rainfall. At the beginning of the rainy season most of the nitrates were concentrated in the top 15 cm of the soil. Three weeks later, after

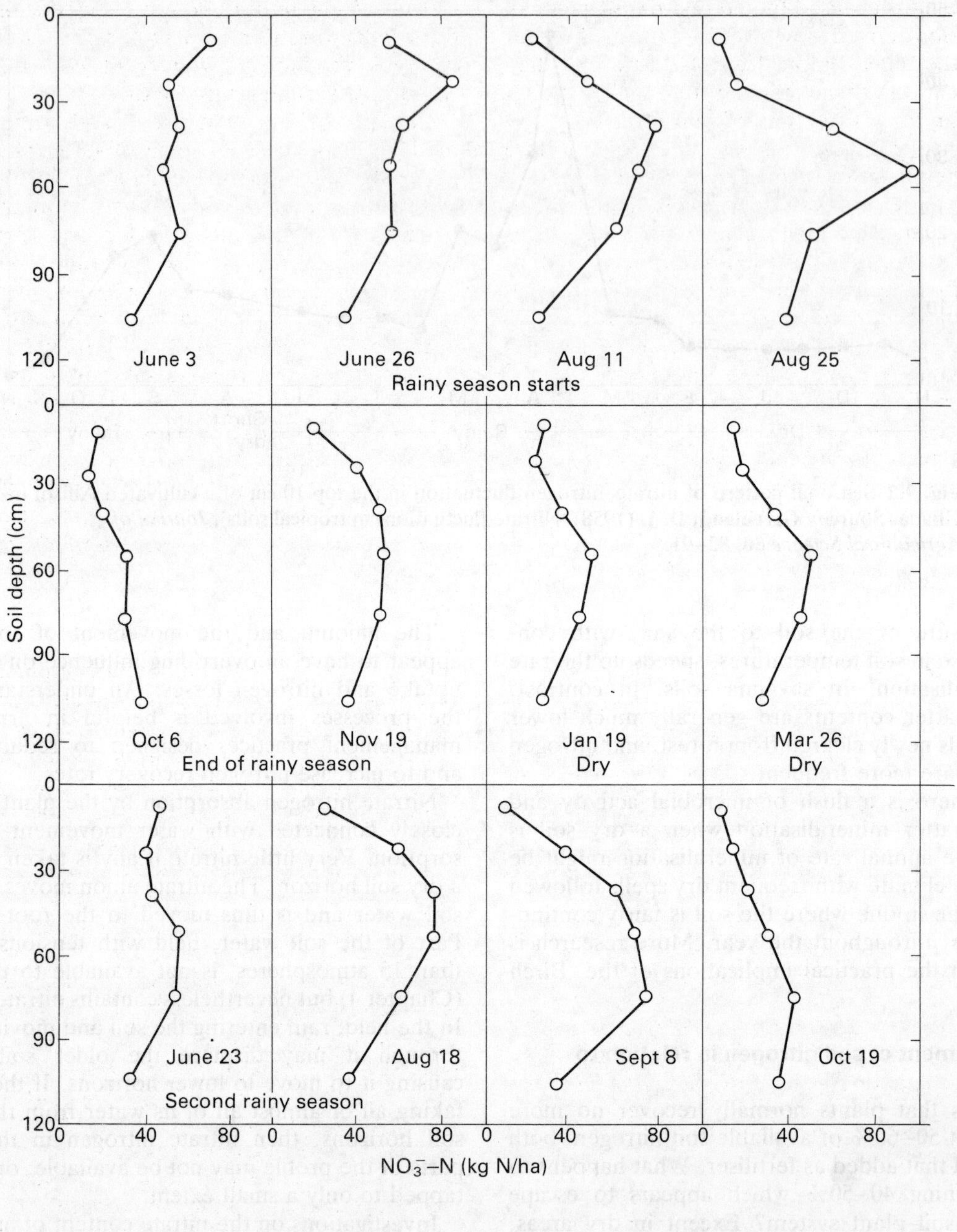

Fig. 9.3 Nitrate fluctuations in a northern Nigerian Alfisol. Source: **Wild, A.** (1972) 'Nitrate leaching under bare fallow at a site in northern Nigeria', *Journal of Soil Science* **23**, 315–24.

the nitrogen flush caused by the first rains, the peak concentration moved down to 15–30 cm. At later stages in the rains it moved to 30–45 cm and then to 45–60 cm. On the average, nitrate moved downwards 1 mm for every 2 mm of rain received.

With a heavy rainfall it is easy to imagine nitrate nitrogen being moved down out of reach of the roots. The pattern of water movement in a soil, and hence the pattern of nitrate movement, varies according to the soil concerned, the rainfall amount and distribution, and the crop. Because rainfall varies from year to year, fertiliser trials for only a single year or for only two to three years may not necessarily be a reliable basis for making recommendations. Even with the same rainfall, two areas with different soils may have quite different water movement patterns. There would be more movement in a soil with a low water content at field capacity than in a heavier soil. With a given rainfall, therefore, water movement tends to be inversely proportional to the amount of water held at field capacity.

The above considerations apply to the relatively mobile nitrate anion. Nitrogen in the form of the ammonium cation behaves very differently, since it is attracted to the cation exchange positions on the soil colloids. The higher the soil cation exchange capacity, the more are ammonium cations likely to be held against movement and leaching. This effect is only temporary since the ammonium cations if not absorbed by the crop or by soil organisms are likely to be fairly rapidly converted in well-drained soils to nitrate anions.

Crop needs for nitrogen

The nitrogen needs of a crop vary within wide limits according to the type of crop (even the variety), the planting density, the stage of growth, the yield, and other factors. The natural supply of nitrogen may or may not be a limiting factor to crop growth, depending on its magnitude and the extent to which growth is also limited by water, other nutrients, or other factors. Where nitrogen is limiting, then the application of fertiliser nitrogen would be expected to raise yields to the extent allowed by other limiting factors, and by the inherent ability of the crop variety to respond to a better nitrogen supply. Many traditional tropical food crop varieties adapted to soils of low fertility do not respond satisfactorily to an added nutrient supply and if it is desired to increase production with fertilisers, then more responsive varieties must be used.

Cereals such as maize, rice and sorghum with modest yield levels of about 1000 kg/ha of grain generally require about 30 kg/ha of nitrogen, and this low requirement is probably more or less in balance with the nitrogen-supplying capacity of many tropical soils. At higher yield levels of 4000–5000 kg/ha, about 100–150 kg/ha of nitrogen are taken up, and this yield level normally requires both high-yielding varieties and fertilisers.

Root crops also remove substantial amounts of nitrogen, with average yields probably taking up 30–50 kg/ha. Nitrogen removal by permanent tree crops is generally relatively low except for tea. The highest extractors of nitrogen are the tropical grasses.

The use of nitrogen fertilisers in the tropics and their effects on the soil are discussed in Chapter 14.

Sulphur

Sulphur in the plant

It is only relatively recently that sulphur has been recognised as a widespread limiting factor in tropical soils. The amounts of sulphur taken up by crops are comparable to those of phosphorus. Sulphur is an essential element that is particularly associated with protein synthesis. It is present in some plant oils, including that of the groundnut. Its role in the synthesis of proteins is suggested by the fact that if sulphur is a limiting factor, nitrates may accumulate in the plant tissue instead of being used for the formation of proteins.

Sulphur deficiency symptoms in plants are similar to those of nitrogen. Plants are often uniformly pale in colour, sometimes pale beige or straw coloured, and stunted in appearance.

Plants absorb sulphur mainly as the sulphate anion (SO_4^{2-}) through the roots, but can also absorb larger sulphur-containing molecules, such as some amino acids. Plant leaves can absorb sulphur dioxide gas (SO_2) direct from the atmosphere.

Sulphur requirements

The amounts of sulphur removed by different crops differ within wide limits. Relatively high sulphur requirements are shown by the legumes and by oil-producing crops. The groundnut, both a legume and an oil producer, forms an important example of a crop with relatively high demands. Relatively low needs are associated with most grasses and cereals, but sorghum, with fairly high needs, is an exception.

Plants also differ in the efficiency with which they extract sulphur from the soil. Grasses, with their fibrous root systems, are thought to be better extractors than legumes. In a grass–legume mixture growing in a soil that is low in sulphur, the grasses may take more sulphur than the legumes so that the legumes are deficient in sulphur and are progressively driven out of the mixture.

The amounts of sulphur removed also depend on the yield level. In northern Nigeria it was found that a relatively low yield of groundnuts of 650 kg/ha removed only 1.4 kg/ha of sulphur, but that a high-yielding crop needed 10 kg/ha or more.

The annual additions of sulphur needed to make good any losses will vary according to

1 the nature of the soil and the amounts of sulphur it releases,
2 whether the rainfall results in soil leaching,
3 the amounts removed by the crop, and
4 whether some sulphur is returned in the residues.

Studies on sandy soils in Senegal have shown annual leaching losses of up to 10 kg/ha of sulphur. Sulphur additions of 15–20 kg/ha would seem to be needed to make good losses due to leaching and crop removal. In Equatorial Africa annual additions of 15 kg/ha have been recommended for cotton. In the poor, sulphur-deficient *campo cerrado* soils of Brazil the estimated annual need is 22 kg/ha.

Natural sources of sulphur

The three main natural sources of sulphur are sulphur-containing soil minerals, sulphur in the atmosphere, and sulphur contained in soil organic matter.

Soil minerals containing sulphur include sulphides and sulphates, especially in poorly drained areas and in tidal marshes. Sulphates, particularly calcium sulphate, are often found in areas of low rainfall and in saline soils.

Atmospheric sulphur occurs mainly as sulphur dioxide gas (SO_2). Amounts vary very much from place to place. They are highest near cities and industrial areas where sulphur-containing fuel is burnt. Sulphur occurs near coasts in salt spray, and in hydrogen sulphide (H_2S) produced in marshes. In most of the tropics, where industrial activity is slight, additions of sulphur are relatively low but may nevertheless be enough to sustain low crop yields. In northern Nigeria, sulphur additions for a number of sites were found to average 1.4 kg/ha a year, the amount removed by a poor groundnut crop of 650 kg/ha. Of this amount, one-sixth fell in dust and five-sixths was washed down in rain.

Organic sulphur includes that in fresh organic residues and that in soil humus. In practice, for most soils the soil humus forms the major reservoir of sulphur in the soil, the sulphur being released as the humus mineralises. Nitrogen/sulphur ratios in humus are normally within the range 10:1 to 10:15. The amount of sulphur released from the mineralisation of soil humus to a growing crop will depend on:

1 the amount of organic matter in the soil,
2 the sulphur content of that organic matter, and
3 the rate at which the organic matter mineralises.

In many respects sulphur release from organic matter can be compared with that of nitrogen. Mineralisation rates depend on soil moisture, soil temperature and microbial activity. A flush of humus mineralisation at the beginning of the rains when a dry soil is first wetted releases nitrogen, sulphur, and a range of plant nutrients.

In practice the sulphur-supplying power of most tropical soils is closely linked to their organic matter content. In soils with a good organic matter content, such as those newly cleared of high forest or from a mature fallow, there is usually enough sulphur for most crops. In soils low in organic matter, including savanna soils that are subject to annual dry season burning, sulphur contents and the rate of sulphur release may both be relatively low and insufficient to meet the demands of some crops such as cotton and groundnuts. It is in such cases that responses can be expected to sulphur-containing fertilisers.

Sulphur retention and movement in soils

The sulphate anion (SO_4^{2-}) is retained by the anion exchange capacity of the soil (Chapter 2). This is associated with the clay fraction of the soil, particularly with kaolinitic clays and iron and aluminium oxides. In practice the anion exchange capacity is higher for most tropical soils in the subsoil than in the topsoil. This is particularly true of soils with an argillic horizon, such as Alfisols and Ultisols (Chapter 8), and of the highly weathered Oxisols. As a result there is often more sulphate sulphur in the subsoil than in the topsoil. This in turn influences the supply of sulphate to the crop. A plant with a shallow root may tap only the topsoil sulphate, which may not be enough, while a deeper-rooted plant may draw on better sulphate supplies in the subsoil. For this reason a crop may show sulphur deficiency symptoms when it is young and the root system is still shallow, but may grow out of the symptoms later as the root system develops.

Although the sulphate anion is not quite as mobile in the soil as the nitrate anion, nor so easily leached out of it, it is nevertheless liable to move in the soil with soil water. It can be washed out of the soil if there is enough water to cause through-leaching. Thus the pattern of soil water movement in the soil influences sulphate movements. With a given rainfall, there will be relatively more water movement in lighter-textured soils than in heavier ones. With increasing rainfall and water movement there is greater likelihood of sulphate anions being moved from the topsoil to the subsoil. With more water, sulphate may be washed down to depths beyond the reach of plant roots or right out of the profile. Because anions and cations in downward-moving soil water are more or less in balance, a good supply of cations facilitates the leaching of anions and vice versa, so sulphate losses can be expected to be higher in soils with a good supply of exchangeable cations. Leaching losses are lowest in highly weathered acid soils that are high in kaolinitic clay and in iron and aluminium oxides, since these soils have a relatively high sulphate retention capacity and are also low in exchangeable cations.

Sulphur responses in the tropics

Prior to the late 1950s, little attention was given to sulphur as a possible limiting factor in crop production. Since then sulphur deficiencies and responses have been reported in many tropical soils. Many of these are low in organic matter, including savanna soils subject to annual burning, but even some forest soils newly cleared from forest in Brazil were sulphur-deficient. Reasons for the relatively late awareness of the importance of sulphur in tropical soils probably include the following:

1 Farmers have been correcting sulphur deficiencies without being aware of it by using fertilisers such as ammonium sulphate and single superphosphate.
2 Heavier population pressures and shorter fallows have resulted in lower soil organic matter levels and less sulphur from organic matter mineralisation.
3 More intensive agriculture has increased yields and demands on soil sulphur, while the use of more concentrated fertilisers containing little or no sulphur has resulted in less sulphur being applied.

In the late 1950s groundnuts in northern Ghana and northern Nigeria showed good responses to single superphosphate and this was assumed to be a phosphorus response. When the source of phosphorus was changed from single super, high in sulphur, to triple super, low in sulphur, yields fell, and it was realised that the main response was not to the added phosphorus but to the sulphur in the single super. Also in the 1950s, sulphur deficiencies were recognised in the *campo cerrado* savanna areas of Brazil. Since then, single super has become a standard recommendation for groundnuts, and the sulphur needs of crops have been much more widely studied. Sulphur deficiencies have been reported from Central America, in many savanna areas of Africa, and in areas of very sandy soils, under both forest and savanna, in the Punjab, in Malaysia and in Hawaii.

In general, sulphur deficiencies appear most likely in tropical soils which are light textured, low in organic matter and subject to leaching. The African savanna areas in which deficiencies are widespread are generally low in organic matter, and subject to annual dry season grass burning during which most of the sulphur and nitrogen are lost. The rains of the wet season, even if short, are often enough to cause some leaching at that period, and movement of sulphate out of the profile and into groundwater.

In northern Nigeria farmers scrape up the sandy topsoil into ridges in which the groundnut roots are largely confined. However, such ridging may lessen root penetration of the subsoil, and may therefore contribute to sulphur deficiency. This illustrates among other things the importance of looking at the root system in soil fertility investigations.

10 Phosphorus, silicon and potassium

Phosphorus

Phosphorus and the plant

Phosphorus is required as a small but vital ingredient in cell nuclei, and is essential for cell division and growth. It is concentrated in the fast-growing parts of the plant, particularly in the root tips. It affects the maturing period of crops and is found in quantity in seeds and fruit.

Phosphorus deficiency symptoms include stunted growth and a poorly developed root system. Leaves may show a purplish or bronzed discoloration. Maturity is delayed and fruiting is poor.

Phosphorus is deficient in very large areas of the world's soils and was one of the first nutrients to be widely used as a fertiliser. Agronomists have long regarded phosphate applications as being particularly useful in promoting a good root system.

Plant absorption of phosphorus

Phosphorus is absorbed by plants wholly or mainly in the form of the di- or mono-hydrogen orthophosphate anions, $H_2PO_4^-$ and HPO_4^{2-}. Of these, the mono-hydrogen ion, HPO_4^{2-} is the more important, particularly in acid soils.

The rate at which the plant absorbs phosphate ions is influenced by their concentration in the soil solution. At any one time the amount of phosphate in the soil solution is very small indeed, and usually considerably less than one part per million (ppm), i.e. less than one part of phosphate to a million parts of soil solution. Optimum concentrations of phosphorus in the soil solution are believed for many crops to be between 0.07 and 0.2 parts per million, depending partly on soil texture. This is usually the equivalent of only a few grams of phosphorus in the soil solution in a hectare of soil to a depth of about one metre. Since a growing crop can easily absorb 10 kg/ha or more of phosphorus in a season, the amount in the soil solution at any one time is only a tiny fraction of the annual needs of the crop. The phosphate in the soil solution has therefore to be taken up by the plant and then replaced by fresh phosphate many times during the season if the needs of the plant are to be met. Where other factors are not limiting, the rate of plant growth may depend on the rate at which phosphate ions removed from the soil solution are replaced from the phosphate reserves in the soil.

An agriculturally very important feature of the phosphate anion is that it is relatively immobile in the soil. In normal well-drained soils phosphate ions do not move very far from their place of origin. This is in strong contrast to more mobile nutrient ions such as the nitrate anion. The practical consequence is that the phosphate ion has to be very near the plant root if it is to be absorbed. Since the phosphate ion cannot move far to the root, then in effect the root has to come to the phosphate ion. For this reason adequate phosphate absorption depends very much on the plant developing a good root system. As the root system expands, the growing point of the root moves through the soil, exploring new areas. The most active phosphorus absorption is from the root tips, with a slower rate of absorption from the older parts of the root system. The amount of phosphate taken up by the plant is thus related to the extent of the root system and the volume of soil it exploits, but the development of the root system is itself dependent on an adequate phosphorus supply.

As phosphate ions are removed from the soil solution, further phosphate ions are released and take their place. They may come from the anion exchange positions in the soil (Chapter 2) or be released from organic or inorganic compounds in the soil.

The availability of phosphorus in the soil

Phosphate that is soluble and capable of entering the soil solution and being absorbed by plants is known as 'available phosphorus'. However, most of the total phosphorus in most soils occurs as forms that are either insoluble or nearly so, and this is referred to as 'unavailable phosphorus'. It remains unavailable as long as it remains in the insoluble or sparingly soluble form.

The unavailable phosphorus is of two types: that which occurs as organic compounds, including those in the soil humus, and that which forms inorganic compounds, such as iron or aluminium phosphate.

Unavailable phosphorus in organic compounds becomes available when those compounds are mineralised, i.e. decomposed to release their constituent ions and elements. Most of the soil organic phosphorus is in the soil humus. The composition of soil humus varies somewhat according to local conditions but in very round figures the relative proportions of carbon, nitrogen and phosphorus in soil organic matter are often found to be about 100 to 10 to 1, so that organic phosphorus is about 1% of the organic carbon and 10% of the nitrogen. The organic phosphate is made available as the organic matter mineralises. The orthophosphate ions are released into the soil solution and are then either taken up by plants or soil organisms, or combined to form other soil constituents. A high organic matter content and a good rate of organic matter mineralisation should ensure a rate of release of phosphate ions adequate for crop production, though most of the phosphate released in this way will be in the topsoil. If not immediately taken up by the plant or by soil organisms it will be converted to inorganic forms.

The unavailable inorganic phosphorus compounds consist mainly of insoluble or nearly insoluble compounds of phosphorus with other elements, particularly with iron, aluminium and calcium.

In acid soils the phosphate is most likely to combine with free iron or free aluminium to form iron III phosphates and aluminium phosphates. These are relatively insoluble.

In alkaline soils phosphate ions are most likely to combine with calcium ions to form calcium phosphate, or they may react with particles of calcium carbonate in the soil.

The various compounds of calcium and phosphate found both in the soil and in fertilisers differ in

1 the relative proportions of calcium and phosphate they contain, and
2 their solubility in water and hence availability to the plant.

It is necessary to distinguish between the following:

1 monocalcium phosphate ($Ca(H_2PO_4)_2$) which is water soluble, and in which there is one calcium ion to two orthophosphate ions;
2 dicalcium phosphate ($CaHPO_4$) which is relatively insoluble and in which there is one calcium ion to one of orthophosphate; and
3 tricalcium phosphate ($Ca_3(PO_4)_2$) which is the least soluble form and in which there is less phosphate in proportion to calcium than the other two forms since there are three calcium ions to two of orthophosphate.

Of the three, only monocalcium phosphate is readily soluble but monocalcium phosphate can change (revert) to one of the other forms and so become insoluble.

The original source of soil and plant phosphorus is apatite. Apatites are a group of minerals with a structure like some silicates which are compounds of calcium, phosphorus and a third component which might be hydroxyl (OH), fluorine (F) or chlorine (Cl). When the third component is fluorine, a fluorapatite is formed; this is the most important in soils, and the main constituent of most natural rock phosphate deposits (Chapter 14).

Phosphorus fixation in the soil

'Fixation' is a broad term referring to soil processes whereby available elements becomes unavailable to plants. The term has no exact definition, and the processes are not irreversible. In relation to phosphorus, essentially it consists of transforming available phosphate into relatively insoluble calcium, aluminium and iron phosphates. Even these relatively insoluble forms release some phosphate slowly over a period of years.

In calcareous soils the phosphate ions form relatively insoluble compounds of calcium and magnesium. In soils containing free calcium carbonate, dicalcium phosphate or apatite may be deposited on the outer surfaces of calcium carbonate particles or concretions, particularly if the calcium carbonate is finely divided.

In tropical soils aluminium and iron oxides are relatively abundant and react with phosphorus to form aluminium and iron phosphates. These compounds when formed in the soil are precipitated as small particles or films, or are adsorbed on the surfaces of iron and aluminium oxides or on the surfaces of clay particles.

Iron and aluminium oxides are intimately associated with the kaolinitic clay fraction of the soil, sometimes as coatings on the clay, and in such situations phosphorus fixation is likely to increase with increasing clay content of the soil. Fixation is thus related to soil texture and would be expected to be greater in clay soils than in light-textured ones. This applies particularly when the clay is kaolinitic: montmorillonitic soils generally have a much lower fixation capacity. The amorphous forms of iron and aluminium oxides fix more phosphate than the crystalline forms such as goethite and gibbsite because the former have a greater surface area. The end product of these fixation processes is the formation of aluminium and iron oxides, but with time the aluminium phosphates, the less insoluble form, tend to be gradually transformed into iron phosphates.

In acid soils with a pH below 5.5 the presence of exchangeable aluminium provides an additional phosphate fixation mechanism. If soluble monocalcium phosphate is added to the soil, as in superphosphate fertiliser, it reacts with the exchangeable aluminium to form aluminium phosphates with the general formula $Al(OH)_2H_2PO_4$.

Some of the phosphates that are 'fixed' can to some extent be expected to become available very slowly over a long period. In very acid soils some phosphorus can often be made available by moderate liming (Chapter 11), and phosphate is also released when a soil is flooded, as in rice cultivation. The latter is discussed further in Chapter 20 in the section on the effects of flooding.

From the above it follows that phosphorus fixation in tropical soils is more likely in clay soils than in sandy soils, particularly

1 in soils containing large amounts of iron and aluminium oxides, especially if they are in an amorphous form, and
2 in very acid soils (below pH 5.5) containing exchangeable aluminium.

Phosphate forms in tropical soils

Normally measurement of the total amount of phosphorus in a soil has no value since the great bulk of the total phosphate is not available to plants, even if some of it may become available very slowly. The combined phosphate in the soil is in the following forms:

1 Organic phosphorus in the soil humus which is not available until the humus is mineralised.
2 The three so-called active fractions, the calcium-bonded phosphates (Ca-P), the iron-bonded phosphates (Fe-P) and the aluminium-bonded phosphates (Al-P), all of which except monocalcium phosphate are relatively insoluble.
3 The two so-called inactive fractions, the occluded and the reductant soluble phosphorus.

The occluded phosphorus ('occluded' means hidden) is the phosphorus, usually in iron and aluminium compounds, that is surrounded by inert material which prevents any reaction with the soil solution. Reductant soluble phosphorus is similarly protected by a coat of material that is inert under aerobic conditions, but may be exposed when the coat is removed by being reduced under anaerobic conditions.

In many tropical soils much of the total phosphorus is in the occluded form, as in the iron oxide concretions which are a feature of many concretionary soils widely described as laterites (Chapter 7). The fact that some is in the reductant soluble form accounts for phosphate release on flooding.

The proportions of the different inorganic forms of phosphorus are broadly related to the degree of weathering of the soil and to its pH. With increasing soil weathering the proportion of calcium phosphate (Ca-P) gets less and the proportion of iron phosphate (Fe-P) increases. This is related to the fact that Fe-P forms are more insolube than Al-P forms, which in turn are more insoluble than Ca-P. Transformations from one form to another occur and are influenced by soil pH. In acid soils the activity of iron and aluminium increases and Ca-P is likely to be slowly converted to Al-P and Fe-P.

Total phosphorus is usually higher in highly weathered soils such as Oxisols and Ultisols than in less weathered soils.

Phosphate requirements of tropical crops

The amounts of phosphates removed by tropical crops vary between wide limits, depending on the crop and even the crop variety concerned, the soil, the climate, and the yield level. The phosphate uptake by cereals and grain legumes is probably less than 10 kg/ha in the case of low yields of about 1000 kg/ha and possibly about 15–35 kg/ha for heavy yields of 4–8 t/ha. Of the root crops, cassava has fairly high phosphorus demands and a good yield of 8 t/ha might take up 35–40 kg/ha, but sweet potatoes and yams have a much lower need for phosphate. Bananas have a relatively low demand, usually less than 10 kg/ha being removed in the bunches, and most of the tree crops remove only low to moderate amounts of less than 2 kg/ha in their harvested parts, though coconuts have higher demands. Sugarcane and intensively managed pastures, on the other hand, may remove 20–70 kg/ha, depending on yield levels. Some crops, such as forage legumes, need fairly high phosphate levels during establishment but less thereafter.

The fact that different varieties of the same crop may differ considerably in their phosphorus needs has been taken advantage of by screening different varieties for their tolerance of low phosphate soils. This has been done systematically on a very wide range of rice varieties and cultivars by the International Rice Research Institute (IRRI).

Phosphate responses in the tropics

Responses to added phosphate are related to the following:

1 The phosphate status of the soil before application, i.e. the concentration of phosphate ions in the soil solution and the rate at which those removed are replaced.

2 The extent to which added phosphate increases the ease with which plants take up phosphate by increasing the concentration in the soil solution, and maintaining it at that higher level. This in turn depends on the type of phosphate fertiliser added, the way it is added, and the nature of the soil, particularly its phosphate-fixing capacity.

3 The nature of the plant and its responsiveness to a better phosphorus supply, and the extent to which it is held back by other factors such as the lack of another nutrient or water.

The residual effects of a phosphatic fertiliser, i.e. the effect on crop yields in the second and following year or season after application, vary considerably according to the amounts given and the fixation capacity of the soil. It used to be thought that most tropical soils fix phosphate so rapidly that residual effects were relatively rare, but more numerous and widespread studies have frequently demonstrated residual effects, and the current view is that most tropical soils can show residual effects lasting a number of years. Of the phosphate added in fertiliser, only a small proportion is taken by the crop. Even after four or five years, perhaps only 20–30% of the added fertiliser has been used by the crop. The use of phosphatic fertilisers over a period increases the total phosphate in the soil and can be expected to reduce the phosphate-fixing capacity. It may result eventually in a soil that is phosphate responsive ceasing to respond to phosphatic fertilisers after sufficient phosphate has been added and accumulated over a period.

Responses to phosphatic fertilisers in the tropics are very widespread, both with traditional farming and with more intensive systems. Thousands of trials were conducted on smallholder plots by the FAO Freedom From Hunger Campaign in the 1960s. These were not rigorously controlled scientific experiments but merely simple trials and demonstrations, but some phosphorus response was thought to be observed at about 90% of the locations. Increases in West Africa were particularly marked for yams, and in northern Latin America for potatoes, while maize responses were very widespread in both areas. Other widespread trials have demonstrated such frequent responses that it is probably true to say that the majority of tropical soils are deficient in phosphorus for the majority of crops, but much depends on the previous history of cultivation and on the yield level envisaged. For low yields of crops grown on soils cleared of original forest or from long fallows, phosphate responses are unlikely. They become more frequent as soils are exhausted by long periods of cultivation, and as more intensive agricultural systems are introduced. Even then, optimum application rates are often quite low. Fertiliser trials on maize in the Ivory Coast, for example, showed that in almost every case, 50 kg/ha of P_2O_5 was enough to obtain maximum yield. Numerous examples of responses to phosphatic fertilisers are given in Chapter 15 for annual crops and in Chapter 16 for perennials.

Phosphatic fertilisers and their use are discussed in Chapter 14.

Silicon

Silicon is not considered an essential element but is of interest because applications of silicon have sometimes produced yield increases. Two different mechanisms appear to be involved. First, silicon applications sometimes decrease phosphorus fixation and increase the supply of phosphate. Secondly, there have been responses to silicon as a nutrient in certain soils and with certain crops, particularly rice and sugarcane.

The effects of silicate applications on phosphorus fixation

Silicate applications can sometimes block some of the phosphorus fixation sites in soils and thus decrease the phosphate fixation capacity of the soil. Silcate ions may also replace phosphate ions held in fixation sites and thus release phosphate to the soil. Applications are usually given as sodium silicate, calcium silicate, or basic slag. Much of this experimental work has been carried out on soils in Hawaii with a high phosphate fixation capacity.

Responses to silicon as a nutrient

Responses to silicon as a nutrient have been obtained with sugarcane grown on Oxisols and Ultisols developed over basalt in Hawaii and Mauritius, and are widespread in rice soils in Japan, Korea and Sri Lanka. Responses are most likely from intensive cultivation of high-yielding crops grown on highly leached soils having low contents of soluble silicon.

Research work on silica for rice has been carried out mainly in Japan where one-third of the 3 million ha devoted to rice are regarded as responding to silica, and calcium silicate slag is applied at the rate of about 1 million t a year.

Sugarcane absorbs large amounts of silica. Sugarcane leaves are often analysed to diagnose fertiliser needs, and the considerable information available on silica contents of the leaf sheath indicate that on low-silica soils only 1% silica may be found instead of the 2.5% silica associated with the better soils. In the former case there was more manganese than in the latter, and one of the possible mechanisms involved is that adequate silica reduces potentially harmful manganese intakes sometimes associated with 'freckling' leaf symptoms. In Hawaii and Mauritius, applications of 5–7.5 t/ha of calcium silicate gave yield increases of 12–25 t/ha of cane.

Potassium

Potassium and the plant

Potassium is often a limiting factor in crop growth and is thus frequently applied as fertiliser. It is essential for the formation and transfer of carbohydrates in plants, and for photosynthesis and protein synthesis. Plants which synthesise and store large amounts of starch usually have relatively high potassium needs. Potassium occurs particularly in the growing points, fruits and seeds of plants.

Potassium is absorbed by the plant as the potassium cation, K^+. This cation is held in the soil, together with other exchangeable cations, by the soil colloids, i.e. by the clay and organic matter in the soil. The root hair absorbs potassium cations from the soil solution or direct from the soil colloid.

Potassium is absorbed by the plant in relatively large quantities, and probably second in amount in many cases only to nitrogen.

Potassium deficiency, particularly when nitrogen is also low, results in stunted plant growth. Leaves are small, sometimes ash-grey in colour, and die prematurely, first as the tips and then along the outer edges. Fruit and seeds are smaller than usual in size, quantity and weight.

Sources of potassium and forms in the soil

The main original sources of potassium in soils are the potassium-containing minerals, principally potassium feldspar, a widespread constituent of igneous rocks, and both black mica (biotite) and white mica (sericite, muscovite). The potassium in these minerals is released when they are weathered chemically. On weathering the potassium is liberated as the potassium ion and is then either:

1 taken up by plants or other living organisms in the soil, or
2 lost in the water draining through the soil, or
3 held by the cation exchange positions of the soil colloids, or
4 converted to less available forms.

Investigators often find it convenient to distinguish between three types of potassium in the soil, potassium that is

1 available.
2 unavailable, and
3 slowly available.

The **available potassium** is that potassium which can be taken up immediately and easily by plants, and this is the potassium in the ionic K^+ form. It includes the potassium cations held by the soil colloids, the potassium ions in the soil solution, and potassium in soluble form. In practice it is usually measured in soils by determining the exchangeable potassium.

The **unavailable potassium** is that potassium in the soil in the form of organic or inorganic compounds that are not soluble. This potassium is not available until the organic matter is mineralised or until the inorganic forms are weathered. Normally the most important soil reserve of unavailable potassium is in unweathered minerals, particularly feldspars and micas. In young soils over rocks containing these minerals there may be a good supply of little-weathered but gradually weathering feldspars and micas in the root zone. In older, deeper and more weathered soils the supply may be very much less in the upper part of the soil profile but may increase with depth as the weathering rock is approached. Biotite (black mica) and the feldspars weather relatively easily in hot, wet environments. White mica is more resistant to weathering and may

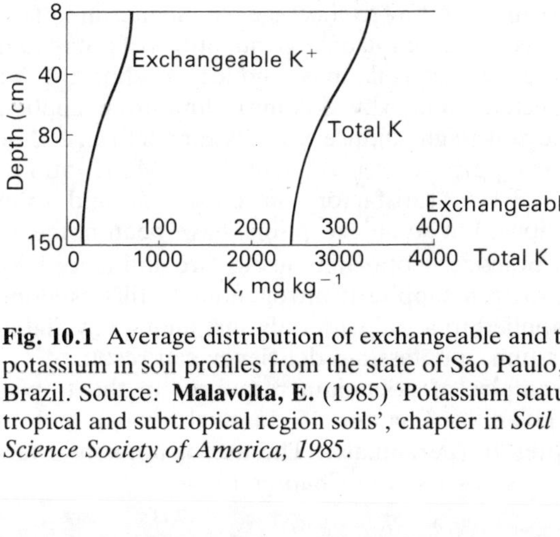

Fig. 10.1 Average distribution of exchangeable and total potassium in soil profiles from the state of São Paulo, Brazil. Source: **Malavolta, E.** (1985) 'Potassium status of tropical and subtropical region soils', chapter in *Soil Science Society of America, 1985*.

be found in older and more weathered soils, and because it is light and easily transported may occur in alluvium. These feldspar-containing weatherable minerals constitute the main inorganic reserves in the soil and vary considerably between soils, depending on the nature of the parent material and the age of the soil and degree of weathering. An example of a particularly potassium-rich parent material is afforded by the volcanic lavas and ashes of East Africa. These are derived from a particularly potassium-rich magma, and soils developed from them are high in potassium.

In addition to the unavailable potassium in weatherable minerals, there are further potassium reserves in the organic fractions of the soil. These reserves are released, together with a range of other nutrients, as the organic fraction is mineralised. Amounts of organic matter, and therefore of unavailable potassium, also vary very much between soils, depending on climate, vegetation, soil texture and history of cultivation. In soils with very low reserves in the mineral fraction of the soil, the organic potassium in the soil humus may constitute the main reserve of this element in the soil.

The **slowly available potassium** is the potassium that is intermediate in availability between the available and the unavailable forms discussed above. This includes potassium that is alternately fixed and released by 2:1 lattice clays in the soil, particularly illites (Chapter 2). Some clays are similar in structure to micas and can change to a mica by absorption of potassium cations, just as micas weather to clays with the loss of these ions. In addition, illites are able to fix potassium cations temporarily in their silica sheets. Drying of the soil sometimes has the effect of releasing some of the slowly available potassium held in this way, particularly in the case of soils that are low in potassium.

The distinction between the available, slowly available and non-available forms of potassium is not rigid, and there is a tendency for potassium to change from one form to another in such a way as to move towards an equilibrium. What this means in practice is that if plants remove readily available potassium, there is a tendency for potassium to move from the slowly available to the available form to restore the equilibrium. Similarly potassium moves from the unavailable to the slowly available and available forms as these are depleted by plant uptake or by other causes. Thus in a soil in which the plant uptake is heavy and there is a steady drawing on the potassium reserves, there would be a general movement from the unavailable to the available forms. If, conversely, large quantities of potassium-containing fertiliser were added to the soil, in excess of that used by the crop and lost by leaching, the extra potassium in the available form would be converted to slowly available and perhaps to unavailable forms.

In most soils only a small proportion of the total potassium is in the available form. Measuring the total potassium in the soil gives an indication of reserves but does not necessarily give much indication of what is immediately available to the plant. Often only 1–2% of the total may be readily available, i.e. exchangeable or soluble.

Factors affecting potassium availability

The ease with which plants absorb readily available potassium in ionic K^+ form varies according to soil characteristics and other cations present in the soil.

Plant uptake of exchangeable potassium is influenced by the proportion of the soil cation exchange capacity occupied by potassium cations. This is sometimes referred to as the potassium saturation percentage. A given amount of exchangeable potassium will be more available in a soil with a low cation exchange capacity than in a soil with a high one, because in the former case the potassium saturation percentage is higher. If 0.1 m.e. of exchangeable potassium is held by a soil with a cation exchange capacity of 10 m.e./100 g of soil, then the percentage saturation is only 1% and the soil is probably deficient in potassium. If the same 0.1 m.e. of exchangeable potassium is held by a soil which has half the cation exchange capacity of the first soil, i.e. only 5 m.e./100 g, then the potassium saturation percentage is 2%, or twice as high, and potassium is probably adequately available to the crop.

In practice it has been found from a limited number of investigations that the potassium supply is adequate for most crops when the potassium saturation percentage is above about 1.5–2%. On oilpalm in south-eastern Nigeria, for example, potassium deficiency symptoms were shown when the potassium saturation dropped below 1.5%. Other investigations have suggested that an optimum level of potassium saturation in Africa is 2–3%, except under very intensive cultivation when it should be a little higher, perhaps as much as 4–5%.

Other research workers have considered the ratio of exchangeable potassium to the sum of the exchangeable bases (calcium, magnesium and potassium) as a practical measure of potassium availability. It appears that under many cropping conditions exchangeable potassium should be at least 2% of total exchangeable bases if potassium uptake by the crop is to be adequate, but this will be influenced by the nature of the soil, and by both the need and the potassium-extracting ability of the crop.

Widespread investigations have also led to the conclusion that in addition to the need for the exchangeable potassium to be sufficiently high in proportion to the cation exchange capacity, or to the total exchangeable bases, a certain minimum amount must also be present. Various studies in French-speaking tropical Africa, in Nigeria and in Brazil have indicated that potassium deficiency symptoms begin to appear on most crops when the exchangeable potassium falls below 0.1 m.e.

In addition, uptake of potassium is influenced by the relative amounts of exchangeable potassium, calcium and magnesium in the soil, and in acid soils is influenced by exchangeable aluminium. There is an antagonism between potassium and calcium, and between potassium and magnesium. The Mg/K ratio – the ratio of exchangeable magnesium to exchangeable potassium – has long been studied by agronomists. Too high leads to shortages of potassium, too low to shortages of magnesium. This explains why addition of large quantities of potassium fertiliser to a soil may induce a shortage of magnesium, and why dolomitic limestone applied to the soil might induce a potassium shortage. For practical purposes it is thought that a Mg/K ratio of about 3:1 is satisfactory in most soil and crop situations, but some exceptions have been noted.

In practice, potassium needs are influenced by the nitrogen supply. If nitrogenous fertilisers alone are applied to a soil with only just enough available potassium, a potassium deficiency can result.

Examples of potassium responses in the tropics are given in Chapter 15 (annual crops) and in Chapter 16 (perennials). The use of potassium fertilisers is discussed in Chapter 14.

11 Calcium and magnesium, soil acidity and liming

Both calcium and magnesium are needed by plants for their nutrition, but these two elements also influence the pH of the soil and thereby have important indirect effects. Calcium, or calcium and magnesium, are often added to soils not so much to supply these nutrients to the plant, but in order to make the soil less acid and therefore to improve the uptake by plants of other nutrients. In tropical soils liming has to be carried out with great caution. Frequently only very moderate lime applications are advisable, in order to bring the soil pH up to about 5.5. Heavier applications may do more harm than good.

Calcium

Calcium is an essential element for all higher plants and is found in relatively large quantities in plant leaves, but plants differ widely in the amounts of calcium they need.

Deficiencies of calcium result in the terminal buds and root tips becoming stunted and failing to develop normally.

Calcium is absorbed by the plant as the divalent cation Ca^{2+}. Calcium is normally the dominant exchangeable cation in most soils but the amounts become progressively less in more acid soils. Plants such as rubber which need a heavy rainfall and tolerate or prefer acid soils generally have relatively low calcium requirements.

Some acid soils in the tropics are deficient in calcium and in those cases liming with calcium carbonate is required to improve the calcium supply for such crops as sugarcane that are not tolerant of low calcium levels.

Magnesium

Magnesium is a constituent of every chlorophyll molecule and therefore is essential for photosynthesis; it also occurs in plant seeds. Green plants cannot exist without it. It is connected with phosphorus metabolism and with the activation of enzymes affecting carbohydrate metabolism and the synthesis of plant oils.

Magnesium is mobile within the plant, being translocated from old leaves to young ones if necessary, and deficiency symptoms first appear on the older leaves. These may take the form of a chlorosis or whitening of the tissue between the veins. Leafy crops such as tobacco are sensitive to magnesium deficiency. In the tropics deficiencies have been observed in oilpalms where they have been held responsible for orange frond disease.

Magnesium is taken up as the divalent cation, Mg^{2+}. In most tropical soils the dominant exchangeable cation is calcium, followed by magnesium. The ease with which the plant absorbs the magnesium cations present is influenced by the amounts and species of other cations present, and in particular by the ratio of exchangeable magnesium to exchangeable potassium, as discussed in the section on potassium (Chapter 10).

When soils are subjected to increasing amounts of leaching, both exchangeable calcium and exchangeable magnesium are removed. In most cases more calcium is removed than magnesium so that the Ca/Mg ratio narrows in the more acid soils. The effects of leaching are thus to reduce the amounts of both calcium and magnesium held in the exchangeable form, to alter the relative proportions of exchangeable calcium and magnesium to each other, and to decrease the two relative to potassium and other cations.

Where magnesium deficiencies are diagnosed or suspected, they can be corrected by the use of dolomite, which contains magnesium carbonate, or of dolomitic limestone, which contains both magnesium and calcium carbonate.

Certain Oxisols of the *campo cerrado* areas of Brazil are very low in magnesium and respond directly to magnesium fertilisers. Of the important tropical crops, magnesium deficiencies have been associated mainly with oilpalm (Chapter 16).

Soil acidity and liming

The majority of tropical soils are more acid in reaction than pH 6.0, though many of the more productive soils selected for intensive agriculture are not acid. Acid soils are particularly extensive in large areas of South America, including the *campo cerrado* of Brazil and the *llanos* of Colombia and Venezuela. In many areas of the tropics there is a general correlation between rainfall and soil acidity. The acid soils are associated with heavier rainfall, though in some cases soils have been made acid by the continued use of acidifying fertilisers (Chapter 14).

The nature of soil acidity and its effects on plants

Soil acidity is a complex subject. It is broadly related to the extent to which basic cations, particularly exchangeable calcium and magnesium, occupy the exchange positions on the soil colloids, but is also influenced by organic matter, by the nature of the clay mineral, by partial pressure of carbon dioxide, and by the occasional presence in certain soils of sulphuric acid. In general terms the removal of calcium and magnesium from the soil, either by organisms or by leaching, lead to the soil becoming progressively more acid. Soils of very high rainfall areas are generally acid in reaction owing to leaching.

Up to the 1950s it was believed that soil acidity was connected with the presence of hydrogen ions (H^+) and that acid soils were more or less hydrogen saturated. Extensive research on soil acidity and its effects since then have shown that in soils with a pH below 5.5, the dominant exchangeable cation is aluminium. In acid mineral soils, exchangeable hydrogen is found in only very small amounts or is virtually absent, though it may be higher in organic soils.

Exchangeable aluminium has to occupy most of the cation exchange sites, probably over 60%, before exchangeable aluminium is found in the soil solution at levels of over 1%. Exchangeable aluminium in the soil solution at more than these levels can cause yield reductions by direct injury to the root system, though plants differ widely in their aluminium tolerance. By accumulating in plant roots, aluminium may impede the uptake and translocation of calcium and phosphorus.

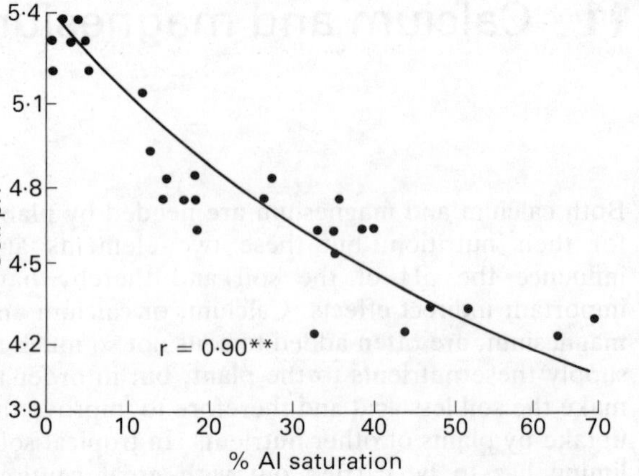

Fig. 11.1 Relationship between soil pH and aluminium saturation in eight Ultisols and Oxisols of Puerto Rico. Source: **Abruna, F. R., Pearson, W. and Perez-Escolar, R.** (1975) 'Lime responses of corn and beans grown on typical Ultisols and Oxisols of Puerto Rico', in **Bornemisza, E. and Alvarado, A.** (1975) pp. 261–81.

The infertility of acid soils may also be due to manganese toxicity, as well as to shortages of calcium and magnesium. Manganese toxicity can occur when the soil pH is below 5.5. Manganese is made more soluble in poorly drained and flooded soils when reducing conditions convert the manganese IV ion (Mn^{4+}) to the manganese II form (Mn^{2+}).

The extent to which crops can grow in acid soils depends on their tolerances of aluminium and manganese, and on their calcium and magnesium demands. Important differences in these respects have been found between crops and between crop varieties. Some species have evolved in acid conditions and are tolerant of high aluminium levels.

Legumes in general are not tolerant of acid soils because of a high calcium requirement, but some tropical pasture legumes are tolerant of high acidity and high levels of aluminium in the soil solution, as discussed in Chapter 21. Some varieties of cassava are tolerant of acid conditions, as are rubber, oil-palm, and coffee, and some plants, such as tea and the pineapple, actually prefer acid soils. In the case of tea (Chapter 16) this is associated with a high aluminium requirement and high aluminium levels in the leaves. Most major tropical food crops include varieties that are tolerant of pH values at least as low as 5.5, so that in these cases there is no point in liming to bring soils above that level.

Liming tropical soils

The purpose of liming is primarily to neutralise the exchangeable aluminium when aluminium-sensitive crops are being grown. This is normally achieved by bringing the soil up to pH 5.5. When manganese toxicity is an additional factor, then the pH should be raised to about 6.0.

The amounts of lime that need to be applied to neutralise the exchangeable aluminium more or less completely, can be simply calculated if the exchangeable aluminium content is determined in a soil laboratory. Normal recommendations on acid soils in South America are of the order of about 1.6 t/ha of calcium carbonate ($CaCO_3$) for each milliequivalent of exchangeable aluminium present. For many soils this works out at about 1.5–5 t/ha of $CaCO_3$. These recommendations are only a fraction of the rates that were applied in the past with very mixed results.

Overliming can result in harmful effects. Any application that raises the soil pH to levels of above 5.5–6.0 is probably quite unnecessary and could reduce yields. A common harmful effect of overliming is to reduce the availability of trace elements, since all trace elements except molybdenum are more available in acid soils than in neutral or alkaline ones (Chapter 12).

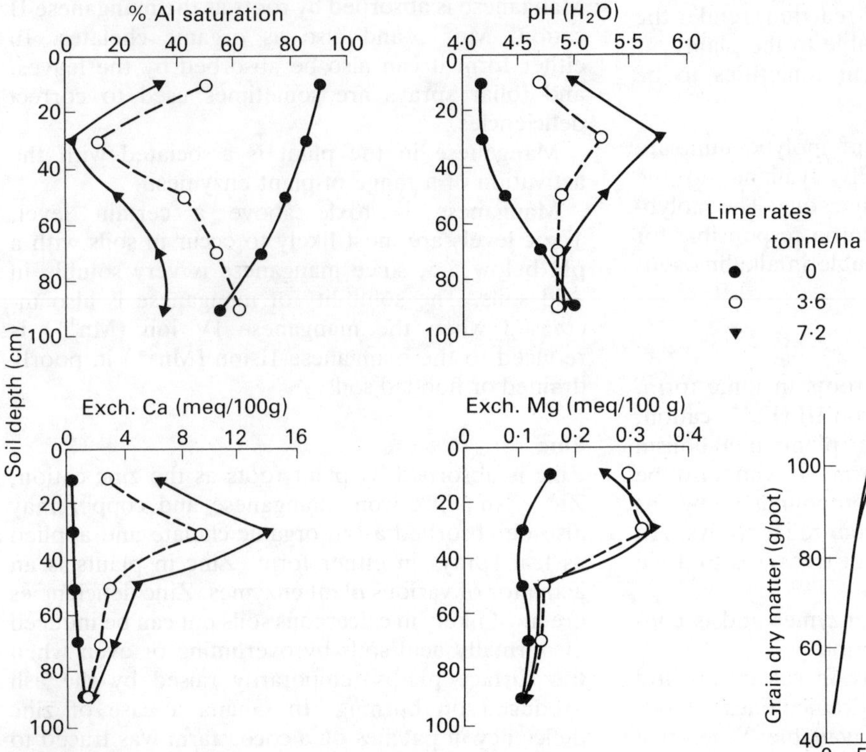

Fig. 11.2 Effects of lime applied four years before sampling on the profile properties of an Oxic Dystrandept from Costa Rica. Source: Sanchez, 1976, calculated from data in **Morelli, M., Igue, K. and Fuentes, R.** (1971) 'Effect of liming on the exchange complex and on the movement of Ca and Mg', *Turrialba* **21**, 317–22.

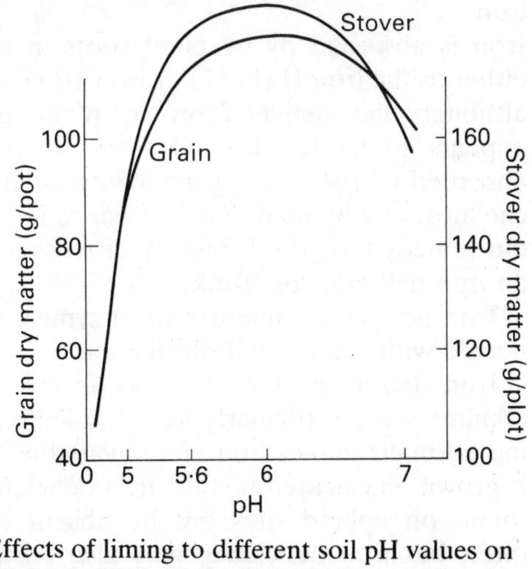

Fig. 11.3 Effects of liming to different soil pH values on maize growth in an Ultisol from Hawaii. Source: **Younge, O. R. and Plucknett, D. I.** (1964) 'Liming materials', *Hawaii Farm Science* **13** (3), 7–8.

12 Trace elements

Trace elements and the plant

Iron, manganese, zinc, copper, boron, chlorine and molybdenum are described as micronutrients or trace elements because they are needed by plants only in very small amounts. With the exception of iron and sometimes of manganese they are present in soils in only minor quantities, or traces. In some cases they may be toxic to plants if present in large amounts.

Trace elements may limit plant growth if

1 the amounts present in the soil are too low to satisfy even the very small requirements of the plant;
2 soil conditions, generally soil reaction, render the trace elements present unavailable to the plant;
3 they are present in sufficient quantities to be toxic.

All the micronutrients except molybdenum are more mobile and more readily available to the plant in acid soils than in alkaline soils. The molybdate anion, needed by the bacteria responsible for nitrogen fixation, is more available in alkaline soils than in acid ones.

Iron

Iron is absorbed by by plant roots in ionic form, either as the iron II (Fe^{2+}) or iron III (Fe^{3+}) cation, although the active form in plant metabolism appears to be the iron II form. It can also be absorbed by roots as organic compounds known as chelates, and by plant leaves from foliar sprays. An old remedy for iron deficiency of trees was to drive an iron nail into the trunk.

Iron activates a number of enzymes and is connected with chlorophyll production.

Iron deficiency is most likely in calcareous and alkaline soils, particularly well-drained ones. Flooding normally makes iron more available. When rice is grown on calcareous soils, iron deficiency may be found on upland sites but be absent on flooded ones, but iron deficiency may also occur in some flooded soils in which the organic matter content is too low to cause soil reduction.

Iron deficiencies are widespread in central and northern India and in the Philippines.

Iron toxicity can occur when certain soils are flooded, giving toxic concentrations of iron II ions in the soil solution and in rice plants. The normal practical solution is to drain the soil to oxidise the iron II to the iron III form. It is also sometimes possible to pre-flood the field and wait for the ferrous iron peak to subside before sowing or transplanting rice.

Manganese

Manganese is absorbed by roots as the manganese II cation, Mn^{2+}, and also as organic chelates. In either form it can also be absorbed by the leaves, and foliar sprays are sometimes used to correct deficiencies.

Manganese in the plant is associated with the activation of a range of plant enzymes.

Manganese is toxic above a certain level. Toxic levels are most likely to occur in soils with a pH below 5.5, since manganese is very soluble in acid soils. The solubility of manganese is also increased when the manganese IV ion (Mn^{4+}) is reduced to the manganese II ion (Mn^{2+}) in poorly drained or flooded soils.

Zinc

Zinc is absorbed by plant roots as the zinc cation, Zn^{2+}, and like iron, manganese and copper may also be absorbed as an organic chelate and applied as leaf sprays in either form. Zinc in plants is an activator of various plant enzymes. Zinc deficiencies are most likely in calcareous soils but can be induced in normally acid soils by overliming or even when the surface pH is temporarily raised by the ash produced on burning. In Ghana a case of zinc deficiency in patches on a cocoa farm was traced to a localised increase in soil pH due to dumps of rotting cocoa pods which were high in calcium and magnesium. Zinc deficiency has been reported in several crops in Malaysia, and in Puerto Rico severe zinc deficiency was found in maize grown on two different Oxisols with pH values of 5.0 and 6.0.

Zinc deficiency is a widespread micronutrient disorder in tropical rice. Under lowland conditions it occurs in parts of India, Bangladesh, the Philippines and Colombia, though some varieties are much more tolerant of low zinc levels that others. Under upland conditions it occurs in the *campo cerrado* of Brazil. Although the *campo cerrado* soils are acid, and zinc solubility is highest at low pH levels, zinc is deficient because these soils contain unusually low amounts.

Zinc can be toxic to plants if present in more than very small quantities.

Copper

Copper is absorbed both by roots and by leaves as the copper II cation, Cu^{2+}, and as copper chelates, and may be applied as foliar sprays. The proportion of copper to the other trace elements is important.

Copper is concerned with the activation of plant enzymes.

Copper deficiencies are most likely to occur in organic soils, including mucks and peats. In Kenya, copper deficiencies in wheat in the Njoro area were traced to soils developed from a particular volcanic ash shower that are very low in this element.

Boron

Boron is absorbed by plants in a number of anionic forms, including $B_4O_7^{2-}$, $H_2BO_3^-$, HBO_3^{2-}, and BO_3^{3-}. Its function in plants is not fully understood. It is widely deficient in Africa and is regularly applied to cotton in the Ivory Coast and elsewhere. Boron deficiencies have been reported in cotton, sisal, wattle, pines and coffee in mainland Tanzania. In Malaysia, boron deficiencies were found in several crops. In South America boron has been reported deficient in the *campo cerrado* area of Brazil and in some soils planted to cocoa in Ecuador.

Molybdenum

Molybdenum is absorbed as the molybdate anion, MoO_4^{2-}. Molybdenum differs from all the other trace elements listed above in that it is more available in alkaline soils than in acid ones.

Molybdenum is required by the plant in very small quantities for nitrogen reduction and therefore for protein synthesis. Molybdenum is needed by *Rhizobium* bacteria for nitrogen fixation, and is sometimes given to legumes on acid soils for this reason.

A spectacular case of molybdenum deficiency in Australia, which rendered large areas unproductive, was cured by adding a trace of molybdic oxide to the phosphatic fertiliser used. Molybdenum deficiency has been found in high-altitude pastures in Kenya. Molybdenum responses have been obtained with groundnuts in northern Senegal, using ammonium molybdate at 28 g/ha, and with the same crop in Ghana and on Latosols in Java.

Other elements absorbed by plants

Chlorine is now known to be another essential element, required in very small quantities, though its functions require further investigation. Excess chlorine has harmful effects on some crops, particularly tobacco and potatoes.

Other elements found in plants are believed to fulfil useful functions even though they are not classified as essential in the sense that the plant cannot grow without them.

Cobalt is not essential to the plant, though it is needed by *Rhizobium* bacteria, and, since it is an ingredient of vitamin B_{12}, by animals, so that its presence in pasture and forage is important.

Vanadium has not proved to be essential to higher plants, though some crops have responded to applications. It has been shown to be able to

Fig. 12.1 Zinc and sulphur trials on rice in Bangladesh.

replace molybdenum to some extent in the nutrition of *Azotobacter*.

Sodium is taken up by plants (as the cation Na^+) and a number of crops are thought to benefit from sodium applications. In some circumstances it can substitute for some of the potassium needed.

Silicon is taken up by some plants, including grasses and sedges, in substantial amounts, and is frequently applied to rice, as discussed in Chapter 10.

Aluminium, iodine, nickel and tungsten are other elements thought by some investigators to fulfil certain functions in plants.

Trace element deficiencies in tropical soils

The absolute amounts of micronutrients in tropical soils are closely related to the soil parent material. They are highest in soils over basic rocks, and generally adequate in young soils and soils developed in recent alluvium which still contain some primary minerals. They are lowest in soils derived from poor siliceous parent materials such as sandstones, and in extremely highly weathered soils of old land surfaces.

The *campo cerrado* area of the central plateau of Brazil, with a savanna vegetation, forms a good example of highly weathered, leached acid soils developed over poor sediments that are low in a range of micronutrients. In some cases good responses have been obtained to applications of zinc, boron and molybdenum but in other *campo cerrado* areas of apparently similar soils the same micronutrients gave no response, so that there was no clear-cut general pattern.

In local cases micronutrient deficiencies have been induced in soils in which the micronutrient supply is normally satisfactory, by raising the pH by overliming or other means.

To some extent, however, the various trace element deficiencies and responses outlined above form exceptional cases, since many widespread and comprehensive trace element tests in various parts of the tropics have often failed to obtain any responses to the trace elements applied. With increasing pressure on the land and with further studies, this picture can be expected to change, and more examples of trace element deficiencies and responses are likely to come to light.

In the past, trace elements have been added, sometimes inadvertently, in fertilisers. The increasing use of higher-analysis chemical fertilisers containing fewer micronutrient impurities appears likely to increase the need for supplementary micronutrients. Organic fertilisers, which are good sources of trace elements, are also used less and less in commercial agriculture.

With the introduction of more intensive farming systems in the tropics resulting in higher yields, increased pressure is being put on micronutrient supplies, and soils that have not shown deficiencies in the past may respond to micronutrient applications. In shifting cultivation systems, greater pressure on the land resulting in shorter fallows and longer cultivation periods is progressively reducing the organic matter content of soils in many areas, and with it the natural supply of trace elements.

Trace element deficiencies and their correction are discussed further in Chapter 14.

13 Soil fertility evaluation

The evaluation of soil fertility involves the assessment of the ability of the soil to supply nutrients to crops. 'Soil productivity', on the other hand, is a wider term which in addition to the nutritional aspects is concerned with soil physical conditions and other soil factors affecting yields.

Soil fertility is in effect evaluated empirically every time a crop is grown and the yields assessed. Farmers have been evaluating soil fertility by practical trial and error ever since agriculture began. The Romans produced numerous books on husbandry, such as Columella's *De re rustica* – 'On farming matters' – which described soils and soil improvement methods adapted to specific soil conditions. In the tropics peasant farmers usually know their soils intimately, and advice is handed down from father to son on how to use them and what crops to plant where and when. Shifting cultivators often assess soil fertility by looking at the vegetation it supports, and consider certain species indicators of fertility or particular soil conditions. The twentieth-century agronomist and soil scientist would be wise to treat this accumulated experience with the respect it deserves. Nevertheless we can bring to the solution of old problems a host of modern techniques, and modern soil fertility evaluation in the tropics makes use of four main approaches:

1 Soil testing.
2 Plant analysis.
3 Pot experiments.
4 Fertiliser field trials.

Results obtained from these approaches have to be carefully assessed in relation to soil conditions, to climate, to the crops grown, and to the level of yields aimed at.

The extent to which these fertility evaluation procedures are used varies considerably within the tropics regionally, and according to the type of agriculture. In general they are more likely to be used by large commercial producers and least likely to be available to the small farmer, though in some countries there may be limited access to some soil testing and other services through government extension services. Regionally, soil fertility evaluation facilities exist in many Latin American countries and in India, but are relatively few and far between in most of Africa, where fertilisers are as yet used only on a very small scale, and in South-east Asia where traditional flooded rice culture is widespread.

Fertility evaluation based on soil testing

Soil testing involves at least three major steps:

1 Soil sampling.
2 Laboratory analysis of the sample.
3 Interpretation of the results and provision of recommendations to the farmer.

Soil sampling

Before a soil can be analysed it has to be sampled. No amount of careful laboratory work is any use at all if the sample is not well taken. Poor sampling is in practice a major source of error and of misleading results. Many agronomists have little or no idea how to sample a soil, and happily bring back to the laboratory soil samples that have been wrongly taken, with insufficient supporting information.

There are two major types of soil sampling. The first is the sampling of the horizons of a profile pit by soil scientists who wish to examine the whole soil profile and to characterise it by careful description and analysis. This type of work is appropriate for soil surveys, and can also be used by the farmer to show what different types of soil are being dealt with. Because it shows the soil only at one specific point, however, it is not appropriate for fertility evaluations designed to provide fertiliser recommendations.

The second type of soil sampling is the bulk sampling of a field or plot in order to characterise the average fertility of the area with the view to providing recommendations regarding the use of fertilisers or amendments. For this, the area sampled has to be fairly uniform as regards slope, drainage, soil colour and texture, and past history of cropping and fertiliser use. If a field includes obviously different areas it must be subdivided for sampling purposes. The soil is then sampled from at least 10 to 20 sites to a standard depth, usually the plough or hoe layer, or the upper 15 cm. The individual cores are then thoroughly mixed to form a composite sample from which a representative subsample of about 500 g is taken for sending to the laboratory. The number of core samples used to make the composite sample is very important if the final sample is to be representative of the area sampled. Even if the area is very small, the number of samples should not be reduced, since soil variations occur over short distances and there can be as much variation in a small field as in a large one. Areas not representative of the field as a whole, such as fence rows, cattle kraals and manure heaps, and termite mounds, should not be included in the sample. For deep-rooted crops, such as tea or sugarcane, a subsoil sample should also be taken. All samples must be clearly labelled and information on site, depth and history of cropping and fertiliser use provided.

Laboratory analysis of the sample

When soil samples reach the laboratory they are ground to pass through a 2 mm sieve, thus separating any coarse fraction (over 2 mm) that may be present. In many laboratories routine analytical procedures are now semi-automatic, enabling a single technician to handle large numbers of samples a day. In Latin America, for example, there are about 50 large soil laboratories which analyse 100–500 samples a day, and which together handle over 400 000 samples a year.

The value of soil tests varies considerably between elements. To be useful and acceptable a soil test must

1 extract all or a regular proportion of the available nutrients in soils with different properties;
2 be rapid and accurate; and
3 give results that can be correlated with crop growth and fertiliser response in the field under practical growing conditions.

Considered in relation to these criteria, current soil testing methods are relatively good for phosphorus, potassium, calcium, magnesium and lime requirements and for the assessment of micronutrient needs. Tests for phosphorus, however, are only reliable for normal well-drained soils since they do not predict the phosphorus that will become available on flooding.

Current soil testing methods are less successful and less reliable for assessing nitrogen needs, and field experiments are needed to predict nitrogen fertiliser requirements and crop responses. Laboratory methods for analysing sulphur are still cumbersome.

Nitrogen is routinely determined as total nitrogen and this gives a good indication of nitrogen reserves, though not necessarily of immediate fertiliser responses since this also depends on the rate of mineralisation of the reserves. For local areas, however, soil nitrogen and organic matter contents are a useful way of distinguishing soils that are likely to respond to nitrogen fertilisers from those that are less likely to do so.

A soil sample can be incubated under aerobic or anaerobic conditions to give an indication of mineralisation rates and the amount of nitrogen released. This method gives a good indication of nitrogen availability in the field, but is too slow for routine use.

Inorganic nitrogen present in the soil as nitrate and as ammonium can be monitored by sampling at intervals but amounts are very small and fluctuating, and difficult to correlate with crop growth. The method is too costly, time-consuming and unreliable to be used as a routine soil fertility evaluation process. It is useful however in monitoring the pattern of changes in available nitrogen during a season or year.

Phosphorus is widely assessed by routine analytical procedures. Since determinations of total soil phosphorus are useless for soil fertility evaluations, these aim instead at extracting a small proportion of the total inorganic phosphorus in the soil which is assumed available to the plant or which correlates with plant responses to added soil phosphate. Several different extractants are used, but these extract phosphorus only from some or all of the Ca-P, Al-P and Fe-P forms discussed in Chapter 10, and not from the occluded or reductant-soluble forms, nor do they reflect the inorganic phosphorus present in the soil humus. Of the commonly used extractants, only the Olsen sodium bicarbonate method (0.5 M $NaHCO_3$ at pH 8.5) reflects soil Fe-P satisfactorily in addition to Ca-P and Al-P, and seems the method applicable to the widest range of

Table 13.1 Critical thresholds of phosphorus deficiency as estimated for five groups of soils by four different methods. The highest amounts are extracted by the Olsen sodium bicarbonate method and the lowest by water extraction. Resin extraction figures showed particularly good correlations with crop phosphorus uptake in a wide range of tropical soils.

Soil group	**Extraction method**			
	Olsen	**Bray 2**	**Water**	**Resin**
	(ppm phosphorus)			
Oxisol		10		10–15
Ultisol	30–40	15–20	0.5	10
Vertisol	30–40	10	0.25	17
Entisol	20	10–15	0.2	10
Alfisol	25	15		10

Source: **Roche, P, Grieve, L., Babre, D., Calba, H. and Fallavier, P.** (1980) *Phosphorus in Tropical Soils: Assessing deficiency levels and phosphorus requirements*. IMPHOS World Phosphate Institute: Paris.

soils. The Bray 2 method (0.3 M NH_4F + 0.025 M HCl) and other acid extractants extract both Ca-P and Al-P and are widely used (see Table 13.1).

Potassium available to the plant is that which is exchangeable and soluble, and this is routinely determined in soil laboratories by extracting with neutral normal ammonium acetate at pH 7.0. This extractant also removes the exchangeable calcium and magnesium at the same time. In soil fertility laboratories potassium is sometimes extracted at the same time as phosphorus by using the Olsen method (see above) or the North Carolina (Mehlich) method (0.025 M H_2SO_4 + 0.05 M HCl).

The **lime requirement** has in the past been assessed by many methods, including the determination of soil pH. The best test to assess lime response in acid tropical soils is now considered to be the determination of the exchangeable aluminium with normal potassium chloride.

Sulphur tests involve the determination of sulphate sulphur in the soil. The various extractants used include water and monocalcium phosphate but these methods are relatively time-consuming.

Micronutrient availability is currently assessed by using a variety of methods, all of which remove

Fig. 13.1 A laboratory technician at the former ORSTOM laboratories, Adiopodoume, Ivory Coast, West Africa, assessing structural stability by a standardised sieving test.

part of the total nutrients present. The Olsen method used for phosphorus evaluation can be used to extract copper, iron, manganese and zinc (as well as potassium, calcium and magnesium) at the same time as phosphorus if modified by the addition of a chelating agent EDTA (ethylenediaminetetraacetic acid).

Interpretation of the results and provision of recommendations to the farmer

The analytical results are meaningless unless they are interpreted. Soil analyses have to be correlated with crop responses under practical growing conditions in the field.

Correlation of a soil test with crop growth is very complex because yield is a function of a large number of variables which can be grouped into four main categories of soil, crop, climate and management. Initial correlations may be performed in greenhouse studies so that some of the variables are eliminated, but the ultimate value of a soil test can only be demonstrated in the field. Correlations of soil test results with absolute yields are exceedingly difficult, and in practice interpretation is directed more to relative yields. A realistic aim is for the soil analysis to predict soil–crop situations where there is a good likelihood of a useful fertiliser response, but not actual yields. Interpretation becomes more reliable when there are large numbers of analyses available for a certain area which have been correlated with crop responses in the field. The fewer the results available for an area, the more difficult they are to interpret.

Finally, the recommendations based on soil tests have to be conveyed to the farmer, and this is often the responsibility of extension services. For some specific crops it may be the responsibility of organisations that buy those crops from the farmer and market them. Unless the yield increases obtained are both spectacular and profitable, the small tropical farmer is not likely to accept the fertiliser recommendations given. Good recommendations include not only the type and amount of fertiliser to be used but also directions as to how and when to apply them.

Fertility evaluation based on plant analysis

Plant analysis involves the analysis of plant parts, usually the leaves, in order to assess the nutritional status of the plant. It is particularly useful for perennial tree crops, and also in areas without effective soil testing services. It shows the net result on the plant of various variables: the soil, the plant, the climate, and the level of management. This has been considered by some to be an advantage since the analysis is a measure of nutrient uptake. In practice a result may have several possible causes, including plant disease or management factors not related to soil fertility. For an annual crop, the plant analysis will probably be obtained too late for that particular crop to benefit from it, though succeeding crops may do so. For perennial crops, nutritional disorders diagnosed can sometimes be corrected before they have an adverse effect on yields. The oilpalm is a good example of a tree crop on which leaf analyses have been used profitably.

Plant sampling and critical values

The analysis of leaves or other plant parts, if it is to be useful, must be related to

1 standardised sampling techniques, and
2 knowledge of critical nutrient levels.

The content of essential elements differs between different parts of the plant, and in the same part at different stages of growth as elements are translocated from older leaves to younger ones and to growing points. Carefully standardised sampling is essential if the analytical results are to be meaningful. In the case of coconut palms, for example, the sampling directions might specify taking the middle third of the central leaflets of frond number 4, 9 or 14 according to the age of the palm.

The interpretation of the results requires a knowledge of the critical levels of the nutrient element considered, above which the crop is sufficiently supplied, and below which it is deficient. Toxicity levels are also needed, above which nutrients such as boron, iron or manganese are present in harmful amounts.

Leaf analyses have been successfully used on coconuts, oilpalm, coffee, sugarcane and other crops. A system of frequent analysis of sugarcane leaf blades and leaf sheaths during the first six months of growth, termed 'crop logging', has been used in Hawaii and elsewhere to increase the efficiency of fertiliser and irrigation practices.

Fertility evaluation based on pot experiments

Pot experiments under controlled conditions in greenhouses have long been used to identify nutrient deficiencies and responses in soils. They involve growing a test plant such as millet in pots containing different soils subjected to various treatments. In the missing-element technique all elements except the one being tested are supplied in adequate amounts, and successive cuttings of the test plant when weighed and analysed give information on growth and the progressive exhaustion of the soil as regards the nutrient tested. Alternatively, various fertiliser dressings can be correlated with soil analyses and plant responses as a preliminary stage before field trials are carried out. Sometimes

Fig. 13.2 Simple fertiliser trials of phosphorus on rice and potassium on maize, with descriptive signs allowing local farmers to know what treatments are being tested (South China).

only small quantities of soil are used; for example, a soil volume of 150 ml might be planted to sorghum grown for 6–8 weeks in small containers.

Pot experiments have been widely used by research workers in francophone Africa. Using missing-element techniques, they consider a nutrient severely deficient if the yield without it is less than 40% of that obtained with the complete fertiliser treatment. Field trials with the elements diagnosed as deficient then follow, and recommendations are based on these.

Possible difficulties attached to pot experiments are that they normally involve a small quantity of soil, not the natural soil profile, and only the early stages of plant growth.

Fertility evaluation involving field trials

Careful, replicated fertiliser experiments under more or less controlled conditions on agricultural stations have long formed a feature of agricultural research. More recently these have been supplemented to an increasing extent by often relatively simple experiments on farmers' fields and plots, sometimes farmer managed, and with varying degrees of supervision. These on-farm experiments, designed to test crop responses under actual conditions in the field, may involve spacing, variety, time of planting, plant protection and other trials as well as those involving fertiliser applications.

In classical field experiments, great attention is devoted to the design of the trial and the statistical analysis of the results. Fertility trials are usually concerned with elucidating response curves and interactions between nutrients. Response curves show yields in relation to increasing quantities of a nutrient and thus the amount which produces the greatest return at the least cost. Interactions between nutrients indicate how the results of applying one nutrient are affected by different levels of other nutrients applied at the same time. Trials on the same site for several years allow residual fertiliser effects to be assessed, and the monitoring of soil fertility changes due to prolonged treatments. Station trials thus give information that is difficult to obtain with comparable accuracy in other ways. Careful supervision by agronomists, and detailed observation and recording at all stages, is possible.

The applicability of the results of field trials depends among other things on the extent to which the experimental site is representative as regards soils and climate of the area it was chosen to serve. Since this type of agricultural research is relatively costly, the number of sites is often small in relation to the need to serve different ecological areas. The fact that the agronomists on research stations appear to have greater access to labour and other inputs often means that local small farmers may regard the results as not being relevant to their situation.

The present tendency is to supplement relatively few carefully controlled but expensive trials on agricultural stations with relatively large numbers of small simple trials away from the stations on farmers' fields and plots. The careful supervision that is given to station experiments is not possible, but to some extent the trials by their greater numbers may be more useful. They represent results under local conditions of climate and management (or mismanagement), and are thus more indicative of the difficulties likely to be encountered and the results achieved by the farmers to whom the recommendations are directed. The innovations are tested in relation to the local cropping system, and the reactions of the farmers themselves give an indication as to whether the recommendations tested are likely to prove acceptable. The theory and practice of carrying out research on agronomic improvements on farmers' fields is now increasingly referred to as 'on-farm research' or OFR, and often embraces the study of the whole cropping system or farming system.

In the 1960s and 1970s the Food and Agriculture Organisation of the United Nations (FAO) carried out a very large number of simple fertiliser trials and demonstrations in many tropical countries, with the two-fold purpose of finding out more about yield responses on small farms under local conditions, and of introducing third world farmers to the practice of applying fertilisers. The underlying assumption of the programme was that fertiliser responses must be evaluated in farmers' fields and not under controlled conditions at agricultural experimental stations. Large numbers of farmers' plots were chosen more or less at random without prior analysis, with individual trials on seemingly similar soils being classed as replicates. The experiments were usually simple, involving two levels of an NPK fertiliser or 2 × 2 × 2 factorial designs, with low rates of application. Only in a small proportion of cases were results correlated with soil tests. No account was taken of local soil variability, and the recommendations that emerged were therefore very broad and general and not site specific. They do nevertheless indicate for broad areas which nutrients appeared most limiting, and showed in a general way that fertilisers can improve yields and are often profitable. The average yield response with the modest fertiliser rates used was 60% of the control plots, and the average profit/cost ratio was three, implying that the value of the increased yield was three times the net fertiliser cost. However, the value varied very much with the crop and local selling prices; in West Africa, for example, yield increases obtained with yams were profitable but those for maize were not.

14 Manures, fertilisers and their use

Composition and use of animal manures

Animal manures are most used in maintaining the fertility of cultivated soils in those areas where livestock and arable farming are carried on side by side or at least in the same area. Their use in the tropics varies considerably from area to area. For example, there are large areas of Africa where the tsetse fly makes it impossible to keep cattle, whereas, in India large numbers of cattle are kept for religious reasons. In some African pastoral societies the dung is burnt as household fuel rather than used as manure.

In the tropics most animals graze natural vegetation (Chapter 21), and concentrates and other foods are rarely given. Where manure is collected and applied to a cultivated field it thus represents a removal of nutrients from the grazed area to benefit the cropped land.

Animal manures are bulky in relation to their nutrient content, and the labour and expenses of transporting them often discourages their use. In the savanna areas of West Africa, for example, cattle are frequently kept in kraals at night and let out to graze during part of the day but the manure that accumulates in the kraals is used only rarely.

In the use of fertilisers on high-yielding grasses and improved grass–legume mixtures discussed in Chapter 21 it is noted that there is a considerable difference in practice between pastures that are cut and fed to animals elsewhere and those that are grazed. Animals return in their dung and urine over 80% of the nitrogen, phosphorus and potassium they take in by grazing, so the net removal is very much less when grass is grazed and the bulk of nutrients are returned to the same area. What happens to the over 80% of nutrients removed which are excreted then depends on whether the excreta are used or wasted, and if used, how they are stored and subsequently applied, since important losses can occur in storage.

Animal manures include dung and urine, and animal by-products such as bonemeal and dried blood. Farmyard manure (FYM) is cattle dung and urine mixed with variable amounts of litter, usually straw, and its composition varies within wide limits.

Dung contains the undigested portions of the animal's food. Its content of plant nutrients and hence usefulness varies very considerably according to a number of factors, including the type of animal, the quality and composition of its food supply, and the age and health of the animal. In much of the tropics the largest volume of available manure is cow manure, which is relatively bulky and low in plant nutrients in comparison with the manure of some other animals. Moreover, in areas of the seasonal tropics where the amount and quality of grazing vary widely between wet and dry seasons, there are corresponding fluctuations in dung composition. If in the dry season the cow feeds on poor grasses with a low protein and a high silica content, its manure will be correspondingly high in undigestible residues of low nutrient value.

Sheep and goat manure are more concentrated than cattle and horse manure. That of pig is also relatively concentrated but its composition varies within very wide limits according to what the pigs are fed on. The richest and most concentrated manure normally available on a farm is poultry manure, and because of its more concentrated nature more care is needed in its application. The fact that the urine and solid excreta are mixed contributes to its higher nitrogen and phosphorus content. One estimate gives the annual manure production of a well-fed hen as about 14 kg of manure a year containing about 200 g of nitrogen, 150 g of phosphorus pentoxide (P_2O_5), 75 g of potassium oxide (K_2O) and appreciable quantities of calcium, magnesium and boron.

Where manure of any sort is produced, the main

cost of its use, apart sometimes from storage costs, is that of the labour and transport involved in its application. Although the very variable quality and nutrient content of FYM makes it difficult to know except within wide limits what is added by a dressing of, say, 10 t/ha, numerous agricultural experiments in the tropics have attested to the long-lasting effect of FYM applications. Residual effects have often been recorded for 10 years or more. Many agronomists believe that part of the beneficial effects lies in the stimulation of bacterial activity and the build-up of soil organic matter content slowly obtained by the use of FYM, and the resultant improvement, in some cases, of soil structure. In some long-term experiments, heavy dressings of up to 50 t/ha of FYM were found to help maintain yields over several years, though in many areas of the tropics such applications would be impracticable. In other experiments equally good results were obtained simply by applying the equivalent amounts of chemical fertilisers.

The effects of FYM in any given situation are partly site specific, depending on the nature of the soil, its cropping history, and the crop grown, but in very general terms it can be stated that FYM always has some beneficial effect in as much as it brings a range of nutrients, including micronutrients, in balanced amounts which become available slowly over a long period. Animal manures do not require much technical knowledge in their application. If they are well weathered and free of pathogens there are few hazards involved, and if not used they would be wasted. Their use is sometimes profitably combined with quicker-acting chemical fertilisers. Phosphate, in particular, is mixed with animal manures to reinforce their generally low phosphate content and to reduce nitrogen losses as ammonia.

Green manures

Green manuring refers to the practice of ploughing in a quick-growing leafy crop in order to increase production of succeeding crops. Ploughing in is best done when the crop is still at a green, fast-growing stage before seed sets, and when there is sufficient soil moisture to allow decomposition to begin before the next crop is planted.

When residues with a low C/N ratio decompose, nitrates and a range of other nutrients are released, and soil bacterial activity is stimulated. In addition to the nutrients released, some organic matter is supplied to the soil. However, the main disadvantage of green cropping is that the effect is usually very short lived, and farmers are often not willing to set aside land and invest their effort for such ephemeral and often doubtful economic results. In some areas with a two-season rainfall pattern it has been found useful in the first wet season to plough in green crops to benefit crops growing in the second season. It has been found beneficial as an occasional measure to maintain productivity in long rotations. In other cases better results have been obtained by applying the small amount of chemical fertilisers represented by the green crop, or at least those elements considered limiting.

Chemical fertilisers and their application

With the development of chemical manufacturing techniques in the last 100 years, combined with recent advances in agricultural and soil chemistry, has come an increasing use of chemical compounds which are applied to the soil and sometimes directly to the plant in order to increase yields by supplying nutrients in available form.

Important characteristics of chemical fertilisers include the following:

1 Compared with organic manures, including animal manure, they are relatively concentrated. Per unit of nutrient, this makes them cheaper to transport and to store.

2 Unlike animal manures, their composition is usually known fairly exactly (and is usually stated on the bag).

3 Again unlike animal manures, most chemical fertilisers in wide use are immediately available to the plant, or nearly so, though slow-release fertilisers are also available.

4 Most chemical fertilisers are of relatively simply composition and low in impurities. They contrast with organic manures which contain a whole range of elements, including trace elements. With chemical fertilisers only the element or elements thought to

be required need be added. They can be supplied in carefully controlled amounts at the time when they are most needed by the plant. They do not contain the pathogens, weed seeds and insects that might be found in animal manures.

5 Some technical knowledge or sound advice is needed to use chemical fertilisers to avoid possibly costly mistakes and waste, since unwisely used fertilisers can fail to give an effect, or even depress yields.

As with many innovations, the initial use of fertilisers was treated with some suspicion, and they were considered inferior to 'natural' manures. Today, doubts about the increasing use of agricultural chemicals in general, fuelled by occasional mistakes, appear to be on the increase in some circles, though not among most practising agronomists. Two facts should be borne in mind. The first is that a nutrient ion, such as an ammonium ion, NH_4^+, is exactly the same whether it is released by the decomposition of cow manure or applied in a bag labelled 'sulphate of ammonia, 21% N'. If a crop plant cannot distinguish between the two sources, then why should we? Secondly, current levels of food production, involving as they do the annual consumption of millions of tons of fertilisers, would be quite impossible without chemical fertilisers, so that if we wish to go back to nineteenth-century methods of food production we have also somehow to reduce the world's population to a fraction of its present number. The answer is not to try to ignore or do without chemical fertilisers, but to study their use and effects with more care, and in relation to an increasing body of knowledge about the world's soils, so that their use becomes more efficient and effective, and mistakes and waste are progressively eliminated.

In the industrialised temperate countries of the world where gigantic quantities of fertilisers are used, a vast amount of experience has been gained of their use under local conditions. Advice is often available to farmers from government agricultural advisers, in publications, from agricultural stations, and from the manufacturers and sellers of fertilisers who have a natural interest in showing farmers how to use their products and in publicising the results of successful fertiliser use.

In the tropics in general, particularly where agriculture is not highly commercialised, fertilisers are used to a small and sometimes negligible extent. This is indicated by the fertiliser import and consumption statistics for tropical countries, available from the FAO, the World Bank and other sources. The relatively very small quantities of fertilisers that are imported into and sometimes made in tropical countries are used mainly by large plantations growing export crops, and on agricultural stations. In much of the tropics, particularly in South America and in Africa, fertiliser use by the small subsistence farmer has hardly begun.

Table 14.1 Very approximate amounts of N, $P_2O^0{}_5$, and K_2O, in kg/ha removed by some important tropical crops.

The amounts removed refer to the yield figures given in the second column, which themselves vary widely. It should not be assumed that nutrients added in fertiliser should be in the same proportion as they are thought to be removed by the plant, because the type and amount of fertiliser required will vary with local soil conditions and will be affected by what is in the soil already, and other factors. The table does indicate, however, that different crops require nutrients in different proportions and amounts, as can be seen by comparing, for example, coconuts with groundnuts.

Crop	Yield (kg/ha)	N	P_2O_5	K_2O
Maize	3 000	72	36	54
Cassava	25 000	161	39	136
Sugarcane	25 000	30	20	60
Groundnuts	1 500	105	15	42
Tobacco	1 700	90	22	129
Coffee (fresh berries)	5 000	35	7	40
Oil palm (bunches)	15 000	90	20	135
Cocoa (dry pods)	1 000	20	10	12
Coconuts	1 500	13	4	12
Rice – grain	1 600	35	9	7
– straw	2 750	11	3	29

Adapted from *Fertilizers and Their Use*, FAO, Rome, 1965.

Methods of fertiliser application

There are four main ways of applying fertiliser to the soil or the plant:

1 Solid fertilisers are applied to the soil surface.
2 Solid fertilisers are placed in the soil.

3 Nutrients are applied as leaf sprays, or by stem injection to trees, so that the plant takes them up directly, thus avoiding possible nutrient transformations within the soil.
4 Solid fertilisers are dissolved in irrigation water ('fertirrigation').

For trace elements, additional methods such as dusting the seed with fertiliser or soaking it in nutrient solution may also be used.

Of the various ways of applying solid fertilisers to the soil, it is helpful to distinguish between those that merely apply the fertiliser to the surface soil, with or without incorporation, and those that place the fertiliser in the soil in a more or less precise position in relation to the roots of the growing crop. Which broad method is used depends partly on the mobility of the fertiliser elements applied.

Fertilisers that are very soluble and then move freely in the soil, including most nitrogen-containing fertilisers such as ammonium sulphate or ammonium nitrate, can be applied to the surface of the soil and will normally dissolve rapidly and move in the soil solution to the plant root. For this reason no particular care is needed as to exactly where they are applied in relation to the plant, and they can also be applied to the surface soil of a growing crop.

In the case of relatively immobile nutrients, particularly the phosphate anion which hardly moves in the soil from its place of origin, it is important to place the fertiliser carefully in such a way within the soil that the growing root can come into close contact with the added fertiliser. In the case of soluble phosphatic fertilisers, placing the fertiliser in well-defined bands or pockets has the additional advantage of limiting contact between fertiliser and soil and thus reducing the possibility of phosphorus fixation (Chapter 10).

Potassium fertilisers generally occupy a place intermediate between the very soluble and mobile nitrogen fertilisers and the relatively immobile phosphatic fertilisers.

Fertiliser application to the soil surface The application of fertiliser to the soil surface, with or without some superficial working in of the fertiliser, includes the following methods:

Broadcasting consists of applying the fertiliser more or less evenly to the whole area, usually before the planting or germination. It can be done by hand or with mechanical spreaders. The fertiliser may be left on the soil surface or lightly worked into the soil.

Top dressing consists of broadcasting fertilizer onto a growing crop. Some of the fertiliser will fall onto plant leaves but will generally do no harm, and be washed down onto the soil by later rain or irrigation water.

Side dressing is the application of fertiliser to the soil surface in band or strips along the rows of plants, rather than to the whole of the cropped area.

Fertiliser placement in the soil Fertiliser placement involves the positioning of fertiliser within the soil in a specific location in relation to the plant roots.

Row placement (also known as 'sidebanding') involves the positioning of the fertiliser, usually by machines and often at the same time as the seed is sown, in a line 5–8 cm to one side of the seed and a little below it, on one side or on both sides of the row.

For tree crops, fertiliser is often placed in the planting hole, below the seedling. For established bushes and trees, fertiliser may be placed in pockets near the tree, or in a circular furrow round the tree, or in a line along the tree rows at a distance from the trunks.

Foliar sprays Nutrients are sometimes applied as foliar sprays, the nutrients then being absorbed through the leaves. This avoids contact of the nutrient with the soil, and possible transformations and the immobilisation of nutrients in the soil. Foliar sprays are best applied in not too dry atmospheric conditions which allow a water film to remain on the leaves. Spray dripping off the leaves is not necessarily lost since it may be absorbed by roots. The advantages of foliar sprays include the fact that the nutrient is quickly absorbed and immediately available to the plant, and the nutrient can thus be given at precisely the time it is most needed. Foliar sprays usually give a high recovery rate of the nutrients supplied, and are often used for micronutrients. Only relatively small quantities of nutrients can be supplied in this way, and leaf corrosion can be caused by solutions that are too strong. Often 1–400 l/ha of solution are sprayed on in fine droplets. Young seedlings are not usually sprayed. Foliar applications may be combined with the application of other agricultural chemicals such as insecticides.

Fertirrigation The dissolving of fertilisers in irrigation water can be an economical and efficient way of distributing fertiliser and getting it to the plant root.

The choice of an appropriate fertiliser application method

The best way of applying fertilisers to the soil depends on

1 the type of fertiliser used,
2 the crop, and
3 the soil and the local climate.

Broadcasting is used where there is a dense stand of crops not arranged in rows, and where the plant roots permeate the whole volume of the soil. It is used particularly when large applications of fertilisers are made, and when the fertilisers are readily soluble. Top-dressing or side-dressing are well suited to the application of nitrogenous fertilisers as split applications to a growing crop at various stages after initial planting and germination. Broadcasting and subsequent incorporation into the soil is also used in the case of relatively insoluble rock phosphates where it is desired to maximise contact with the soil in order to increase the amount dissolved.

Placement of fertilisers is used in the careful distribution of small quantities of fertiliser, and for plants with a poorly developed root system or where there are large spaces between plants or plant rows. It is used for phosphatic and sometimes potassium fertilisers where there is a danger of fixation, and contact with the soil is to be minimised. Placement of chemical fertilisers in the planting hole, often with some organic manure, is often used for tree crops.

Foliar sprays are particularly suitable when nutrients such as magnesium or the micronutrients are absorbed with difficulty or made unavailable if applied to the soil, and when nutrients are required only in very small quantities.

The type of crop and its arrangement in the field will often dictate the appropriate fertiliser application method.

Fig. 14.1 Careful hand application of small quantities of fertiliser applied as a side-dressing to market-garden crops grown on volcanic soils in Java, Indonesia. Note the clean weeding to help maximise returns.

For **row crops** the placement of fertiliser 5–8 cm to one side and a little below the seed line is often carried out at the same time as the seed is drilled in, particularly for phosphatic fertilisers, with or without potassium and with a starter dose of nitrogen. More nitrogen may be applied later by top or side-dressing. In the case of small grains, low-analysis fertilisers in moderate amounts (less than 300 kg/ha) can be mixed with the seed and drilled in with it without causing serious germination injury.

Table 14.2 Approximate composition of some of the commoner fertilisers, with their equivalent acidity or basicity

Fertiliser and formula	% N	% P_2O_5	% K_2O	Equivalent acidity	Equivalent basicity
Nitrogen fertilisers					
Sodium nitrate $NaNO_3$	16				29
Ammonium sulphate $(NH_4)_2SO_4$	21			110	
Ammonium nitrate NH_4NO_3	33.5			60	
Calcium nitrate $Ca(NO_3)_2$	15.5				21
Ammonium nitrate – limestone (including nitrochalk and similar mixtures of NH_4NO_3 and $CaCO_3$)	15–26				
Urea $CO(NH_2)_2$	45			80	
Calcium cyanamide $CaCN_2$	21				63
Ammonium chloride NH_4Cl	24			128	
Ammonium sulphate nitrate $(NH_4)_2SO_4$ and NH_4No_3	26			93	
Phosphate fertilisers					
Single superphosphate $Ca(H_2PO_4)_2$ and $CaSO_4$		16–20			
Triple superphosphate $Ca(H_2PO_4)_2$		43–52			
Di-calcium phosphate $Ca(HPO_4)$		35			25
Rock phosphate		20–40			XX
Basic slag		16–20			XX
Potassium fertilisers					
Muriate of potash, or potassium chloride KCl			50–60		
Potassium sulphate K_2SO_4			50		
Potassium magnesium sulphate $K_2SO_4.2MgSO_4$			21		
Potassium nitrate KNO_3	13		44		23

The *equivalent acidity* is the number of kg of calcium carbonate ($CaCO_3$) required to neutralise 100 kg of the fertiliser; the *equivalent basicity* shows the acid neutralising capacity, expressed as kg of $CaCO_3$, of 100 kg of the fertiliser; for example 100 kg of calcium cyanamide have the same neutralising effect as 63 kg of $CaCO_3$. The symbol xx indicates that the compound is basic, but the strength varies with the composition, since both rock phosphate and basic slag have no fixed composition. The acidifying effects of fertilisers such as ammonium sulphate are likely to be important only on soils with a low buffer capacity (i.e. soils low in clay and humus whose pH is relatively easily changed) and only if applications are repeated for several years running.

Pastures can be fertilised by broadcasting prior to seeding, followed by thorough incorporation of the fertiliser into the soil. Once established, top-dressings can be applied so as not to injure foliage or root crowns.

Bushes and trees are often treated individually, and can be fertilised in a relatively large number of ways. Before planting, a basal application of chemical fertilisers with or without manures is placed in the planting hole. Once established, fertiliser may be placed in a circular furrow away from the trunk, or in lines parallel to tree rows, at distances related to the age and root distribution of the trees, or fertiliser may be applied to the surface soil and then worked in. In the so-called 'perforation method', small pockets are dug around trees within the outer half of the branch-spread zone, and deep enough to extend beyond the topsoil into the upper subsoil

Fig. 14.2 Studying root development of bananas in relation to fertiliser applications, Panama.

(often to 20–30 cm or more). Fertiliser is placed in the holes which are then filled again. Any cover crops grown in association with the trees should be treated separately, with the fertiliser broadcast and then worked in prior to seeding, followed by later top-dressings at intervals.

The time of fertiliser application

Correct timing of fertiliser applications is of great practical importance and is connected with the type of fertilisers used and the method of application. Broadcasting and row or sideband placement are carried out before or during seeding. With fertilisers such as the phosphates which do not move far in the soil, and with less available fertilisers which become available over a period, placement in the soil at or before planting is normally recommended. Soluble, mobile fertilisers, such as the nitrates, can be added during the growing season as top or side-dressings, or dissolved in irrigation waters.

The application of the same fertiliser more than once to the crop is known as a 'split application'. Split applications are useful in the case of soluble and mobile fertilisers in areas of high rainfall because they tend to reduce the amount of the added fertiliser that is lost by leaching, and thus to increase the proportion of added nutrient taken up by the plant. Split applications are widely used when nitrogen-containing fertilisers are applied to maize and other crops which have been shown to benefit from extra nitrogen at a particular stage in their development.

In the case of perennial crops grown in seasonal rainfall areas, fertilisers are often applied before the main rains.

Single and mixed fertilisers, and fertiliser grades

In the following sections, the more common commercial fertilisers are described. Some contain only one major nutrient element, and are referred to, somewhat loosely, as **single fertilisers**. Examples of single fertilisers include ammonium sulphate and single superphosphate, because the first contains the major element nitrogen, and the second the major element phosphorus. Both compounds, however, also contain sulphur, and single super additionally contains calcium. These additional elements may in some cases be beneficial, and increase yields. The term 'single fertiliser' can thus be misleading, and often agronomists require to know the chemical composition of the fertiliser and not merely which major elements it contains.

Chemical fertilisers that contain two or more major elements are conventionally referred to as **mixed fertilisers**. For example, the sulphate of ammonia and single superphosphate mentioned above could be mixed to make a mixed NP fertiliser, or, with the addition of potassium chloride, an NPK fertiliser. In other cases, so-called mixed or compound fertilisers consist of a single chemical compound such as ammonium phosphate which provides two major elements, and are thus not technically a mixture.

The conventional way to designate the nutrient content or grade of mixed fertilisers is to give the percentage by weight they contain of nitrogen, P_2O_5 (phosphorus pentoxide) and K_2O (potassium oxide). Thus a mixture labelled 10–15–5 would contain

10% by weight of elemental nitrogen, 15% by weight of P_2O_5 and 5% by weight of K_2O. If the amounts of elemental phosphorus or potassium are required, rather than the conventional P_2O_5 and K_2O, then these figures have to be multiplied by the conversion figures given on pages 158 and 163.

The grade of fertiliser thus indicates the relative proportions of the major elements (expressed as N, P_2O_5 and K_2O) and also the concentration of the fertiliser. A 10–10–10 mixture has the same proportions as a 15–15–15 mixture, but the latter is more concentrated since it contains 50% more nutrients in the same weight of fertiliser. More concentrated fertilisers save transport and handling costs per kilo of major nutrient supplied, but in some cases the additional elements in the low-analysis fertilisers may be of value too. The more concentrated commercial mixed fertilisers are made possible by using ingredients containing higher percentages of the nutrient elements desired, such as ammonium nitrate or ammonium phosphate instead of ammonium sulphate and single superphosphate.

The use of mixed fertilisers has several advantages over the use of single fertilisers, particularly in unskilled hands. Mixed fertilisers are less likely to cause an unbalanced supply of nutrients in the soil than the uninformed use of single fertilisers. They may assure a more even distribution of nutrients in the field. They require less technical knowledge in their use, and applying several nutrients at once may be less costly than applying them separately.

The use of single fertilisers also has advantages which may be important. Single (or straight) fertilisers containing as they do only one major nutrient allow the different nutrients to be applied at different times and in different ways, as when phosphatic fertilisers are applied in concentrated pockets or bands at planting time and nitrogen-containing fertilisers are broadcast or side-dressed later in the season. Soluble fertilisers, such as most potassium and nitrogen-containing fertilisers, are likely, in wet areas, to have their effects increased by being given as split applications. In some cases a crop needs only one nutrient, and others supplied in a mixture would be wasted.

Whether single fertilisers or mixed fertilisers are used, the farmer and research worker interested in interpreting fertiliser responses must be fully aware of the chemical composition of added fertilisers, and not merely of the NPK content. We do not and we cannot add nitrogen, phosphorus or potassium to the soil; we can only add these elements in combination with others. If a response is obtained, this may be due to the presence of the additional sulphur, calcium or other elements contained in the fertiliser but not indicated in the N–P_2O_5–K_2O grade given on the bag. Failure to remember this has led to some wrong conclusions in the past.

Nitrogen-containing fertilisers

Nitrogenous fertilisers are used by farmers in greater quantity than any other. Animal manures and green manures are examples of ways in which nitrogen can be supplied to the soil, but most of the nitrogen now given to crops by commercial farmers in the more developed countries is in the form of chemical compounds. The nitrogen fertiliser manufacturing industry is based mainly on the production of ammonia (NH_3) formed by reacting nitrogen and hydrogen gases.

Nitrogen in fertilisers is simply calculated as the percentage nitrogen by weight. Thus ammonium sulphate with the formula $(NH_4)_2SO_4$ has a molecular weight of 132, of which the two nitrogen atoms together count for 28, so that the nitrogen in the compound is 28/132, or about 21%. The sulphate anion is relatively heavy. If it is replaced by a nitrate anion as in ammonium nitrate, then a compound with a higher nitrogen percentage is obtained, and the fertiliser is more concentrated. The percentage nitrogen in some of the commoner nitrogenous fertilisers is given in Table 14.2.

Fertiliser nitrogen can be supplied either as a cation, as ammonium (NH_4^+), or as an anion, as nitrate (NO_3^-), or as both. Plants can absorb nitrogen in either form, and usually with equal ease, but there are sometimes advantages in applying nitrogen as the cation since in this form it is held in the soil by the cation exchange capacity of the soil colloids (Chapter 2) and is thus less readily leached downwards or out of the soil than is the nitrate anion. Unless it is quickly taken up by plants or immobilised in other ways, however, the ammonium cation is likely to be nitrified by soil bacteria to the nitrate form (Chapter 9). The nitrification of NH_4^+ to NO_3^- releases H^+ ions and contributes to the acidifying effect of ammonium-containing fertilisers (Table 14.2)

In the following paragraphs the more common nitrogen-containing chemical fertilisers are grouped into those containing ammonium nitrogen (or forming ammonia when in the soil), those containing nitrate nitrogen, and those containing other forms of nitrogen.

Ammonium-containing nitrogenous fertilisers

The ammonia-derived nitrogenous fertilisers include ammonium sulphate, which is widely used in the tropics, and other forms such as ammonium nitrate, ammonium chloride and anhydrous ammonia. Urea is also included in this section: it does not contain ammonia when manufactured but produces ammonia when allowed to react with the soil. Ammonium phosphate is dealt with in a later section on phosphatic fertilisers.

Ammonium sulphate, or sulphate of ammonia, is one of the oldest of the nitrogen-containing fertilisers. It is a white crystalline substance containing 21.2% nitrogen and 24.2% sulphur by weight. It has long been popular in the tropics because of its good handling and storage properties and its relatively low cost. It is often mixed with other ingredients such as potassium chloride and single or triple superphosphate to make mixed fertilisers. The sulphur content may have beneficial effects when applied to soils deficient in this element (Chapter 9).

Ammonium sulphate, like all ammonium containing fertilisers, can lead in some circumstances to acidification of the soil, as discussed further below.

Ammonium nitrate (NH_4NO_3) contains nitrogen in both the cation (NH_4^+) and anion (NO_3^-) forms. The nitrate anion is probably the more immediately absorbed but also the more likely to be leached during wet periods, whereas the ammonium cations are likely to be held against leaching until taken up by the plant or converted by micro-organisms to the nitrate form. Ammonium nitrate contains 32–33.5% nitrogen, depending on the purity of the commercial fertiliser. It needs care in handling and storage because it is hygroscopic (i.e. it picks up moisture from the air) and it can form an explosive mixture if mixed with carbonaceous substances such as oil.

CAN, or calcium ammonium nitrate ($NH_4NO_3.CaCO_3$), also known as **nitrochalk**, is a mechanical mixture of ammonium nitrate with calcium carbonate, sometimes standardised to give a nitrogen content of 26%. The admixture of calcium carbonate has two effects. First it neutralises the equivalent acidity of the ammonium nitrate so that it has little or no effect on soil pH. Secondly, it improves the storage and handling properties of the ammonium nitrate. CAN is widely used in East Africa and elsewhere where a nitrogen-containing fertiliser is needed that does not acidify the soil. Like ammonium nitrate (above) the nitrogen is provided in both the ammonium and nitrate forms.

Ammonium chloride (NH_4Cl) contains about 26% nitrogen and is used mainly on rice paddies, in preference to ammonium sulphate, where the reducing conditions in the flooded paddy fields would reduce the sulphate in ammonium sulphate to hydrogen sulphide (H_2S), thought harmful to rice.

Anhydrous ammonia (NH_3), unlike the other fertilisers listed here, is a liquid which volatilises when applied to the soil to ammonia gas. It is the cheapest source of nitrogen, much used in the United States, but since it requires special equipment for its application it has as yet been little used in the tropics. It can also be dissolved in irrigation water.

Urea ($CO(NH_2)_2$) is an important nitrogen-containing fertiliser with the relatively high nitrogen content of 45%. It is a reasonably cheap source of nitrogen produced by reacting carbon dioxide with ammonia. It does not contain ammonia as such, but when added to the soil it rapidly reacts with water to form first ammonium carbonate, which then decomposes to ammonia and carbon dioxide. It is therefore grouped here with other ammonium-containing fertilisers. In most cases the hydrolysis of the urea, once added to the soil, is rapid. If applied to the soil surface or if mixed into the soil only to a shallow depth, some of the ammonia released may be lost as ammonia gas, particularly in alkaline soils. Too much urea, as in the case of other ammonia-releasing fertilisers, may, if placed too near the seed or seedling, cause injury through excess ammonia, but placement at a distance from the seed avoids injurious effects. Urea can also be supplied in leaf sprays. It acts then directly as a nutrient, and is not likely to cause leaf corrosion.

Biuret is a toxic by-product of urea manufacture sometimes produced on drying. Amounts contained in commercial urea vary, and in some countries are limited by law. The biuret content should be less than 0.5% for urea applied to the soil and less than 0.25% for that used in leaf sprays. The sensitivity of plants to biuret varies between crops, pineapple being particularly sensitive.

Recent developments include the manufacture of extra-large granules of urea, usually of 2–4 g each, known as urea supergranules, which are suitable for individual placement, for example in paddy rice.

Urea is sometimes mixed with other fertilisers to give such combinations as **urea–sulphur** (40% nitrogen and 10% elemental sulphur), to be used where sulphur is needed as well as nitrogen, and **urea–phosphate** (29% nitrogen, and 29% P_2O_5, equivalent to 12.7% phosphorus).

The acidifying effect of ammonium-containing fertilisers is due to two mechanisms:

1 The ammonium cation is absorbed on to exchange sites on the soil colloids (Chapter 2) and displaces other cations such as calcium and magnesium into the soil solution. These cations may then be leached out of the soil, thus lowering the soil pH.
2 If the ammonium cations are not quickly absorbed by plants or other organisms, they are likely to be nitrified by soil bacteria to nitrate (NO_3^-). The conversion from NH_4^+ to NO_3^- releases hydrogen ions which contribute to soil acidity and a lowering of soil pH.

The extent to which the application of ammonium-containing fertilisers lowers the pH of the soil depends on two factors:

1 The amounts and frequency of application of fertilisers that contain ammonium or which release ammonium on contact with the soil.
2 The buffer capacity of the soil, i.e. its capacity to resist a change in reaction. This in turn depends on its content of colloids (clay and humus) and its cation exchange capacity. The higher the cation exchange capacity, the greater the buffer capacity, resulting in smaller pH changes in response to a given application of ammonium.

In practice, small and irregular applications of ammonium-containing fertilisers have little effect on the pH of clay soils or of soils high in organic matter. Sandy soils are more easily affected. Regular and heavy applications of ammonium-containing fertilisers to light-textured soils can result in substantial pH changes. Ammonium-containing fertilisers are sometimes used to deliberately acidify some soils, for example those used for crops such as tea or pineapples which prefer acid soil conditions. Where it is thought that acidification of a soil is undesirable, then a fertiliser such as calcium ammonium nitrate with little or no equivalent acidity can be used instead.

The **equivalent acidity of a fertiliser** (Table 14.2) is the number of units of calcium carbonate needed to neutralise 100 units of the fertiliser. Ammonium sulphate has an equivalent acidity of 110, indicating that it requires 110 kg of calcium carbonate to neutralise the acidifying effects of 100 kg of ammonium sulphate. Ammonium nitrate and urea have somewhat lower equivalent acidities (Table 14.2). Other fertilisers have an acid-neutralising rather than an acid-producing tendency which is expressed as their **equivalent basicity**, i.e. as the number of units of calcium carbonate equivalent to 100 units of the fertiliser.

Nitrate-containing nitrogenous fertilisers

Nitrate-containing fertilisers are very soluble. Once dissolved, the nitrate ion is very mobile in the soil and in a moist soil usually finds its way quickly to the plant root where it is easily absorbed. The nitrate anion, in heavy rainfall conditions, is also relatively easily moved downwards in the soil or leached out of the soil profile.

Ammonium nitrate, considered in the previous section, is the most widely used nitrate-containing fertiliser. Sodium nitrate, calcium nitrate and potassium nitrate are all used to a lesser extent. These three metallic nitrates are not, as is ammonium nitrate, acid forming. In practice the nitrate anion is often absorbed more rapidly than is the sodium, calcium or potassium cation, and the addition of these cations to the soil may in fact have a slight tendency to raise the pH rather than to lower it.

Sodium nitrate (Na_2NO_3) occurs in natural deposits in the dry areas of Chile. Chilean sodium nitrate was once more important as a fertiliser than it is now. It contains about 16% nitrogen when pure, and the natural deposits contain trace elements, particularly boron, though a purer synthetic variety is also marketed.

Potassium nitrate (KNO_3) contains 13.8% nitrogen and 36.5% potassium, so that it is higher in potassium than in nitrogen. It needs care in handling and storage because of its explosive properties.

Calcium nitrate ($Ca(NO_3)_2$) contains 15.5% nitrogen and 19.5% calcium when pure, but is sometimes sold as a mixture of 82% calcium nitrate and 5% ammonium nitrate to give a composition of about 16% nitrogen. It is hygroscopic, and easily soluble, and has a net alkaline effect on the soil. Pure calcium nitrate has an equivalent basicity of about 21% (Table 14.2).

Other nitrogen-containing fertilisers

Some minor nitrogen-containing fertilisers either do not fall into the two broad classes considered above or are best considered separately because of their slow-release properties. These include calcium cyanamide and a number of slowly available nitrogen-containing compounds deliberately designed to release their nitrogen over longer periods than the generally very soluble fertilisers considered above.

Calcium cyanamide ($CaCN_2$) contains 21–22% nitrogen, and is also used as a weedkiller. When added to the soil it reacts with soil water to form urea, but it should be mixed thoroughly with the soil to allow the reaction to take place in order to avoid the formation of toxic by-products, preferably two to three weeks before planting. Calcium cyanamide is very little used in the tropics.

Slowly available nitrogenous fertilisers are of particular interest in areas of heavy rainfall where much of the nitrogen added in quickly available forms is likely to be washed out of the soil, and in those cases where it is desired that an initial application should be relatively long lasting, as when a basal dressing is put into the soil before a pasture is established or into a planting hole for tree crops. Fertilisers that have been developed to meet these needs include the urea–formaldehyde compounds, which vary from soluble to quite insoluble depending on the proportion of formaldehyde they contain, and such metallic ammonium phosphates as magnesium ammonium phosphate (about 9% nitrogen) which releases ammonia slowly. These compounds are relatively expensive per unit of nitrogen they contain but slow release might result in a greater proportion of the nitrogen they contain finding its way into the plant than in the case of the conventional use of soluble fertilisers.

The recovery of fertiliser nitrogen

The percentage of the added fertiliser nitrogen recovered by the crop varies widely according to conditions, but is usually less than 50–60%. Applied nitrogen is lost in various ways. Apart from fertiliser nitrogen lost to weeds, some may be lost as nitrogen gas, some as ammonia gas, and some by fixation of the ammonium by some clays, but the main loss, particularly in wet areas, is by the leaching out of the soil of nitrates.

Phosphate-containing fertilisers

The very large phosphate fertiliser manufacturing industry is based mainly on the treatment in various ways of naturally occurring deposits of rock phosphate of varying composition, many of which are found in Africa, in order to make them more soluble. Finely ground rock phosphate can also be used, particularly in high rainfall areas with acid soils.

Classes and specifications of phosphate-containing fertilisers

The amount of phosphorus in a fertiliser has for long been conventionally calculated and referred to commercially as the content of P_2O_5 (phosphorus pentoxide), even though the phosphorus does not occur in this form in the fertiliser and could not be absorbed by the plant if it were (Chapter 9). Thus if a fertiliser contained 10% elemental phosphorus by weight, this would be considered as equivalent to 22.9% of P_2O_5, since the P_2O_5 molecule weighs 2.29 times as much as the phosphorus atom, and this would be the percentage or grade marked on the bag. The use of the P_2O_5 content is a rather artificial convention. Recently there has been an increasing tendency among agronomists and, more reluctantly, among manufacturers and suppliers to express the grade of the fertiliser as the content of elemental phosphorus. This makes the grading comparable to that of nitrogen, which is also as a percentage of the element itself by weight (see page 152). The old and the new gradings are easily converted to each other by means of the following conversion factors:

P (elemental phosphorus) $= P_2O_5 \times 0.436$
P_2O_5 (phosphorus pentoxide) $= P \times 2.29$

Not all the phosphorus contained in a fertiliser is immediately soluble and available to the plant, so that in addition to the grade (i.e. the percentage by weight of phosphorus in the fertiliser) the user also needs to know what percentage is available. The phosphate in the fertiliser may or may not be insoluble in water. Of that which is not water-soluble, some may be soluble in dilute acids, such as citric acid. Both the water-soluble and the citric acid soluble phosphate in a fertiliser is conventionally considered to be available to plants. We can therefore distinguish between the following classes of phosphate:

Water-soluble phosphorus is that phosphorus which can be dissolved in water. It is expressed as a percentage by weight (either as P_2O_5 or as P) of the total fertiliser. This is the phosphorus that is most quickly and easily available to the plant.
Citrate-soluble phosphorus is that phosphorus which is not soluble in water but which is soluble in a solution of neutral normal ammonium citrate. The citrate soluble phosphorus is also expressed as a percentage by weight (of P_2O_5 or of P) of the total fertiliser. It is assumed that the citrate-soluble phosphorus is also available to plants, though it may not be as immediately soluble and available as the water-soluble forms.
The available phosphorus is all that phosphorus which is either water or citric acid soluble. The available phosphorus in a fertiliser is therefore the sum of the water-soluble and citric acid soluble percentages.
The total phosphorus is the total phosphorus percentage in the fertiliser, whether it is available or not.

Water-soluble phosphatic fertilisers

The more important water-soluble phosphatic fertilisers are single superphosphate, triple (or double) superphosphate, and the ammonium phosphates. Others are listed in Table 14.3.

Single superphosphate is formed by mixing rock phosphate with an equal weight of sulphuric acid. The reaction of the two gives off considerable heat. The heat helps in drying the product, which sets hard and has to be ground up. The resulting fertiliser material is a mixture of calcium sulphate (gypsum) and calcium phosphate, though the exact composition depends on the composition of the original rock phosphate treated, which, being a natural product, varies somewhat according to the nature of the deposits mined. Most of the calcium phosphate is monocalcium phosphate (Chapter 10) which is water soluble. Single superphosphate contains 16–20% P_2O_5, equivalent to 7–9.5% P, of which almost all is water soluble. Its average composition is approximately as follows:

	%
Calcium sulphate	48
Monocalcium phosphate	30
Dicalcium phosphate	9
Tricalcium phosphate	2
Iron and aluminium oxides, silica	9
Moisture	2
	100

Single superphosphate (often abbreviated to 'single super') has been used successfully in many parts of the tropics, and is still widely used. The sulphur content may also be beneficial in those areas where sulphur is deficient (Chapter 9), such as the savanna areas of West Africa and parts of Brazil. It is often the standard recommendation for groundnuts and other plants having a relatively high sulphur requirement (Chapter 16). Single superphosphate is often mixed with other ingredients such as sulphate of ammonia and muriate of potash to make mixed fertilisers. Where the sulphur is not needed and where transport costs are high,

Table 14.3 Composition of some of the more common phosphatic fertilisers. The percentage of the phosphorus available refers to that which is removed by neutral normal ammonium citrate. The extent to which the phosphorus is water soluble is indicated by (1), (2), and (3), where (1) means all or almost all is water soluble, (2) means that some or most is water soluble, and (3) means that very little or none is water soluble.

Fertiliser	% P	% P_2O_5 *(a)*	% *(b)*	Other constituents available				
				% N	% K	% S	% Ca	% Mg
Single superphosphate	7–9.5	16–21.8	97–100 (1)	..	..	11–12	18–21	..
Triple superphosphate	19–23	43–52	96–99 (1)	..	..	0–1	12–14	..
Enriched superphosphate	11–13	25–30	96–99 (1)	..	..	7–9	16–18	..
Ammoniated single superphosphate	6.1–8.7	14–20	96–98 (2)	2–5	..	10–12	17–21	..
Ammoniated triple superphosphate	19–21	43.5–48	96–99 (2)	4–6	..	0–1	12–14	..
Dicalcium phosphate	23	52.7	98 (3)	..	..	..	29	..
Phosphoric acid	23	52.7	100 (1)	..	..	0–2	..	..
Superphosphoric acid	34	78	100 (1)	..	..	..	..	..
Potassium phosphate	18–22	41–50	100 (1)	..	29–45	..	..	..
Ammonium phosphate–nitrate (c)	4	9	100 (1)	30	..	..	..	..
Ammonium polyphosphate	25	57	.. (1)	15	..	..	..	..
Magnesium ammonium phosphate	17	40	.. (3)	8	..	..	..	14
Raw rock phosphate	11–17	25–39	14–65 (3)	..	..	..	33–36	..
Heated (defluorinated) rock phosphate	9	20.6	85 (3)	..	..	..	20	..
Rhenania phosphate	12	27.5	97 (3)	..	..	..	30	0.3
Basic slag	3.5–8	8–18	62–94 (3)	..	..	0.2	32	3
Ammonium phosphates % N; % P_2O_5; % K_2O								
21-53-0	23	52.7	100 (1)	21	..	..	..	..
21-61-0	27	61.8	100 (1)	21	..	..	..	..
11-48-0	21	48	100 (1)	11	..	0–2	..	..
16-48-0	21	48	100 (1)	16	..	0–2	..	..
18-46-0	20	45.8	100 (1)	18	..	0–2	..	..
16-20-0	8.7	20	100 (1)	16	..	15	..	..

(a) % P×2.29 *(b)* Ammonium citrate method. *(c)* Essentially a mixture of ammonium phosphate and ammonium nitrate.

Adapted from **Tisdale, S. L. and Nelson, W. L.** *Soil Fertility and Fertilisers* (2nd edn). Macmillan: New York.

savings may be effected by using more concentrated (higher analysis) fertilisers such as triple superphosphate or one of the ammonium phosphates.

Triple superphosphate is formed when rock phosphate reacts with phosphoric acid instead of with sulphuric acid. The resulting product has between two and three times the phosphorus content of single superphosphate, and is referred to as double or as triple superphosphate. Neither term is quite correct, since triple superphosphate contains 43–52% P_2O_5, or 19–23% P, which, on average, is more like 2.5 times as much as single super. Nearly all the phosphorus in triple super consists of monocalcium phosphate, and is water soluble. Triple super lacks the substantial sulphur content of single super, though it does contain up to 1% sulphur, and may therefore be less suitable for sulphur-deficient areas unless combined with another sulphur source such as ammonium sulphate. The higher phosphorus content gives triple super an advantage over single super where it has to be transported long distances or at considerable cost.

Ammonium phosphates are made by allowing ammonia gas to react either with phosphoric acid or with a mixture of phosphoric and sulphuric acids. The most important ammonium phosphates are **monoammonium phosphate ($NH_4H_2PO_4$)** or **MAP**, which commonly has grades of about 11-48-0 to 11-53-0 (higher when pure), equivalent to 21–23% P, and **diammonium phosphate ($(NH_4)_2HPO_4$)**, or **DAP** which has grades of 16-48-0 or 18-46-0, equivalent to 20–21% P. Ammonium phosphate–sulphate (16-20-0, with 8.6% P) is also made. The ammonium phosphates are relatively high-analysis fertilisers which have the advantage of being completely water soluble and of having good handling qualities. Their use in the tropics is widespread, and is increasing more rapidly than any other class of fertilisers. They form an excellent starter fertiliser to be placed at planting since they combine a substantial content of phosphorus with a useful starter dose of nitrogen, which can be supplemented if necessary later in the growing season by additional top-dressings of nitrogen fertiliser when needed. They are usually applied as such but are also often used in NPK mixtures. Because of the ammonia they contain, care must be taken not to place ammonium phosphates, and particularly DAP, too near seeds or seedlings. The ammonium in both MAP and DAP has an ultimate effect of lowering soil pH. The initial reaction of DAP with the soil is alkaline, but that of MAP is acid; in calcareous soils MAP is to be preferred to DAP which is likely to form a greater proportion of insoluble reaction products.

Polyphosphates with particularly high phosphorus contents are made by reacting various elements or ions with superphosphoric acid, a more concentrated acid containing condensed phosphate radicals, which contains 34% P (79% P_2O_5) as against the 24% P (55% P_2O_5) of ordinary phosphoric acid. Polyphosphates include ammonium and calcium polyphosphates.

Nitric phosphates vary considerably in composition, as well as in the percentage of the contained phosphate which is water soluble, which ranges between 0 and 70%. Those with low percentages of water-soluble phosphate belong more to the next section on water-insoluble phosphatic fertilisers.

Water-insoluble phosphatic fertilisers

Phosphatic fertilisers in which the phosphorus is all or mainly water insoluble include raw rock phosphate, basic slag and some other fertilisers of minor importance.

Raw (untreated) **rock phosphate** is obtained by mining deposits of rock phosphate. Phosphate deposits are of various types but most consist mainly of apatite, a naturally occurring mineral which contains calcium, phosphate, and a third constituent which in the case of rock phosphates is often fluorine, though normally there are various degrees and types of impurities also present. The phosphate is usually entirely insoluble in water, but some of it is citric acid soluble, depending on the nature of the deposit concerned. Raw rock phosphate is relatively cheap as compared with the more soluble and concentrated treated fertilisers discussed in the previous section, and is often applied in relatively large quantities. Under some conditions the higher quantities of total phosphate applied can have beneficial effects comparable to those achieved with smaller quantities of more soluble fertilisers. This is most likely to be the case where

1 the rock phosphate is finely ground so as to increase its surface area and speed of reaction, and is mixed well with the soil;
2 the rock phosphate is applied to acid soils.

Most rock phosphates are relatively ineffective in neutral or alkaline soils and in dry areas, but deposits differ in their availability. The relatively soluble deposits of Tilemsi in Mali, for example, though little exploited as yet, react relatively quickly when applied to the soil.

Many different types of rock phosphate are found in the tropics, and because they are bulky in relation to their nutrient content their use in untreated form is particularly likely to be important in areas near their place of origin. The International Fertiliser Development Corporation (IFDC) of Muscle Shoals, Alabama, has completed detailed investigations of many tropical deposits, some of which are not yet mined.

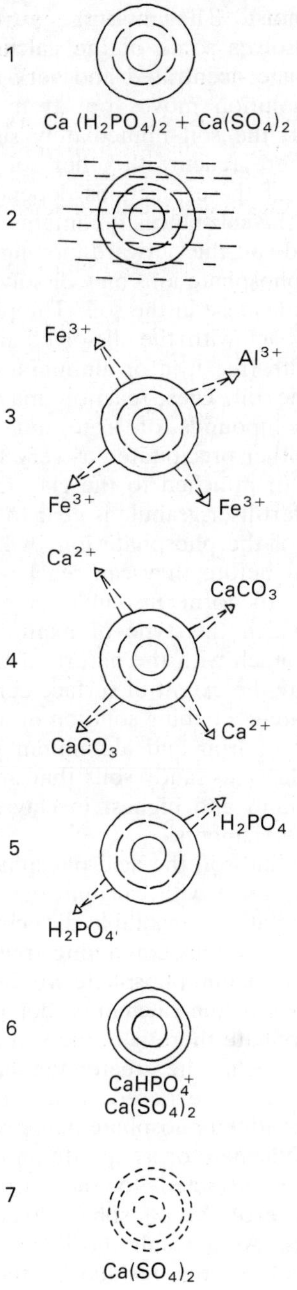

1 A fresh granule of single superphosphate added to the soil. The granule consists of a mixture of calcium sulphate and monocalcium phosphate.

2 The granule attracts moisture. The monocalcium phosphate is dissolved and moves into the surrounding soil . . .

3 In acid soils some of the phosphate reacts with iron and/or aluminium ions in the soil to form insoluble ironIII or aluminium phosphates. These are precipitated, or attached to the clay minerals.

4 In an alkali soil rich in calcium ions the phosphate may be precipitated as dicalcium phosphate. If there is free calcium carbonate in the soil, dicalcium phosphate or apatite may be precipitated on to this.

5 As the granule dissolves, more phosphate than calcium is lost . . .

6 . . . so that the residue becomes relatively rich in calcium and reverts from monocalcium phosphate, which is soluble, to the much less soluble dicalcium phosphate.

7 Eventually even the dicalcium phosphate may be dissolved in the soil solution and only a porous shell of calcium sulphate (gypsum) remains.

Fig. 14.3 What happens to a granule of single superphosphate in the soil.

Because of their relative cheapness, raw, finely ground rock phosphates have been recommended in the form of a relatively large basal application which can be topped up at the beginning and at later intervals by small applications of soluble phosphatic fertilisers. A massive initial application may also be useful in order to progressively quench the phosphate fixation capacity of some soils where this is known to be high.

Apart from its cheapness rock phosphate has the advantage that since it dissolves only slowly it has a long-lasting effect and may produce responses over many years, so that is represents a relatively long-term investment in the soil. It appears to be most useful in heavy rainfall areas of acid soils, and has been used effectively in parts of Malaysia and tropical Africa.

Heat-treated rock phosphates are heated to high temperatures. Heating drives off the fluorine in the apatite and increases the citrate-soluble phosphorus content to about 8% P (18% citrate-soluble P_2O_5). Rhenania phosphate (12% citrate-soluble P, or 27.5% citrate-soluble P_2O_5) is another heat-treated rock phosphate made by calcining a mixture of soda ash, silica and rock phosphate; it is effective on acid soils.

Basic slag, a by-product of the open-hearth steel-making process, is another source of phosphorus which has been effectively used in the tropics, particularly on acid soils. The phosphorus content is very variable, from 3.3 to 8% P (8–18% P_2O_5), of which over half is citric acid soluble. Because of the basic nature of this product it has a high acid-neutralising capacity and acts like lime when applied to acid soils. It has been found effective for rubber grown on acid soils in Malaysia.

What happens when phosphate is applied to the soil

The fate of applied phosphatic fertilisers depends on the type of fertiliser concerned and the nature of the soil. Climate, particularly rainfall, also exerts an influence.

When a granule of a water-soluble phosphate, such as single super (a mixture of calcium phosphate and calcium sulphate) or triple super (mainly calcium phosphate), is added to the soil, it first attracts moisture to itself from the surrounding soil. If the soil is dry, little reaction occurs until it becomes moist. The moisture attracted to the granule dissolves some of the calcium phosphate and forms a concentrated and very acid solution. The acid solution moves out from the fertiliser granule into the soil immediately surrounding it, and may there react with other soil constituents. The nature of these reactions depends on the soil pH and on the substances it contains.

In an acid soil the outward-moving acid solution containing phosphate ions may dissolve iron, aluminium or manganese in the soil. The phosphate ions may then react with the dissolved iron or aluminium, or with free iron or aluminium ions already present in the soil. These reactions may then produce insoluble compounds of iron and/or aluminium which are either precipitated as very small particles of crystals, or attached to the clay fraction of the soil. If the fertiliser granule is near to plant roots, a proportion of the phosphate ions will be taken up by the plant before they can react with other soil constituents to form insoluble compounds. The extent to which this type of fixation takes place varies very much with the nature of the soil, being influenced by the extent of surface contact between the phosphate-containing solution in the soil and by the amount of iron and aluminium present in it. Fixation is least in sandy soils that are low in iron and aluminium, and highest in clay soils that are high in these elements.

In an alkaline soil the acid phosphate-containing solution may react with calcium ions in the soil to precipitate relatively insoluble dicalcium phosphate (Chapter 10). In soils containing free calcium carbonate the dicalcium phosphate will be precipitated on the surface of the calcium carbonate. The more calcium carbonate there is in the soil and the more finely divided it is, the greater the surface area of calcium carbonate exposed, and the greater the likelihood of added phosphate being precipitated as dicalcium phosphate or as apatite on this surface.

The water attracted to the fertiliser granule removes more of the phosphate ions than it does calcium ions. As a result the ratio of calcium to phosphate left in the residue of the granule increases, and its composition changes from monocalcium phosphate, which is soluble, to dicalcium phosphate, which is relatively insoluble. If the original granule was of single superphosphate, containing originally monocalcium phosphate and calcium sulphate, it might later change or revert to a mixture of dicalcium phosphate and calcium sulphate. Eventually the sparingly soluble dicalcium phosphate may also be slowly dissolved, leaving only a porous gypsum (calcium sulphate) shell of the original granule.

The reaction between water-insoluble phosphatic fertilisers and the soil solution is relatively slow. In some cases rock phosphate added to a soil gave no responses initially but increased yields some years later. Calcium phosphates which are insoluble in pure water are to a limited extent soluble in the soil solution, which though very weak contains a range of substances dissolved in it, or the phosphate in the fertiliser may be released by reacting with other soil constituents, including plant root secretions and organic matter decomposition products. The interaction between soil and added rock phosphate is speeded up when the rock phosphate is finely ground in order to increase its surface area. Whereas in the case of water-soluble fertilisers it is best to reduce contact between the fertiliser and the soil to a minimum, in the case of water-insoluble fertilisers this contact is beneficial because it increases their slow availability.

Applying phosphatic fertilisers

The most appropriate way to apply phosphorus-containing fertilisers depends on the fertiliser selected, on the soil, on the climate and to some extent on the crop involved. Sometimes the choice is influenced by whether the plant also needs other nutrients such as sulphur contained in some phosphatic fertilisers.

Normally the placing of phosphorus in the soil, if practicable, is far better than merely applying a dressing to the soil surface. The placement is best done at or before planting time. With tree crops, phosphate can be applied in the planting hole below the roots. With annuals, placement is often done a little below and a little to one side of the seed, often 5 cm below and 5 cm to one side in the case of maize and beans. With mature trees the phosphate can be placed in holes or in shallow trenches dug round the tree or between rows approximately below the outer edge of the crown. Placement is necessary because phosphate does not move far in the soil and must therefore be near the growing root. At the distances mentioned there is little danger in the case of most phosphatic fertilisers that the fertiliser will harm the root. In the case of ammonium-containing phosphates, closer proximity must be avoided because they may, particularly in alkaline soils, release too much free ammonia near the roots and thus injure them.

Where water-soluble fertilisers are used, applying the fertiliser in large granules or in coated pellets slows down the release of the phosphate and lessens the percentage lost through fixation. Fixation is lessened by minimising the contact between applied soluble fertilisers and the soil. The water-insoluble fertilisers, in contrast, are best ground up finely and mixed thoroughly with the soil to give greater, more intimate contact between added fertiliser, the soil and the roots. In low rainfall areas the more soluble fertilisers should normally be used.

Potassium-containing fertilisers

In contrast to phosphatic fertilisers, the potassium in all the widely used potassium-containing fertilisers is water soluble. For this and other reasons, the use of these fertilisers is simpler than is the use of phosphate fertilisers. Potassium fertilisers are derived from the mining of very large deposits of soluble potassium salts in various parts of the world, some of them of a high degree of purity.

Specifications of potassium-containing fertilisers

The potassium content of fertilisers is at present usually calculated as the percentage by weight of K_2O (potassium oxide), even though it is not present in this form. As in the case of phosphatic fertilisers (see above), there is an increasing tendency to express potassium not in terms of its oxide but in terms of the percentage of the element itself. This can be done using the following conversion factors:

$$\% \text{ K (elemental potassium)} = \% \text{ K}_2\text{O} \times 0.83$$
$$\% \text{ K}_2\text{O (potassium oxide)} = \% \text{ K} \times 1.2$$

The principal potassium-containing fertilisers

Potassium fertilisers are mainly simple compounds of potassium with chloride, sulphate or nitrate.

Potassium chloride (KCl) is by far the most widely used potassium-containing fertiliser, both by itself or in mixtures, and this is the potassium salt mined in the largest quantities. The natural salt is often purified to increase its KCl content to 83.5 or 96% equivalent to 50 or 60% K_2O (41.5% or 49.8% K). This product is often referred to commercially as **muriate of potash**.

Potassium sulphate (K_2SO_4) contains about 48–53% K_2O, or about 40–44% K. It is used mainly on crops such as potatoes and tobacco which are sensitive to large quantities of chlorides. It is prepared by reacting potassium chloride with other salts or with sulphuric acid. It supplies sulphur to the plant, which is more frequently needed than is chlorine, and is particularly recommended for drier areas.

Potassium magnesium sulphate (K_2SO_4. $MgSO_4$) is a naturally occurring double salt, a mixture of potassium sulphate and magnesium sulphate. It contains 22% K_2O (18% K), 18% MgO (11% Mg) and 18% sulphur. It is of value where soils are deficient in sulphur and magnesium as well as in potassium.

Potassium nitrate (KNO_3), also known as saltpetre, contains 44% K_2O (37% K) and 13% nitrogen. It is mined from natural deposits and is one of the oldest fertilisers known, being mentioned by Roman authors and in the Bible, but is now also produced by reacting KCl with nitric acid. It is a good source of both potassium and nitrogen and is very soluble, but its relatively high cost in most areas in the past has restricted its use mainly to high-value crops, such as tobacco and citrus crops, on which it is used as a side-dressing.

Applying potassium fertilisers

Agronomic experiments have suggested that of the above fertilisers, one form is as good as another as a source of potassium, and usually the cheapest form per unit of potassium is preferable, taking into account the content of magnesium and/or sulphate where these are needed when fertilisers containing these elements are used. In the case of chloride-sensitive crops such as tobacco, sweet potato and citrus, care must be taken in the use of large quantities of potassium chloride.

All these fertilisers are soluble, so that where moisture is not lacking the potassium is quickly available to the plant. Potassium is not as mobile in the soil or as easily leached as nitrate, but is more mobile than phosphorus. Because of this intermediate position between nitrate and phosphate as regards mobility, potassium-containing fertilisers can be applied either by placing in the soil, often as bands to the side and below the seed or plant, or by being broadcast. Placement is advisable in very heavy soils, or where rainfall is light. In most tropical soils where potassium fixation by the clay fraction (Chapter 2) is relatively unimportant, broadcasting potassium fertilisers is not likely to result in significant losses through fixation, but in soils that are high in micas and illite, which are known to fix potassium, banding is preferable to reduce fixation. A well-known example of an illite-containing soil that fixes applied potassium occurs in Trinidad, but most highly weathered tropical soils do not fix potassium.

The potassium cation is held by the exchange complex so that leaching losses in wet areas are not as likely as are losses of nitrate, particularly where the soil has a good exchange capacity. Leaching losses would be expected to be highest from sandy soils that are low in humus in areas of high rainfall. In some cases potassium is washed down from light-textured upper horizons of a soil to be held by subsoil clays.

Calcium, magnesium and sulphur

Liming of the soil, i.e. the use of calcium-containing compounds to change the pH of a soil, is discussed in Chapter 11, and calcium is usually supplied to soils mainly for this purpose. In some cases, calcium and magnesium give yield increases not by giving rise to a more suitable soil pH, but because of a direct need for these elements by the plant. In practice, calcium and magnesium are often added to a soil when fertilisers containing a major element are applied. Calcium occurs in the calcium phosphate and calcium sulphate of single superphosphate, and in the calcium phosphate of all the remaining major phosphatic fertilisers. Magnesium is given as magnesium sulphate or as potassium magnesium sulphate. Magnesium applications are necessary when the use of potassium fertilisers has resulted in too high a potassium/magnesium ratio in the soil, causing an induced deficiency of the latter.

Sulphur (Chapter 9) as a nutrient has received greatly increased attention in recent years. It is conveniently supplied to the soil as the sulphate anion, the form usually taken up by the plant. Sulphate is contained in the calcium sulphate of single superphosphate, in ammonium sulphate, in potassium sulphate, in magnesium sulphate and in a range of other fertilisers. Of these, the first two are the most important sources. In areas of heavy rainfall the sulphate ion can be leached out of the soil, or leached down to the subsoil where the anion exchange capacity of many tropical soils is typically higher than in the topsoil (Chapter 9).

Sulphur can also be supplied to the soil as elemental sulphur but as such has a marked effect on soil acidity. It is often used for the purpose of deliberately acidifying (or acidulating) soils and reclaiming alkaline soils. Elemental sulphur in the soil is oxidised to sulphate in the soil by sulphur-oxidising bacteria of the genus *Thiobacillus*, and sulphuric acid is formed. Conversion is quickest when the sulphur is finely ground.

Correcting trace element deficiencies

The correction of trace element deficiencies has in some cases resulted in considerable yield increases (Chapters 15 and 16) and trace element deficiencies are now more widely recognised in tropical soils than formerly (Chapter 12). Since the availability of trace elements is influenced by soil reaction, trace element deficiencies are sometimes cured by modifying the soil pH. More commonly they are corrected by applying small amounts of the missing element, either to the soil or direct to the plant, as in foliar sprays.

For application both to the soil and as leaf sprays the metals iron, copper, manganese and zinc are commonly chelated. The chelates (also known as sequestering agents) are organic compounds that are capable of attaching themselves to a metallic atom with various degrees of tenacity. The metal cation is bound on several sides by chelating agents (ligands) like clamps and is thus surrounded. The term 'chelate' is derived from the Greek *chele*, a crab's claw. Chelates are soluble in water, but when dissolved the metallic cation and the chelate are dissociated only to a small degree and the plant can absorb and then decompose the complete molecule to metabolise the metal in it. The metal is thus protected in the soil against being fixed or immobilised by forming inorganic combinations which would render it insoluble.

Iron deficiencies are most likely in calcareous soils, particularly in drier areas, and in soils high in phosphate; they are not likely in most acid tropical soils except pale sands. Iron is absorbed by the plant in both the iron II (Fe^{2+}) and iron III (Fe^{3+}) forms, but the iron II form is more soluble and mobile in the soil than the iron III form. The solubilities of both forms fall with increasing pH. Many tropical soils have high total iron contents, though because most of the iron occurs as very sparingly soluble iron III oxides, total iron in the soil is not necessarily any indication of availability to the plant, unless it is unusually low. Iron deficiencies are sometimes associated with imbalances between ions such as copper, iron and manganese, and can be induced by particularly high copper or, more commonly, manganese levels in soils. High concentrations of phosphorus have been shown to interfere with iron uptake by the plant.

Iron deficiencies are most often corrected by foliar sprays, either of iron II sulphate (20% Fe) or of Fe-chelates. Two important Fe chelates are Fe-EDTA, which contains iron in the divalent, iron II form and is more effective on acid soils, and Fe-EDDHA, containing iron in the trivalent, iron III form, which remains longer available in calcareous soils. As stated previously, iron deficiencies in trees have been cured simply by driving an iron nail into the trunk.

Manganese in tropical soils varies considerably in amount, but manganese deficiencies, like most iron deficiencies, are in most cases not due to low absolute amounts in the soil but to fixation in insoluble forms. These insoluble manganese compounds occur mainly in neutral to alkaline soils, and may be induced by too much liming. Manganese is made available by reducing conditions in the soil and is more available in wet, poorly drained soils than in well-aerated ones. Dry conditions, on the other hand, increase manganese fixation in insoluble forms. The available manganese in soils can be increased by acidifying alkaline soils (as with the use of ammonium-containing fertilisers), by preventing soils drying out, and by applying easily decomposed organic matter. The amounts of manganese removed from the soil by crops are generally low, less than 0.5 kg/ha. For the quick correction of deficiencies in a growing crop, the crop can be sprayed with manganese sulphate ($MnSO_4.4H_2O$) at rates of 2–4 kg/ha dissolved in 400 l of water, or with manganese chelates.

Zinc is less available in neutral and alkaline soils than in acid ones, but some acid soils may exhibit zinc deficiences due to low total amounts in the soil. Deficiencies have also sometimes been correlated with high phosphorus levels in soils or with the presence of calcium and magnesium carbonates, and have been induced by over-liming, and are often most marked during dry periods. The amounts of zinc removed from the soil by crops is very small, usually less than 0.4 kg/ha. Plants vary in their zinc requirements and in their ability to extract zinc from the soil. Crops that are particularly sensitive to zinc deficiencies include maize, citrus, cotton and legumes. Sugarcane has relatively low zinc requirements.

Zinc deficiencies are most commonly corrected by the use of zinc sulphate (e.g. $ZnSO_4.7H_2O$, though the amount of combined water and hence the zinc content varies with different salts). As such it can be applied to the soil, e.g. at rates of 10–20 kg/ha, or mixed with other fertilisers, or used to lightly dust seeds. As a leaf spray, zinc sulphate is acidic and can cause corrosion damage to leaves unless applied in very weak solutions (up to 0.5 kg zinc sulphate in 400 1 water). For this reason, zinc chelates are better as a foliar spray than is zinc sulphate. Too much zinc in the soil (above about 500 ppm) can be toxic to plants.

Copper is most likely to be limiting in organic soils, probably owing to the formation of unavailable copper–humus complexes which reduce the amount of copper in the soil solution. Deficiencies may also occur in alkaline soils, though copper is less pH dependent than are manganese and zinc. Copper deficiency may be induced by heavy phosphate applications, or by high levels in the soil of zinc or iron. The cereals, including wheat and maize, are particularly sensitive to copper deficiencies. High copper levels have been found to reduce the latex quality of rubber trees without affecting growth or yields.

Copper has long been supplied to soils and plants as copper sulphate (also used as a fungicide). The copper content varies according to the water or crystallisation (e.g. $CuSO_4.7H_2O$, with 25% Cu, and $CuSO_4.H_2O$, with 36% Cu). As a foliar spray, copper sulphate can cause leaf corrosion unless applied in weak solutions (0.2–0.5 kg in 400 l of water), but copper sulphate can be applied direct to the soil at rates of up to 5–10 kg/ha, preferably mixed with other fertilisers used in larger quantities in order to get better distribution in the field. Once applied, copper is relatively very immobile in the soil.

Boron is widely applied as a trace element but the fact that plants differ widely in their needs may pose problems, since a satisfactory level for plants with high boron requirements (such as cotton, sisal and legumes) may be toxic for plants with low boron needs (rubber, cereals and grasses). Amounts of boron removed by a crop from the soil range from about 500 g/ha/year for plants in the first group to perhaps one-tenth of this amount for those in the second, and in practice only the first group are usually fertilised. Boron deficiencies are most commonly found in light-textured acid soils that have been leached by high rainfall or high levels of irrigation, or which have been limed, and boron is the trace element most often analysed in commercial soil surveys. Boron levels in soils are generally less than 3 ppm in the humid tropics but are often higher in arid and semi-arid areas, rising to 10–40 ppm. At these levels boron may be toxic. Boron is likely to be toxic in saline–sodic soils. Levels below 0.5 ppm indicate a probable boron deficiency.

The most commonly used boron fertiliser is borax ($Na_2B_4O_7.10H_2O$, 11% B), both as a soil fertiliser and as a foliar spray. It is best added at rates of up to 2–4 kg/ha to other fertilisers used in larger quantities. For cotton-growing areas of the Ivory Coast, for example, the use of a 10-18-18 fertiliser containing 5% S and 0.4% B is standard practice. Boric acid (H_3BO_3, 18% B), a white crystalline powder, is also used, especially as a foliar nutrient, but is very toxic to humans. Borax and boric acid are very soluble and can easily be leached from soils, especially light-textured soils, and in those cases it may be preferable to use less soluble sources of boron which are released slowly over a period and which also reduce the risk of toxicity. These include the natural boron-containing mineral colemanite, and glass frits which dissolve slowly to release boron.

Molybdenum is present in particularly small amounts in soils, the average soil content being only 2 ppm. Unlike all the other trace elements considered above, molybdenum in the soil is more available to the plant in alkaline soils than in acid soils. Deficiencies are normally found only in acid soils.

Molybdenum deficiencies are normally corrected by adding very small amounts of a molybdenum-containing compound at rates of 0.1–1.0 kg/ha to normal NPK or phosphatic fertilisers. Compounds used for this purpose are sodium and ammonium molybdate, both of which are soluble and can also be used as leaf sprays. In some cases deficiencies are cured by soaking seeds for eight hours in a 0.5% sodium molybdate solution.

15 Some tropical crops in relation to soils and fertiliser use: annuals

General aspects of fertiliser use on annual crops

Responses of a crop to added nutrients supplied in fertilisers depend on the needs of the crop in relation to the existing fertility of the soil, on the climate, on the ability of the particular crop variety to respond to a better nutrient supply, and on management factors.

Management factors include correct time of planting, appropriate planting densities, protection against pests and diseases, and adequate and timely weeding.

For these reasons responses are to some extent specific to the crop, crop variety and site. For the same crop on the same soil, responses may vary from year to year if rainfall amounts and distribution, or other factors, change. In areas where there are two rainy seasons a year, a response may be obtained in one but not in the other.

The importance of the general level of management and of appropriate agronomic practices has been stressed by many agronomists.

In the Kitale district of Kenya, for example, studies of fertiliser use on maize showed that yields were less if fertilisers were used but management was poor, than if no fertilisers were used but management was adequate. Adequate management in this case was defined in relation to correct time of planting, correct planting densities, and early weeding.

In Senegal it is emphasised that the recommended fertiliser dressings on groundnuts give their full benefit only if the following conditions are met:

1 Correct selected seed strains.
2 Use of a fungicide/insecticide dressing on the seeds.
3 A high planting density of at least 110 000 seeds/ha (corresponding to a spacing of 60 × 15 cm).
4 Early sowing: every day that sowing is delayed after the optimum planting date results in a yield decrease of about 1%.
5 Adequate weeding.

The choice of an appropriate cropping system involves fitting crop varieties with an appropriate length of growing season to the local rainfall patterns, and taking account of rainfall confidence limits and the likely start and end of the growing season. Much attention has been given, particularly in Africa, to optimum planting dates. These are related mainly to rainfall distribution, but may also be affected by day length and by soil factors such as soil temperature and the nitrogen flush that commonly occurs at the beginning of the rains. Yields are lowered if crops are sown after the optimum date, and this effect often cannot be compensated for by added fertiliser.

Two additional common causes of disappointing responses to fertiliser applications are too low planting densities, and the use of a traditional crop variety which may have been selected over the years because of its ability to do well under local conditions, including its ability to grow on poor soils, but which does not respond satisfactorily to an increase in the nutrient supply.

Dry periods within the growing period may retard crop growth and limit fertiliser responses. In general the cereals, particularly maize and upland rice, are relatively sensitive to dry spells, particularly in the early stages. The root crops, including yams and cassava, are less affected and thus have an advantage over cereals in those areas where rainfall is erratic or poorly distributed.

A crop cannot respond adequately to fertiliser nutrients if there is strong competition for water and nutrients from weeds, or if the plant is attacked by pests and diseases.

The effect of a fertiliser depends not only on the type and amount of nutrient, but also on the form of fertiliser, the way it is applied, and the timing of the application (see Chapter 14).

Applications to annual crops have to take into account possible residual effects from the preceding crop and those affecting the following one, and must thus consider not only the current crop but the crop rotation. The levels of fertiliser application should be decided in relation to the yield levels aimed at. In addition to the immediate responses expected from the current crop, attention should be given to the long-term maintenance of the fertility of the soil.

Cereals: maize, sorghum, millet

Maize (*Zea mays*)

Maize is the most important of the cereals after wheat and rice, and is very widely distributed. It is one of the highest-yielding grain crops. Maize originated in the New World, where it was grown over a wide range of latitude and altitude from what are now Chile to southern Canada, and from sea level to 1000 metres. After the discovery of America by Europeans in 1492, maize spread rapidly in the tropics, subtropics and warm temperate areas of the world. In areas of Africa that are suited to the crop it largely replaced traditional grains such as sorghums and millets. In Asia it became particularly important in areas that are too dry for rice.

Climate and soil requirements Different varieties of maize are adapted to a range of climates and growing seasons. In recent years new varieties have been bred which have further enlarged the geographic area of the crop. These include varieties that require less sunshine than usual, which are grown in areas formerly considered unsuitable.

Growing seasons are mainly within the range 90–200 days. There is a general increase in growing season of about 20 days for each 100 m increase in altitude above 1500 m, or for each 0.5 °C fall in mean temperature below 20 °C. Maize in some high-altitude areas of the tropics, as in some areas of the Andes, may take more than a year to mature. Typically, however, maize is a crop of warm areas and adequate rainfall.

Most maize is grown in areas with temperatures of 21–30 °C at the tasselling period. The optimum temperature at germination is considered to be 18–21 °C. Maize has a C4 photosynthetic pathway; photosynthetic rates peak at 30–40° C, and rates of leaf emergence at about 30° C. However, production can be expected to be greatest when nights are cool and there is a marked diurnal range of temperature.

Rainfall requirements vary with the variety, and range from 200 to 900 mm in the growing season. Maize is an important component of cropping systems in the tropics where rainfall is in the range 750–1750 mm/year. In drier areas the growing season may be too short or the incidence of within-season dry spells too frequent, and in Africa and South-east Asia maize is then replaced by sorghum and millet.

The seed is large, and relatively large amounts of water are required at imbibition. Maximum moisture demand is from the tasselling to the 'hard dough' stages, after which water needs decline. Maize is relatively sensitive to periods of low rainfall and water stress. It has long been recognised that maize is particularly sensitive to water deficit at flowering. Water stress reduces leaf area, and lowers the leaf photosynthetic rate during the stress period, though this may recover afterwards. It delays silking and reduces grain numbers. Maize yields in the tropics are generally lower than those in temperate regions. This appears to be related in part to above-optimum temperatures and to sensitivity to stress when water is limiting.

Maize is susceptible to flooding. Experiments in Nigeria showed that flooding for 48 hours to a depth of 5–15 cm reduced yields by 31–63%.

Maize grows well on a wide variety of soils, but best on intermediate textures (sandy loams to clay loams) with a good structure and aeration. Poor soil structure, inadequate aeration, and soil compaction restrict root development and lower yields. On gravelly soils in western Nigeria with low total porosity and water-holding capacity, root volume and root elongation rates were found to be low, rendering the crop more susceptible to the adverse effects of dry periods. Drought resistance and nutrition are improved where soil conditions allow unrestricted development of the fibrous root system which under favourable conditions can penetrate to a depth of over 2 m, though most water uptake is usually from the upper 1 m.

At the International Institute for Tropical Agriculture (IITA), Ibadan, Nigeria, emphasis has been on minimum soil tillage techniques. These techniques involve clearing by the use of weedicides and the planting of maize in small holes between the dead weed mulch, with a minimum of soil disturbance or cultivation. This technique has been recommended on the medium-textured Alfisols of the area in which root proliferation appears adequate without soil cultivation. In contrast, on sandy soils in Senegal, French workers have found that deep cultivation has a marked positive effect on the rooting of cereal crops and on yields, and on some soils deep cultivation resulted in maize root densities increasing by 300% and yields almost doubling.

Soil compaction induced by heavy machinery results in higher soil bulk densities and poorer aeration, and limits yields because of the adverse effect on root development and on root metabolism and nutrient uptake. The formation of surface crusts may seriously reduce seedling emergence.

The normal soil pH range for maize is 5.0–8.0, with a pH of 6.0–7.0 probably being optimal for most varieties. On soils more acid than pH 5.0, aluminium toxicity can seriously limit yields.

Nutrient uptake by maize and deficiency symptoms Maize is a heavy feeder with a high nitrogen requirement. It is also demanding for phosphorus, and sensitive to a low phosphate supply, particularly at the early stages of growth. Because of this it is sometimes used as a test crop to assess phosphorus deficiencies.

The better-known deficiency symptoms include the following:

Nitrogen: pale green or yellow colour; yellowing and dying along midribs, with a later 'firing' of lower leaves.
Phosphorus: stunted growth, delayed ripening, dark green or purplish leaf colour, especially at leaf tips.
Potassium: poor root system and stalk breakages; small whitish-yellow leaf spots followed by scorching and browning of leaf tips and edges.
Magnesium: whitish or yellowish striping between leaf veins.
Zinc: chlorotic leaf fading to give whitish areas, especially towards the leaf centre and base.

Leaf analyses have been widely used to denote deficiencies, using either a particular leaf or the whole plant.

Maize fertiliser responses and practices

Nitrogen Nitrogen is the most important nutrient for maize and the one that most often limits yields. Nitrogen increases vegetative growth and the photosynthetic capacity of the plant. Of the total nitrogen taken up, about two-thirds has normally accumulated in the grain at maturity.

The responses of high-yielding maize varieties to nitrogen are often spectacular. Nitrogen applications are commonly in the range 40–80 kg/ha, but much higher applications can be profitable with responsive varieties and with adequately high plant densities. In the Philippines, responses were obtained

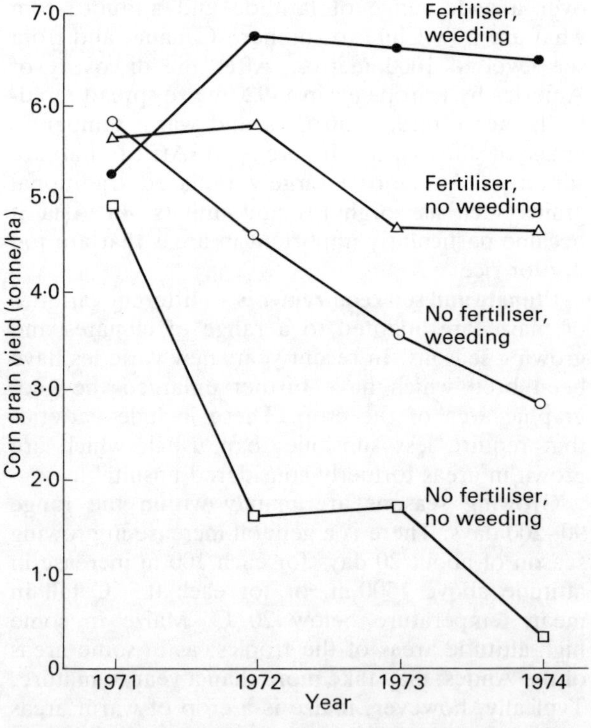

Fig. 15.1 Effects of fertiliser and weeding on rainy season maize yields in a continuously cropped experiment on an Alfisol near Ibadan, Nigeria. Fertiliser was applied at rates of 120 kg K/ha, 40 kg P/ha and 50 kg/ha per crop. Source: International Institute of Tropical Agricultural, Ibadan.

up to 180 kg N/ha with good plant stands, but only up to 45 kg N/ha with low densities.

The following yields were obtained in Madagascar with nitrogen applications in the range 40–200 kg/ha:

Kg N/ha	Yield, kg/ha grain
40	1211
80	1561
120	3212
160	4117
200	5218

These results show a linear response over a fairly wide range of application rates, and an average yield increase of 26 kg of grain for each kg of nitrogen applied.

The utilisation by maize of fertiliser nitrogen at various sites in West Africa showed that about 40–45% of the applied fertiliser nitrogen was in the aerial plant parts at harvest, and that succeeding crops continued to take up the fertiliser nitrogen at rates of 4–5% for the second year and 2–3% for the third. Maize response to low rates of application were about 20 kg of grain per kg of nitrogen supplied, and remained linear for most varieties up to 50–100 kg/ha.

A summary prepared in 1979 of numerous nitrogen field experiments in Brazil, Puerto Rico and Ghana found that 95% of maximum yield was obtained in the range 80–120 kg/ha of applied nitrogen, and it was concluded that response curves in the tropics are qualitatively comparable with those in temperate areas, and rates of recovery of applied nitrogen about the same.

Nitrogen responses are not likely on soils that are high in organic matter, such as soils newly cleared from high forest. On a Kikuyu friable clay in Kenya in which the high organic matter content mineralises about 180 kg/ha of nitrogen, no response was obtained to nitrogen at 90 kg/ha. In most savanna soils, and in forest soils in which organic matter levels have declined during cultivation, nitrogen responses can be expected. In the Guinea savanna areas of West Africa, yields of 5000 kg/ha have been obtained with 100–175 kg/ha of nitrogen and 60–150 kg/ha of phosphorus. In the drier Sudan savanna, with a shorter growing season and lower rainfall, 60–120 kg N/ha and 40–100 kg P/ha are considered appropriate for yields of about 3000 kg/ha.

Phosphorus Maize is a relatively demanding crop for phosphorus, but since tropical soils vary very widely in their available phosphorus status and in their ability to fix added phosphate, yield responses are essentially patchy, and amounts needed to obtain near-maximum yields have been found to vary within relatively wide limits. A summary in 1979 of extensive trials in Brazil, Puerto Rico and Ghana showed that satisfactory growth was obtained with as little as 18 kg P/ha but in extreme cases as much as 440 kg/P/ha was needed. Although some Ultisols and Oxisols fix phosphate, residual phosphorus effects are often satisfactory on these soils.

As in the case of nitrogen, substantial quantities of phosphate can be liberated if organic matter levels and rates of mineralisation are good. In one study on maize grown after the clearing of secondary forest in Nigeria it was found that 25% of the organic phosphorus in the 0–15 cm layer was mineralised during two seasons, with about three times as much phosphate being released as was taken up by the crops.

Phosphorus uptake is sometimes improved when mycorrhiza are present on the maize roots.

Phosphorus utilisation is best if the phosphate is incorporated into the soil at planting. It is often the practice with intensive maize cultivation to incorporate a mixed fertiliser at planting containing all the phosphate to be given, and one-third to half the nitrogen. This ensures that the phosphate is available to the growing root system. The rest of the nitrogen is then applied as a top-dressing at a later stage, often at tasselling. Nitrogen–phosphorus interactions are frequently recorded.

Farmyard manure has traditionally been widely used for maize, but since it is relatively low in phosphorus in relation to nitrogen and potassium, it is often supplemented with a phosphatic fertiliser or with a mixed fertiliser with a high phosphate component.

Potassium Potassium is taken up in fairly large amounts by maize, but it is a relatively efficient extractor of this element and responses to added potassium in the tropics are patchy, especially if the residues are returned. In the savanna areas of West

Africa it has been estimated that a total dry matter production of 6–8 t/ha involves an uptake by the maize crop of 100–200 kg K/ha. Of this total, only 25–30 kg was in the grain, and the remainder could have been returned to the soil in crop residues. Where maize plants are left to decompose in the field after grain harvest, it is sometimes found that the soil within the rows is markedly enriched with potassium but that between the rows it is not.

Potassium as a fertiliser nutrient is given to maize less frequently than is nitrogen and phosphorus. In India, moderate to fairly heavy dressings of a mixed NPK fertiliser with a ratio of about 6:3:2 are recommended for hybrid maize, but for traditional varieties the recommendation is for nitrogen alone at 40–60 kg/ha.

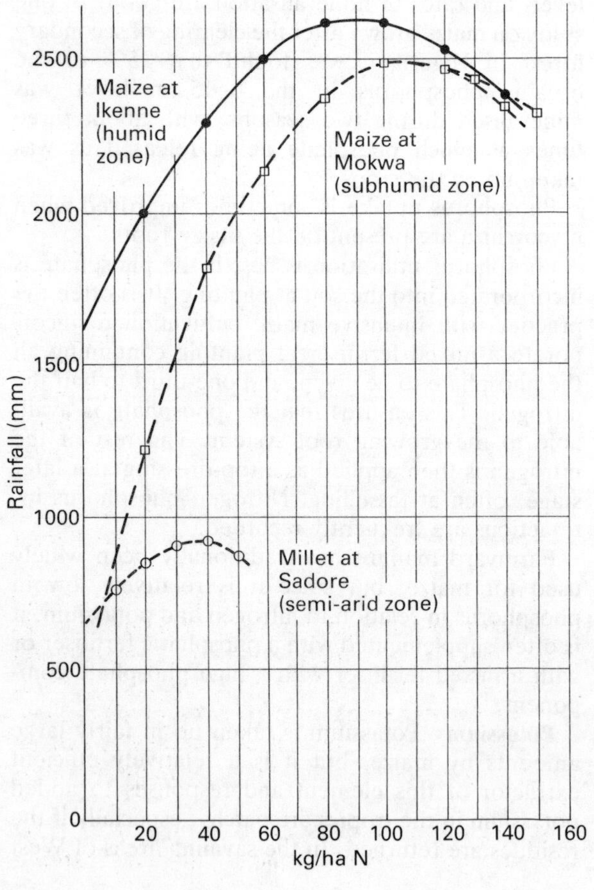

Fig. 15.2 Examples of responses to broadcast urea in West Africa showing yields of millet at the ICRAF station at Sadore, Niger, and yields of maize at two sites in Nigeria, at Ikenne in the humid zone, and at Mokwa, in the subhumid zone.

The results obtained with millet on a poor sandy Inceptisol at Sadore in the semi-arid zone showed that 30 kg/ha of urea applied broadcast raised yields from about 600 kg/ha of grain to 900 kg/ha, and that heavier doses were not worth while. This gave an increase of only about 10 kg grain per kilo of fertiliser applied, which has to be evaluated in terms of the local cost of fertiliser and selling price for millet.

The yield increases obtained with maize in Nigeria were greatest in the subhumid zone since initially low yields of about 500 kg/ha were substantially increased by modest additions of urea. The curve is very steep up to 40 kg N/ha and steep up to 80 kg N/ha, which raised yields to about 2400 kg/ha of grain. This represents an increase of about 24 kg grain per kg of nutrient supplied.

In the humid zone of Nigeria total yields were highest but since yields without urea were about 1500 kg/ha maize the increases obtained were substantially less per kilo of nutrient applied than they were in the subhumid zone.

Tests with banded and placed urea, the latter using urea supergranules, showed that on the sandy soils used with these trials banding and placement were less effective than broadcast urea applications. This was interpreted as being due to increased access to soil exchange sites and hence decreased leaching when the fertiliser was broadcast, but on heavier textured soils elsewhere placement appeared to be the best method of urea application.

The results reproduced here are given merely as examples of the varied and sometimes contradictory responses obtained when fertilisers are applied to tropical crops, particularly in areas of unreliable rainfall. The results should be used and extrapolated to other soils and to other areas with caution, remembering that strictly speaking experimental results are site and season specific and thus valid only for a specific site in the years during which the experiment was carried out. Different responses might have been obtained from other sites in the same zone or at the same sites in other years, or with different levels of management or crop varieties.

Source: **Moghogo, S. K., Batiano, A., Christianson, B. and Vlek, P. L. G.** (1986) 'Management of nitrogen fertilisers for Tropical African soils'. Paper in **Mokwunye, A. U. and Vlek, P. L. G.** (eds), 1986.

Fig. 15.3 Maize field trials showing improved growth of maize which has been weeded, on the right, compared with weed infested maize on the left. The trials illustrate the importance of timely weeding (see also Fig. 15.1).

Micronutrients Micronutrient deficiencies in maize have been mainly of zinc and molybdenum. Zinc deficiencies are most likely on calcareous soils and may be induced by heavy applications of superphosphate. Molybdenum deficiencies are thought not to occur if there is at least 0.08 ppm molybdenum in the seed.

Sorghum (*Sorghum bicolor*)

Sorghum is particularly important, and a staple food crop, in many of the drier areas of the tropics, including those of Africa, India and China. In areas of Africa with a Sudan climate, sorghum and millet are the important cereals. In the wetter areas of the tropics sorghum is usually outyielded by maize, which needs less labour.

Sorghum is thought to have originated in Africa, and was grown in West Africa in neolithic times. From Africa it spread to India and China, and, more recently, to America.

Climatic requirements Sorghum grows in a wide range of ecological conditions. Partly because of its dense, well-branched root system, it is tolerant of periods of drought, and can remain dormant until rains fall again. High potential photosynthetic rates, efficient water use and an ability to tolerate drought conditions result in sorghum being concentrated in areas where the rainfall is 600–1000 mm, but the crop needs at least 200–300 mm of rain during the growing season. Most sorghum varieties prefer warm or hot conditions, with optimum temperatures of about 28 °C, but high-altitude varieties can grow in cooler latitudes and at higher elevations. In the Kigezi highlands of Uganda, sorghum is grown to 2500 m.

In the savanna areas of West africa sorghum is grown mainly in a belt receiving 600–1000 mm of rain. Sorghum gives way to maize in the southern, wetter parts of that belt, and to millet in the drier areas with a shorter growing season to the north, thus occupying a position between the two crops though overlapping with both of them. Sorghum growing periods in this belt vary from 120–135 days in the north to 240 days in the south. Local varieties, sensitive to day length, usually head at the end of the rains so that the seed sets as conditions get drier and matures on residual soil moisture. This avoids the smuts that often attack grain maturing in the rains. Partly for this reason sorghum is not grown in the wetter areas of West Africa, where it is in any case outyielded by maize. Since local cultivars are photoperiodic they mature at about the same time even if planted late, so that delayed planting results in a shorter growing season. Correct date of planting is therefore important, 110–120 days being required to produce good yields, though longer cycles show no yield advantage.

For a map showing areas of South America considered climatically suited to sorghum, see Chapter 1, Fig. 1.4.

Soil requirements Sorghum will tolerate relatively poor soils though it grows best on moderately fertile ones. It is often grown late in a crop rotation after fertility has been lowered by preceding crops. The best soils for sorghum are probably well-drained sandy loams but sorghum is also widely grown on heavy clays. It can stand short periods of waterlogging.

In West Africa sorghum is grown mainly on Alfisols which are typically light to medium-textured in the surface horizons but become a little heavier in the subsoil. In India sorghum is also grown on Alfisols but in addition is widely grown on Vertisols, i.e. on heavy, black, cracking clays, often as a post-rainy season crop on residual soil moisture. Sorghum is also grown on Vertisols in Ethiopia and in the Sudan. A good water storage capacity is an advantage as it allows crops to mature on stored soil moisture after the end of the rains. Crusting soils may adversely affect seedling emergence. Sorghum is considered to be less tolerant of acid soils and aluminium toxicity than maize. On neutral and alkaline soils sorghum appears susceptible to iron deficiency, and iron deficiency has been reported on Vertisols in India that are rich in calcium carbonate.

Sorghum fertiliser responses and application rates In many sorghum areas production is limited more by lack of moisture than by soil fertility, and in such conditions no response can be expected from fertiliser applications, which may even depress yields. Much less attention has been given to fertiliser use on sorghum in proportion to production than to fertiliser applications to maize or rice.

Fig. 15.4 Sorghum trials, ICRISAT, Hyderabad, India.

In West Africa sorghum is grown as a subsistence crop that is mainly consumed by the grower and his family. Sorghum is grown by extensive methods, often in a shifting cultivation system with long fallows, and fertilisers are almost never used. Yields are low, generally averaging 600–750 kg/ha, but there are usually only small variations between good and bad years and even in very poor years there is usually some yield. In the more densely populated areas where fallows have become inadequate, however, yields are declining.

The current view of many agronomists is that in sorghum systems in these areas only low levels of fertiliser are at present to be recommended, with fertility maintenance being helped by basal dressings of local rock phosphate and the incorporation of residues, and fertilisers being applied mainly to the cash crop in the rotation. Residue incorporation appears particularly important in order to maintain potassium levels. Sorghum has been shown to benefit from fertilisers applied to a previous crop of cotton or groundnuts.

In sorghum, nutrient accumulation can be spread out over longer periods than in other cereals, and right through to maturity. Although sorghum and maize may contain similar amounts of total nitrogen at harvest, sorghum normally has a longer growing season and takes up its nutrients at a slower rate, with peak nitrogen demands at about half of those for maize. Since the daily needs of sorghum are lower, it appears better able to be grown with little or no fertiliser.

Nitrogen applied to traditional tall, late-maturing West Africa varieties with fixed flowering date have sometimes caused the plants to grow to enormous heights with little or no grain increase and an increased likelihood of lodging. Where water is limiting, nitrogen can depress yields by leading to vegetative growth early on and to greater water stress. Fertiliser use on short varieties can, however, give good yield increases where water is not limiting. Under these conditions response curves to nitrogen can show a steep yield increase up to 50 kg N/ha, with increases in grain yield averaging about 20 kg of grain per kg of applied nitrogen. Responses to phosphorus have given increases up to 50–80 kg P_2O_5/ha with an immediate increase of 9–20 kg grain per kg P_2O_5 applied, but there are also residual effects over several years which may bring the total increase per kg of P_2O_5 up to 30–40 kg grain. In northern Nigeria, an initial response was found to phosphate, and additional increases to added nitrogen.

In India, hybrid sorghums consistently outyield local varieties, even on the poorer soils. The root and stubble residues have a high C/N ratio and Indian farmers have found that they may cause a nitrogen deficiency in the succeeding crop (Chapter 9). Nitrogen fertilisers are applied to sorghum only where moisture is not likely to be limiting, usually at modest rates of 20–40 kg/ha. For hybrids grown with adequate moisture, applications of up to 70–100 kg N/ha and 35–50 kg P_2O_5/ha are recommended.

Millet (*Pennisetum typhoides, Pennisetum americanum*)

Bullrush or pearl millet probably originated in the savanna areas of West Africa from where it spread to the rest of Africa and to India. It is now a staple food in the drier areas of Africa and also of India where it is grown in areas of low rainfall and on poor soils.

Climatic requirements Millet is well adapted to relatively high temperatures and high solar radiation. Optimum temperatures for net photosynthesis are 30–35 °C.

Millet is associated mainly with an average annual rainfall of 200–800 mm. It shows a wide range of growing periods, from as little as 55 days to 180 days and more. In West Africa where total rainfall is closely correlated with latitude, the 55–65 day millets are found in the northern millet-growing areas near the 200 mm isohyet. Longer-season varieties are found in areas of less low rainfall. In some areas, such as parts of northern Nigeria, two different millet types are grown together, with the shorter season day-length insensitive Gero millets of 75–100 days maturing in the middle of the rains and the longer season day-length sensitive Maiwa millets taking 120–280 days. In India millet is grown mainly as a short season crop of 90 days or less. The International Crops Research Institute for the Semi-Arid Tropics (ICRISAT) has devoted considerable research to millet/groundnut and millet/sorghum intercropping in India and it appears that the yield advantage is greatest under water stress conditions.

Millet has a reputation for tolerating low rainfall. This is connected with its deep and rapid root penetration, usually to 1.2–1.5 m (up to 3.6 m has been recorded), though most of the roots are in the upper 30 cm.

Although millet can grow with a lower total rainfall than sorghum, it does not have the ability that sorghum has of surviving dry spells by becoming dormant, so that a more even rainfall distribution is required. Also unlike sorghum, it cannot tolerate waterlogging.

In West Africa, millet seedlings have often to contend with strong winds blowing sand at ground level. Local millet cultivars are resistant to this, but when ICRISAT introduced non-resistant Indian cultivars to their agronomic station near Niamey in Niger, the seedlings were wiped out for this reason.

Soil requirements Millet is grown on a range of soils, both light and heavy, but light, sandy soils predominate. Millet is known for its ability to grow on relatively infertile soils and in rotations is often grown after other crops that have lowered soil fertility. It can grow on soils that are too poor for sorghum, provided they are well drained.

In West Africa millet is grown mainly on Aridisols, Alfisols and Entisols, mostly young soils with little profile differentiation. Over half are developed in aeolian sands, and many of the remainder in often sandy lake and river alluvium. In India, millet is also grown widely on sandy soils, including deep sands in the Rajasthan desert and the western Punjab which are similar to some of the sands of the Sahel, and on gravelly soils of the Deccan and southern India, but millet is also grown on heavy, black cracking clays (Vertisols).

Soils that crust may restrict seedling emergence. Millet is tolerant of acid soils and high levels of aluminium. It also appears to be relatively tolerant of saline conditions.

Millet fertiliser responses and application rates In West Africa millet is not normally given either manures or fertilisers. Its capacity to grow on soils of low natural fertility appears related to its ability to develop a deep root system rapidly. When given fertilisers or grown on richer soils, however, it can take up large amounts of nutrients. Responses to nitrogen may be linear up to fairly high amounts if water is not limiting. In practice, under West African conditions of limited water availability, fertiliser might promote early vegetative growth and a higher water demand and thus result in lower yield. The recommended rate of application in Samaru, northern Nigeria, is only 13 kg N/ha. Millet has been shown to benefit in its nitrogen nutrition if it is grown after a legume, usually groundnuts.

Under rainfed conditions the demand of millet for phosphorus appears to be modest. Though some responses to this element have been recorded, trials at Samaru showed that 66 kg P/ha depressed yields.

Potassium uptake is relatively heavy but millet appears to be an efficient potassium extractor and responses are not often recorded. A complete NPK fertiliser is recommended for millet in Zimbabwe where it is grown under rainfalls of 500–625 mm.

In India, bullrush millet (*bajra*) is the second most important rainfed crop after sorghum in areas with 400–1000 mm of rainfall. Hybrid varieties are now important. Results of different nitrogen applications to millet in Uttar Pradesh using both local and hybrid varieties were as follows:

Nitrogen (kg/ha)	Yields (kg/ha)	
	Local varieties	Hybrids
0	1856	2154
40	2569	3561
80	3069	4348
120	3806	5645
160	3393	5967

With local varieties yields increased with nitrogen applications up to 120 kg/ha, with an average response of 16 kg of grain per kg nitrogen, though hybrid varieties gave about double this increase per kg over the same range of application rates.

Root crops: cassava, yams, cocoyam, sweet potato

Cassava (*Manihot utilissima, Manihot esculenta*)

Cassava is the most important root crop in the tropics. It can grow in soils that are too poor for many other crops, and is drought resistant. It may be the last crop in a sequence, planted after fertility has been depleted by more demanding crops.

Cassava originated in America and was widely distributed there at the time of Columbus, being taken to Africa by the Portuguese in the late sixteenth century. Recently it has spread widely in Africa, expanding from the traditional forest areas to savanna areas where it can survive long dry periods. It forms a useful reserve crop, almost immune from locust damage, which can remain in the ground for extended periods until it is needed. Cassava cultivation is very widespread in the tropics, but greater quantities are grown in Africa than elsewhere, almost all for human consumption. Cassava as an export crop, largely for animal feeds, has become important in Thailand, Madagascar and Brazil.

Climatic requirements Cassava needs a warm climate, with average temperatures of 25–29 °C. At temperatures above 29 °C yields are reduced. Growth ceases at about 10 °C. Cassava is grown mainly at low altitudes. At altitudes higher than 1000 m growth is slower and yields lower. Cassava is a short-day plant growing best between latitudes 15° N and 15° S.

Cassava is found growing in areas with rainfalls of 500–2500 mm, but 1000–1500 mm is considered the optimum range. Cassava is often grown in areas where the natural vegetation is forest, but its resistance to drought has allowed it to spread to areas of seasonal rainfall and savanna vegetation. Some cultivars mature at 6–10 months and in Papua New Guinea a 3½-month variety is grown, but most take longer than this, typically 12–18 months, and some remain in the ground for two years or more.

Soil requirements Cassava is grown on a very wide range of soils from sands to heavy clays, and also in gravelly lateritic soils and drained organic soils. The crop is grown mainly on Oxisols, Ultisols, Alfisols and Histosols. For best results extremes of texture should be avoided, and the soil should be free of stones and gravel and deep enough to allow adequate rooting volume for the development of the tubers. On heavy soils there is a tendency to produce stems and leaves at the expense of roots. High bulk densities and waterlogging are unfavourable.

Cassava is tolerant of a wide range of soil reactions, from pH 5.0 to 9.0, though yields decline at above pH 8.0 and in saline soils. With soils more acid than pH 5.0, modest liming may be beneficial.

Reasonable yields of 3–6 t/ha can be obtained even on infertile soils on which little else can grow satisfactorily. Solution culture studies have shown that cassava can tolerate relatively low calcium, nitrogen and potassium in the root environment. In the field, its phosphate uptake is often assisted by the presence of mycorrhiza on the roots which thus increase their effective surface area. Mycorrhiza have also been shown to benefit uptake of potassium and sulphur. The root system is typically deep and well developed, though coarse, thick and poorly branched, and thus not well equipped to take up phosphorus without the mycorrhiza. Nutrient uptake is relatively heavy, particularly for potassium, and cassava cultivation can deplete soil fertility. Cassava is often grown at the end of a rotation. It is not as effective at controlling erosion and leaching as are other crops such as the grasses.

Cassava fertiliser responses and application rates Although negative responses have sometimes been obtained to nitrogen, particularly at high rates, frequent responses have been obtained for applications as high as 60–180 kg N/ha. In the derived savanna areas of Nigeria, on sandy soils derived from granites and gneiss that are low in nitrogen, 28 kg N/ha raised yields by over 3000 kg/ha, from about 9000–11 000 kg/ha without nitrogen to 12 000–14 800 kg/ha with nitrogen. This represented an increase of over 100 kg of tubers per kg of applied nitrogen. Responses to phosphorus and potassium were lower.

In Latin America phosphorus appears to limit cassava production more than it does in Africa, particularly on acid Oxisols and Ultisols. In one experiment on acid infertile soils in Colombia, yield increased from about 5 t to over 20 t with 176 kg P/ha. Some cassava cultivars are much better adapted to phosphorus-deficient soils than others. Aluminium toxicity appears to be as important in limiting yields in South America as phosphorus deficiencies, but again seems less of a limiting factor in Africa. Liming may improve phosphorus availability and yields, but rates of only 0.5 t/ha are generally recommended since at higher rates trace-element deficiencies, particularly of zinc, may be induced.

Potassium responses are often obtained, especially on highly weathered soils. In West Africa, however, little response so far has generally been obtained to potassium applications, and nitrogen and phosphorus have given better results in trials, but the low selling price of cassava has not encouraged fertiliser use.

A general recommendation for soils that are moderately deficient in both phosphorus and potassium would be to use an $N:P_2O_5:K_2O$ ratio of about 1:1:2 unless the soil was known to be well supplied with one of these nutrients. When other factors are not limiting, high yields of up to 40–50 t/ha can be obtained with applications of 40–80 kg N/ha, 40–60 kg P_2O_5/ha and 80–150 kg K_2O/ha.

Yams (*Dioscorea* spp.)

Edible yams are grown for their starchy tubers and are widely distributed in the tropics. The main concentration of yams is in West Africa, about two-thirds of the world's yams being grown in the so-called 'yam belt' which extends across West Africa between about 4° N and 10° N through the Ivory Coast, Ghana, Togoland, Dahomey and Nigeria to Cameroon. Yams are also grown in the Caribbean, particularly in Jamaica, in South-east Asia, including Papua New Guinea, and throughout the tropical Pacific. In many areas yam production is declining in favour of other food crops that need less labour.

The main edible yams belong to about ten species. The most important are *Dioscorea alata*, the water yam, greater yam or winged yam, *Dioscorea rotundata*, the white yam, and *Dioscorea cayenensis*, the yellow yam. The first is a native of Asia, and is the main species now grown in Asia and the Pacific. The other two are natives of West Africa, and *D. rotundata* is the main yam of the West African yam belt. Many wild species are collected for food and for diosgenins, which are used in the pharmaceutical industry.

Climatic requirements Most edible yams require high temperatures, between 25 and 30 °C, and since within this range growth increases with temperature, 30 °C is considered optimum. Below 20 °C growth is poor except for some minor varieties that are found further away from the tropics or at higher altitudes.

Rainfall of 1000 mm a year is thought to be minimal, and 1500 mm well spread over all least nine months is better. *D. alata* is said to need a higher rainfall than *D. rotundata*. In the yam belt of West Africa yams are found in the northern forest and southern savanna areas. Both areas have marked dry seasons, which grow longer in the savanna areas with increasing latitude. Yams are traditionally planted before the rains begin, and harvested at the end of the rains or at the beginning of the dry season, though some early-maturing varieties are harvested during the wet season and others are harvested twice, the second harvest being for seed yams. Most yams in West Africa have a growing season of 7–8 months, and they are generally grown in mounds at about 10 000 or more to the hectare.

Soil requirements The best soils for yams are friable, deep, well drained and free of stones. A friable soil is needed to allow the expanding yam tuber to penetrate the soil. Good drainage is necessary because waterlogging causes the tuber to rot. In West Africa yams are grown on individual yam mounds generally 50–100 cm or more high, made by pulling up the generally friable topsoil with a hoe. The yam mound increases the effective rooting depth and drainage, so that the soil in the mound is not waterlogged even during intense downpours, and the working of the soil increases its friability. A piece of sod or some mulch is often placed on the top of the mound to reduce temperature and moisture losses. In mechanised cultivation, yams have been grown on ridges.

Most yam varieties require a fairly fertile soil, and wherever practicable traditional varieties in West Africa are planted immediately after clearing, at the head of the rotation. For maximum yields they need a stake to climb up. In the central area of Ivory Coast, newly introduced heavy-yielding *D. alata* varieties, including the 'Florido', have recently been found to be less demanding as regards soil fertility and these can be planted later in the rotation, even after another yam crop, and the stakes dispensed with.

Yams are thought to be sensitive to acid soils with free aluminium.

Yam responses to fertiliser The main supplementary nutrient need for yams in West Africa appears to be nitrogen, though nutrient needs vary with the species. There is often a lower rate than usual of nitrogen mineralisation in West Africa on clearing a grass fallow in which *Andropogon* species occur. Although various trials have shown profitable yam responses to nitrogen applied as sulphate of ammonia, very little fertiliser is at present applied to yams in West Africa. Farmyard manure and composts have traditionally been used.

In Ghana, recommended dressings are 15–30 kg N/ha. In the derived savanna zone of western Nigeria, on sandy soils derived from granites and gneisses, applications of 56 kg N/ha resulted in large yield increases of 3000 kg/ha, equivalent to over 50 kg of tubers for each kg nitrogen. In 1973 a three-year study of yam responses near Tamale, northern Ghana, showed that the main responses of *D. rotundata* were to nitrogen, which increased yields by 22% at 67 kg N/ha, and to phosphorus, which increased yields by 6% at the low rate of 15 kg P/ha. No responses to potassium were obtained in the Tamale trials, but other trials at Ibadan, Nigeria, showed very substantial responses to this element at 75 kg K/ha. Nitrogen applications in Trinidad got responses only when they were delayed until three months after planting.

Phosphate needs of the yam appear to be only moderate, and uptake is sometimes helped by mycorrhizal infection of the roots. *D. rotundata, D. alata* and *D. esculenta* have all been shown to be efficient at taking up phosphate from low levels of soil solution phosphorus. In West Africa, responses to phosphorus have been obtained, but have generally been negligible or even negative.

Cocoyam (*Xanthosoma sagittifolium* and *Colocasia esculenta*)

The terms 'cocoyam', 'dasheen', 'taro' and 'tannia' are applied to various members of the Araceae which produce edible corms. The two most widely grown are *Colocasia esculenta* and *Xanthosoma sagittifolium*.

Colocasia esculenta originated in Asia. It has a large central corm which is relatively soft when cooked, and is found mainly in poorly drained areas. Known as 'old cocoyam', 'dasheen', 'eddo' or 'taro', *Colocasia* is widely grown in the Pacific and is also popular in the West Indies.

Xanthosoma sagittifolium originated in tropical Central and South America. It produces numerous subsidiary flask-shaped corms, is firmer when cooked than *Colocasia*, and is grown mainly on better-drained soils. It is the preferred cocoyam in much of West Africa and is widely grown in the forest zone.

Climatic requirements Cocoyams are widespread throughout the tropics, and different species and varieties appear to be adapted to a range of climates and altitudes. Moist conditions with rainfalls up to 2500 mm are best, and temperatures of 21–27 °C. In some drier areas the crop is irrigated.

Cocoyam soil requirements and fertiliser responses Cocoyams need a rich soil to give good yields. Deep alluvial loams are well suited to the crop. The optimum pH range is 5.5–6.5.

In Ghana, the crop yields well and is widely grown in the cocoa belt, growing best on non-gravelly soils cleared from high forest. Fertilisers are not applied to the crop.

On less good soils elsewhere, responses have been obtained to nitrogen and, to a lesser extent, to phosphorus and potassium. In Madhya Pradesh, India, the economically optimum application to *Colocasia* proved to be 120 kg N/ha. On acid, highly weathered soils in Puerto Rico, *Xanthosoma* responded to 56 kg N/ha but not to other nutrients. In Hawaii and in Eygpt heavy NPK dressings are recommended for *Colocasia*.

Sweet potato (*Ipomea batatas*)

Sweet potatoes (also known as 'yams' in the United States) are very widely grown in tropical, subtropical and warm temperate areas between 40° N and 32° S, particularly in Asia, and a wide range of cultivars are suited to different soils and climates.

The sweet potato originated in Central or South America, and is a perennial that is generally cultivated as an annual. The first harvest is usually after six months, and the 'tubers' (technically secondary thickened roots) are harvested progressively thereafter.

Climatic requirements Best growth is obtained with temperatures above 24 °C, abundant sunshine and warm nights. Lower temperatures result in slower growth. Annual rainfalls of 750–1000 mm are considered most suitable, with a minimum of 500 mm in the growing season. The crop is sensitiv to drought at the tuber initiation stage 50–60 days after planting. In some areas the crop is irrigated.

Soil requirements Sweet potatoes are grown on a variety of soils, but well-drained light and medium-textured soils with a pH range of 5.0–7.0 are considered best. More acid soils are sometimes limed. Gravelly soils are not very suitable, and heavy clay soils may given low yields of inferior quality, but soils with a high clay content that are friable because of stable micro-aggregation, such as well-aggregated Oxisols, are very favourable. The plant is not tolerant of waterlogging, which may cause tuber rot, and it suffers if aeration is poor.

The sweet potato can grow on soils with very low available phosphate levels. This appears to be due both to the fact that it has a relatively low phosphorus requirement, and that it is a good extractor of phosphorus from soils deficient in this element. Potassium requirements, like those of cassava, are relatively high. The plant is often thought of as growing well on poor, light-textured soils.

The plant quickly develops a complete ground cover formed of a mat of vines, and protects the soil from erosion during cropping.

Sweet potato fertiliser responses A yield of 15 t/ha has been estimated to remove about 70 kg N/ha, 20 kg P_2O_5 and 110 kg K/ha. These amounts are relatively low, especially for phosphorus. Fertiliser experiments have generally confirmed that the phosphorus requirement of sweet potato is low, although yield increases have occasionally been reported.

Nitrogen applications have given variable results. In some experiments they have depressed yields, or merely increased foliage growth without increasing tuber production. On soils that are low in nitrogen, moderate applications have increased yields.

A potassium uptake that is high relative to nitrogen and phosphorus can be associated with the finding that the greatest tuber enlargement is found when the nitrogen/potassium ratio in the fertiliser is low. Responses to potassium are frequent and occasionally high.

In Taiwan, nitrogen applications at rates of 40 kg/ha have generally proved profitable. In Zimbabwe, nitrogen at 67 kg/ha is recommended. In India, recommendations differ between states, generally for a mixed NPK fertiliser with usually in the ranges 56–62 kg N/ha, P_2O_5/ha and 40–50 kg K_2O/ha, though one set of trials resulted in yields being depressed when nitrogen exceeded 56 kg/ha. In São Paulo, Brazil, applications of about 50 kg/ha each of nitrogen, P_2O_5 and K_2O have been advised.

Fibre crops: cotton and jute

Cotton (*Gossypium* spp.)

Cotton includes several species of perennial shrubs which are often grown as annual subshrubs. Commercial cotton cultivars are classified according to the length of the lint, though this is now less important than it was. Commercial cotton production extends well inside the temperate zone. A relatively small proportion of total world production comes from tropical countries, including India, Uganda, Tanzania, the Ivory Coast, Nigeria and Sudan.

Climatic requirements Cotton needs a temperature of 21 °C or more at sowing, and an average growing season temperature of 25–30 °C for optimum growth. It is a sun-loving plant, intolerant of shade.

Rainfed cotton is grown mainly in areas with 1000–1500 mm of rain. In dry areas it is often irrigated. Cotton can withstand a certain amount of rainfall irregularity and dry spells. Ideally conditions should be moderately wet for early vegetative growth, then relatively dry for the first flowering period, then wetter for boll swelling and renewed growth, and finally dry for ripening and harvest. Rain during flowering causes boll shedding. Because of the adverse effects of rain at the wrong time, the highest cotton yields are obtained with carefully controlled irrigation in dry areas.

Soil requirements Cotton is grown on soils with a very wide range of textures from light sandy soils to heavy cracking clays. The Vertisols of the tropics are often referred to as 'black cotton soils' because of their widespread use for cotton in India, but yields are generally better on more friable soils, particularly loams and clay soils. Deep friable soils allow the deep tap root to develop freely, which it cannot do if there is a hard layer in the soil or a heavy concentration of gravel. Cotton is sensitive to waterlogging, and also to aluminium toxicity when grown on acid soils below pH 5.0–5.5. The optimum pH range for most varieties of the crop is 6.5–7.5. If acid soils are effectively limed only in the surface layer, roots may be largely confined to the limed layer.

Cotton fertiliser responses The amounts of nutrients removed in the fibre are very small indeed, being only about 1 kg nitrogen, 0.2 kg phosphorus and 3.0 kg potassium per tonne, but more is removed in the seed, which makes up about two-thirds of the seed cotton weight. A yield of 500 kg of fibre and 1 t of seed have been estimated to remove about 35 kg of nitrogen, 6 kg phosphorus, 13 kg potassium and 4 kg magnesium.

Leaf analysis has proved very useful in monitoring cotton nutrition and as a guide to fertiliser use.

Nitrogen is usually the most important nutrient needed to increase cotton yields, and nitrogen is used on cotton in much greater quantity than any other nutrient, though in very variable amounts depending on local soil and other factors. On lighter soils, applications are often split to reduce leaching losses. If the nitrogen flush with the first rains is important, nitrogen may not be recommended until flowering. Alternatively a mixed fertiliser may be incorporated at planting and additional nitrogen top-dressed later, though to avoid seedling injury fertilisers containing nitrogen applied at sowing should not come into contact with the seed. Banding the fertiliser about 7–8 cm to the side of the seeds and about 5 cm below them is often recommended.

Nitrogen in excess, especially if applied early, causes increased vegetative growth and delays flowering, but when insufficient it leads to boll shedding.

Fertiliser responses depend on adequate plant protection from insects, without which yields can be very low, and on adequate and timely weeding.

Yield responses to nitrogen in Africa and in India in particular depend partly on correct planting dates. Planting at the beginning of the rains is very important and yields fall off if planting is delayed. In trials in Uganda, Kenya and Tanzania, a four-week delay in sowing reduced yields to an average of 60% of optimum sowing date yields, thus causing a reduction averaging 40%.

On the Tana River scheme in eastern Kenya

responses to 250 kg/ha of ammonium sulphate (52 kg/ha of nitrogen) fell from 525 kg/ha for the earliest sowing on 15–21 March to only an uneconomic 109 kg/ha for the last sowing in mid-April. Experiments in Uttar Pradesh, India, found that yields were reduced by about 10 kg/ha for every day that sowing was delayed between 15 March and 15 May, because earlier sowing increased the life-span and thus the yield.

Phosphorus promotes early growth and root development, and early maturity. Phosphorus responses on cotton soils have been patchy but the element is frequently applied with nitrogen at planting, often with an $N:P_2O_5$ ratio of 1:2, or an $N:P_2O_5:K_2O$ ratio of 1:2:1.

Potassium applications of 40–120 kg K_2O/ha are sometimes recommended on soils known to be potassium deficient, but in most tropical cotton-growing areas, present indications are that potassium in most cases is less likely to give responses than nitrogen or phosphorus.

Fig. 15.5 Harvesting cotton on an African peasant farm. For adequate yields, frequent insecticide sprays are usually needed and moderate fertiliser dressings which include nitrogen, sulphur and sometimes boron are widely recommended.

In many part of the tropics, particularly in savanna areas, cotton has responded to sulphur. Sulphur is normally supplied either as ammonium sulphate or single super. Boron applications have been found to give responses in West Africa and elsewhere.

In India, which has the largest area under cotton in the world, cotton is grown on a wide range of soils, though average yields are low. Rainfed cotton yields are often limited by water shortages. The main cotton-growing areas and soils are the alluvial soils of the Indo-Gangetic plain, the Vertisols (black cotton soils) of central India, the medium black soils of the Deccan Plateau, and red lateritic soils of eastern and southern India. Four-fifths of the total area is rainfed, and sown mostly at the beginning of the monsoon rains.

Fertiliser recommendations for cotton in India vary considerably from area to area, but generally emphasise nitrogen and phosphorus, potassium being applied only in certain areas of known deficiency. Nitrogen recommendations are generally of the order of 30–50 kg/ha for rainfed cotton, but about double this for irrigated cotton. Phosphate recommendations are often in the range 25–40 kg P_2O_5/ha for rainfed cotton, increasing to 45–60 kg P_2O_5/ha under irrigation. In the few areas where potassium is known to be needed, recommended application rates are moderate, at 20–60 kg/K_2O/ha. Farmyard manure is widely used and recommended, especially as a basal dressing with some phosphate followed by a moderate top-dressing of nitrogen. For fertilisers, a standard recommendation is to place the phosphorus and half the nitrogen at planting and to top-dress the remaining nitrogen later.

In Africa, cotton is grown widely in francophone West Africa and in Nigeria. It is grown in three separate climatic zones: the Sudano-Sahelian zone, with a single wet season of less than 1000 mm, the Sudan savanna zone, with a single wet season of more than 1000 mm, and the southern Guinea savanna with two wet seasons in summer separated by a short drier period in July–August. In the last zone a common practice is to grow food crops, particularly maize in the first wet season followed by cotton in the second. In the areas with a single wet season, early planting is essential for good yields, as discussed above.

In West Africa in general the nutrients most applied to cotton are nitrogen and phosphorus, often with sulphur and sometimes with boron.

On savanna soils in the Ivory Coast there is often a deficiency of nitrogen and sulphur in the first year, but with further cultivation the deficiencies are mainly nitrogen and phosphorus. Standard fertiliser recommendations for cotton are 50–100 kg/ha of a 10:18:18 mixture also containing sulphur and boron at planting, followed by 50 kg/ha of urea later in the season. Residual effects are usually obtained on food crops grown the following season or year.

In northern Nigeria, farmers who maintain good management standards for cotton are recommended to use 125 kg/ha each of ammonium sulphate and single superphosphate. At Samaru, good responses on cotton were obtained to trace element applications, mainly to molybdenum. In Western Nigeria nitrogen applications to cotton were found to be profitable, but phosphate responses were variable and generally unprofitable.

In East Africa cotton is widely grown. In the lake region of western Tanzania the standard recommendation is 125 kg/ha of ammonium sulphate and the same amount of single superphosphate a year. These applications are said to increase average yields from 700–800 kg/ha seed cotton to 1100–1200 kg/ha. Residual responses to the phosphate are obtained. Farmyard manure and compost are recommended, and combined with fertiliser applications have enabled good yields of up to 2240 kg/ha seed cotton to be obtained after 15 years of continuous cultivation. In the northern and eastern areas of Uganda the same general recommendation of 125 kg/ha each of ammonium sulphate and single superphosphate is reported to give average yield responses of about 250 kg/ha seed cotton, but early planting is essential.

In Malawi, nitrogen alone is recommended on alluvial soils, and both nitrogen and phosphorus on basement complex soils. Sulphur deficiency was noted at the cotton research station at Makoka, and boron deficiency in the central region.

In the irrigated Gezira scheme in the Sudan, on heavy black clays (Vertisols), cotton is the cash crop in a cotton–sorghum–*Dolichos lablab* rotation. Continuous cotton with 120 kg N/ha has given yields of 1700–1800 kg/ha seed cotton, but a *Dolichos*–fallow–cotton rotation gave higher yields of 2700–2800 kg/ha seed cotton without fertiliser, thus demonstrating the effects of the fallow on nitrogen nutrition. Where fertilisers are used they are generally confined to nitrogen, at recommended rates of 90 kg/ha.

In Brazil, cotton is grown mainly in the south where standard fertiliser practice is to apply phosphorus and potassium with a little nitrogen at planting, and the rest of the nitrogen later in the season. Recommended total rates per year are 30–45 kg N/ha, 30–90 kg P_2O_5/ha and 30–70 kg K_2O/ha. The heavier phosphate applications are for soils newly brought into cultivation, and the lower ones for soils that have been previously fertilised with phosphate and in which some reserve has been built up. The heavier potassium applications are recommended on soils that were acid but which have been limed for cotton production. Sulphur is thought to be widely deficient.

In Colombia, most cotton soils are mainly in need of nitrogen. In northern Tolima phosphate is also required, and in central Tolima good responses were obtained to 30 kg K_2O/ha in combination with with 60 kg N/ha.

Jute (*Corchorus* spp.)

Jute produces a fibre that is second in importance to cotton. The main species grown is *Corchorus capsularis*, with *C. olitorius* being of subordinate importance. Jute is made into twine, and woven into coarse cloth for sacks and other uses. About 90% of the world's jute is grown in India and Bangladesh, the main growing area being the Ganges–Brahmaputra delta. Minor quantities are grown on alluvial soils in the Amazon floodplain in Brazil and elsewhere.

Climatic requirements Jute is grown in the tropics at low altitudes with temperatures of 24–35 °C, and rainfalls of over 1000 mm a year, preferably concentrated in the latter part of the growing season, which is 100–135 days. Young jute is sensitive to waterlogging, but mature *C. capsularis* can withstand deep flooding.

Soil requirements Jute grows best on fertile, loamy-textured alluvial soils that are well drained and well aerated. Heavy clay soils, very sandy soils and alkaline soils are all relatively unsuitable.

The main jute soils in India and Bangladesh are the Brahmaputra alluvium and the old red alluvium. The former is well supplied with potassium and has adequate phosphorus; it benefits from annual silt deposits. The latter soil is more acid, low in phosphorus and moderately low in available potassium.

Responses of jute to fertiliser On alluvial soils enriched by annual silt deposits, jute often receives little or no fertiliser. Elsewhere, farmyard manures, wood ash and moderate applications of nitrogen (20–40 kg K/ha) have been used to increase yields.

Trials in India and Bangladesh have mainly emphasised the need for nitrogen. At the Jute Agricultural Research Institute, Barrackpore, 80 kg/ha of nitrogen increased yields by an average of 80%. The general recommendation is 40 kg N/ha for *C. olitorius* and 60 kg N/ha for *C. capsularis* on the Brahmaputra alluvium, with these amounts increased by 50% for the less fertile old red alluvium soils. Phosphorus, at 20–40 kg P_2O_5/ha, and potassium, at 20–60 kg K/ha, are recommended for soils known to be deficient in these elements, including some of the old red alluvium soils.

Sugarcane

The term 'sugarcane' (often also written as two words, 'sugar cane') refers to six species of grasses of the Andropogon tribe which originated in Asia. The most important commercial sugarcanes are cultivars of *Sacharum officinarum*, known as 'the noble cane' because it has thicker stems than other species. Originally grown for chewing, the making of sugar from the cane was discovered in India 2–3000 years ago. Sugarcane has been much improved by breeding from seed since 1888. The most important tropical sugarcane areas are India, the West Indies and Guyana, and Indonesia, but sugarcane is widespread in other parts of the tropics.

Climatic requirements For good growth and production sugarcane requires high temperatures and high levels of sunshine. The optimum temperature for growth is between 25 and 28 °C, but for the germination of the stem cuttings or sets from which cane is grown, somewhat higher temperatures of 32–38 °C are said to be most favourable. Cane growth slows or ceases below 21 °C. A relatively cool dry period is favourable for the final ripening period, when the sucrose content of the juice goes up prior to harvesting. Sugarcane is easily damaged by frost.

Sugarcane is a sun-loving plant. It fixes carbon by the C4 pathway and, depending on the variety, photosynthesis increases with increasing light intensity over a wide range.

For maximum production and fertiliser use efficiency, a good moisture supply is essential. An average rainfall of 1500 mm a year, well distributed, is regarded as the minimum. In areas where rainfall is adequate in total but unreliably distributed, supplementary irrigation to eliminate the harmful effects of dry spells has improved fertiliser responses and given a greater control over production. In some areas a fairly linear relationship has been found between dry matter production and the amount of water used, averaging about 125 g water to 1 g dry matter. Provided that yields are not limited by periods of water stress or other factors, cane yields are about 2–4000 kg/ha for each 2.5 cm of water used in evapotranspiration.

The climatic requirements of sugarcane vary markedly according to variety. In Hawaii cane is grown from below sea level to altitudes of over 1000 m, and in areas with rainfalls from very low to over 7500 mm and with a wide range of cloud cover and solar radiation. Over a dozen varieties are needed for the main ecological zones in which cane is grown on the island.

Most sugarcane is grown for 14 to 18 months for the initial crop, known as the plant crop, and for about 12 months for each of the two to four ratoon crops that normally succeed the plant crop in the production cycle.

Soil requirements Sugarcane is grown successfully on a very wide range of soils, from light sands to heavy, cracking clays. In general, medium to heavy-textured soils are considered best. Very heavy clays can be improved by using filter mud and cane trash. On heavy, poorly drained soils in high rainfall areas, ditches or cambered beds are used to improve soil drainage and aeration.

Sugarcane can tolerate acid soils and low levels of calcium, though some responses to liming have been reported. Filter mud, sometimes applied to cane soils, contains 20 kg $CaCO_3$/t. The usual pH range for cane is given as 5.0–8.0 or over. Below pH 4.5–5.0, aluminium toxicity may occur. In Guyana it has been reported that exchangeable aluminium should be less than 60% of total exchangeable bases throughout the root zone if good cane growth is to be obtained.

Sugarcane is often grown on the same soils year after year. The fibrous root system has a beneficial effect on soil structure.

Sugarcane nutrient uptake and deficiency symptoms Sugarcane has a high nutrient demand. Nutrient removal by crops of 28, 74 and 100 t/ha has been reported as follows:

Cane yield (kg/ha)	Nutrients removed (kg/ha)		
	N	P_2O_5	K_2O
28 000	30	20	60
74 000	107	60	300
100 000	200	85	420

These separate investigations all show a high potassium uptake in relation to nitrogen and phosphorus.

Leaf analysis, at frequent intervals during the first 18 months of growth, is widely used to monitor sugarcane nutrition and to suggest fertiliser applications. The leaves, the leaf sheaths and sometimes the stem are used.

The best-known deficiency symptoms are as follows:

Nitrogen: yellow-green colour of all leaves, drying and dying of old leaves, slower growth and thinner canes.
Phosphorus: shorter, thinner canes, shorter internodes, little tillering; leaves are narrow and greenish-blue, and older leaves dry and die at tips and margins.
Potassium: suppressed growth, yellowing and spotting of older leaves which first become orange-yellow with chlorotic spots and then brown with necrotic centres before dying back from tips and margins, though young leaves remain green.

Sugarcane responses to fertilisers The use of manures and fertilisers on sugarcane varies very considerably with the cultivar, the climate, the soil, whether the crop is irrigated or rainfed, and economic conditions.

Formerly, commercial estates used large quantities of animal manures. Chemical fertilisers have largely taken their place, and yields have steadily increased. For fertiliser use efficiency a good water supply is necessary, and weeds must be kept down.

Sugarcane has a high nutrient demand in the early stages of growth and fertilisers should be applied at planting or soon after.

Nitrogen is particularly important during early growth. Too much nitrogen, or nitrogen applied too late in the growing season, results in delayed crop maturity and ripening and sometimes in a lowering of juice quality and sucrose content. In the ripening period before harvest it is necessary that nitrogen in the plant is low enough to reduce the rate of vegetative growth and force the conversion of reducing sugars to sucrose.

Optimum nitrogen application rates are usually in the range 50–220 kg/ha, applied soon after germination or soon after cutting in the case of a ratoon crop, or split partly at planting and partly when the crop begins to close in. Some varieties respond to higher application rates. Responses by first ratoon crops are frequently better than those obtained from the preceding plant cane, so that optimum ratoon application rates are often considered to be about 50 kg N/ha above those for the plant crop.

With sandy soils in high rainfall areas where nitrogen leaching losses can be expected, it may be preferable to apply nitrogen as a top-dressing within three months rather than in the furrow at planting. In Guyana, 70–90 kg N/ha are applied within two weeks of planting or ratooning, with the subsequent applications being based on foliar analysis.

The amount of nitrogen that it is profitable to apply in the form of fertiliser is influenced by the amount released by mineralisation of soil organic matter (Chapter 9). Under the plant cane, nitrogen mineralisation from organic matter may be higher than under the succeeding ratoon crops, when the rate may stabilise at a lower level.

The average sugar yield increases obtained with a standard application of 56 kg N/ha in 46 trials in 8 Commonwealth countries are summarised in Table 15.1.

Fig. 15.6 Sugarcane trials carried out by the Mauritius Sugar Research Institute, showing plots treated with NP (above) and NPK (below).

Table 15.1 Average sugar yield increases in 46 trials in 8 Commonwealth countries

Country	Yield increase (kg sugar/ha) from 56 kg N/ha		
	Plant cane	First ratoon	Second ratoon
Jamaica – rainfed	890	1267	1054
– irrigated	1317	1655	1969
St Kitts	439	2860	–
Antigua	334	1267	–
Barbados	1104	2108	–
Trinidad	1355	1104	1305
Mauritius – rainfed	477	1317	1332
– irrigated	838	2308	3023
South Africa	389	979	1305
Queensland, Australia	702	1831	2584
Average	784	1670	1796
Average per kg N/ha	16	33	36

Phosphorus requirements of sugarcane appear to vary widely between varieties, but in general the responses to nitrogen described in the previous paragraphs can be expected only where phosphorus levels in the soil are satisfactory. Phosphorus encourages root growth, early and vigorous tillering, improved cane height and thickness, and early ripening. Responses to phosphatic fertilisers have been very variable, depending on the phosphorus status of the soil and on the extent to which fertiliser phosphorus is fixed in the soil (Chapter 10).

Phosphatic fertilisers are normally placed in the furrow at planting. On soils with a high phosphorus fixation capacity, better results may be obtained by mixing the phosphatic fertiliser with farmyard manure or with compost.

Recommended application rates in the West Indies vary between 125 and 500 kg/ha single superphosphate (equivalent to 22–90 kg P_2O_5/ha) at planting for plant cane to about 125–250 kg/ha single superphosphate (22–45 kg P_2O_5/ha) for ratoon cane. In Kenya, responses to phosphate applied to sugarcane are usually low, but 65 kg P_2O_5/ha are used in some estates. In South Africa the recommended dressing is 675 kg/ha single superphosphate (120 kg P_2O_5/ha) in the furrow at planting and one-third of this for each ratoon crop applied on top of the trash layer. In Hawaii, where fertiliser application rates are heavy compared with most sugarcane production areas, phosphate is applied when leaf analyses indicate a need for this nutrient at rates of 112–448 kg P_2O_5/ha.

Potassium requirements of sugarcane are high relative to other nutrients. Potassium is needed for carbohydrate synthesis, and therefore for sucrose production. The potassium in the cane juice is left in the molasses, which contain 3–7% K_2O, though these are not normally returned to the field.

Recommended application rates for muriate of potash (60% KCl) are usually in the range 60–375 kg/ha, equivalent to 36–225 kg K_2O/ha, with recommended rates of 100–200 kg K_2O/ha being the most frequent.

Trace element deficiencies have been reported in many areas. Copper deficiency, causing 'droopy top', has been reported in Australia and South Africa. Iron deficiencies are found, usually on very calcareous soils or induced by lime applications, and are usually cured with chelates. Manganese deficiency has been reported in Hawaii, and zinc deficiency in South Africa and on limestone soils in the Bahamas.

Silicon, although not an essential element, has sometimes given yield responses when applied to sugarcane (Chapter 10).

Groundnuts

The groundnut (*Arachis hypogaea*) originated in South America and was widely distributed there and in the West Indies at the time of the discovery of America in the late fifteenth century. In the sixteenth century groundnuts were taken to Africa, South, South-east and East Asia, and the Pacific, and they are now grown virtually throughout the tropics and subtropics. After soya beans, they are the second largest source of vegetable oil. Large quantities are grown in India, West Africa and South America.

Climatic requirements Groundnuts are grown over a wide range of latitude, from 40° N to 40° S, but it is a warm-season crop, which is killed by frost. During the growing season, usually 3½ to 5 months, a rainfall of at least 500 mm is needed. Adequate moisture is particularly important between flower initiation and flowering, but dry conditions are needed for ripening and harvesting. Under heavy, continuous rainfall, vegetative growth increases, resulting in low yields and longer growing periods.

Soil requirements Light-textured soils are best suited to groundnuts, particularly loose, friable sandy loams that are well supplied with calcium and not low in organic matter. Soils must be well aerated and well drained, and should not harden when dry or form crusts, as this hinders the penetration into the ground of the pegs on which the nuts form. Harvesting is easiest in loose friable soils but becomes more difficult in heavy soils that adhere to the nuts and form hard clods when dry.

Groundnut responses to fertiliser Although groundnuts are generally grown on sandy soils of only moderate to low fertility, fertiliser responses are often low and unpredictable. When grown in rotation, best results have sometimes been obtained by applying fertilisers to the preceding crop, with the groundnuts benefiting from a residual effect.

The standard recommendation for groundnuts in West Africa is single superphosphate, since it supplies three elements needed by the crop – phosphorus, calcium and sulphur. The crop has a relatively high demand for both sulphur and calcium, which increase nodulation. Calcium is necessary for large healthy kernels, and calcium deficiency leads to 'pops', i.e. to empty or poorly filled pods. Calcium

Fig. 15.7 Groundnut fertiliser trials.

appears to be somewhat immobile in the plant, but additional calcium is taken up directly from the soil by the gynophores and young pods.

Although the nitrogen supply of groundnuts is normally augmented by symbiotic *Rhizobium* in the root nodules, nodulation is not effective until three weeks after germination, so the plant does not benefit from nitrogen from this source during this early period. Although nitrogen is released by the microbial flush which often occurs at the beginning of rains, responses to a modest early nitrogen application are frequent.

In Africa, responses to potassium have been uncommon but sulphur deficiencies are widespread, particularly in the savanna areas where the groundnut is grown most extensively. Sulphur in good supply has been credited with causing the pods to be more firmly attached to the plant, so that fewer are left in the soil at harvesting. Responses have been obtained in West Africa to molybdenum, and in India to boron.

In India, the world's largest producer of groundnuts, the standard fertiliser recommendations vary somewhat between states, but are mainly within the ranges 10–30 kg N/ha, 20–60 kg P_2O_5/ha and 0–60 kg K_2O/ha. In some areas a basal dressing of farmyard manure is also recommended.

In West Africa, the largest producers in the past were Nigeria and Senegal, using 90–120 day varieties, but of recent years production has fluctuated. Most groundnut soils in these areas require phosphorus and sulphur, and potassium is deficient in parts of Niger and Senegal. Nitrogen deficiencies occur mainly at the beginning of the cultivation cycle immediately after a grass fallow has been cleared, and on soils that are particularly low in organic matter.

In Nigeria the usual fertiliser recommendation is 60–120 kg/ha single superphosphate. This is conveniently placed in the furrows, and the old ridges than split to make new ones above the placed fertiliser.

In Senegal, sulphur is needed in the central and southern groundnut growing areas, where there is a strong sulphur–phosphorous interaction when sulphur is supplied at 5 kg/ha and dicalcium phosphate at 30 kg/ha. Sulphur–farmyard manure interactions are also obtained.

The standard fertiliser recommendations in Senegal for groundnuts are in the ranges 40–70 kg/ha ammonium sulphate, 0–60 kg/ha dicalcium phosphate and 0–50 kg/ha muriate of potash (60% KCl). These are best applied as a side-dressing rather than broadcast. These recommendations have produced average yield increases of 520 kg/ha in the southern region and 310 kg/ha in the drier central region, and also gave a residual effect, average 15%, on the succeeding millet crop. In some districts, particularly in the Louga area, molybdenum has given responses averaging 80 kg/ha when included in the seed disinfectant at the rate of 15–20 g/ha. Foliar diagnosis has proved useful for groundnuts in Senegal, using the fourth or fifth leaf of the main stem collected 40–45 days after planting.

In the Gambia (now part of Senegal), large responses to single superphosphate have been obtained on sandy loam soils at Yundum agricultural station and elsewhere, the amount of the response depending on soil potassium levels. Potassium needs may be supplied to some extent by grass burning before sowing. A general recommendation for groundnuts in Gambia is 170 kg/ha of an 8–12–8 compound fertiliser.

In Madagascar, phosphorus was also found to be the main nutrient needed. A dressing of 75 kg P_2O_5/ha approximately doubled yields.

16 Some tropical crops in relation to soils and fertiliser use: perennials

General aspects of fertiliser use on perennials

Perennial crops include pastures, considered in Chapter 21, a wide range of tree crops, of which six are considered in this chapter, and crops like sisal which are grown in cycles of several years.

The difference between perennials and annuals is not always clearcut, since some shrubby perennials such as cotton and pigeon pea are treated as annuals. The life of perennials such as tea and coffee, on the other hand, is very long and somewhat indeterminate. Some tea and coffee estates are 100 years old or more and still growing and yielding well, and in Ghana some cocoa farms are over 80 years old and still bearing well. In practice perennials such as tea, coffee and rubber may be uprooted and replaced not because they have ceased production but in order to replant with newer, higher-yielding clones. Oilpalms and coconuts may be replaced after about 30 years partly for this reason and partly because old palms grow so tall that harvesting becomes difficult.

The growth cycle of an annual crop is a matter of months – as little as two months in the case of short-season millets. In tropical areas with seasonal rainfall the annual crop is fitted into the wet season and typically is harvested before the end of the rains or at the beginning of the dry season. The climate during the rest of the year has no direct effect on the crop. For perennial crops, on the other hand, the climate all the year has to be considered. Also, exceptional events which may occur only in occasional years or at long intervals, such as frosts on coffee in Brazil, or years of quite exceptionally low rainfall in West Africa, are of importance and may reduce yields or kill the plant.

A perennial tree crop is a long-term investment, and as such merits even more careful planning and preparation than an annual. The grower may go to great pains not only to select the planting material, but also to dig planting holes and furnish them with topsoil, compost, manure and additional fertilisers. Once established, however, a tree crop may need less labour than a succession of annual crops.

Fertiliser use on tree crops can typically be considered in four stages:

1 in the nursery,
2 at establishment in the field,
3 for young trees before they come into bearing, and
4 for mature trees in full bearing.

Fertiliser requirements may differ markedly between different stages. In the nursery, relatively generous fertiliser dressings may be used, sometimes with mulches. At establishment in the field, manure and fertilisers are usually put in the planting hole. The fertiliser recommendations thereafter for young trees before bearing may differ substantially from those for trees in full bearing, which are actively exporting nutrients from the soil. For coconut palms, for example, nitrogen is emphasised for palms before bearing, but potassium is the main or only need when they are in production and vegetative growth is slower. For young oilpalms nitrogen is also the main nutrient, but phosphorus may receive most attention once the palms begin to yield.

Although leaf analysis as a tool for diagnosis of nutrient deficiencies is widely used for some annual cash crops such as cotton and sugarcane, it has proved particularly appropriate and valuable for perennial tree crops such as coffee, tea, oilpalm and coconuts. In the case of the palms, the continuous leaf production of monocotyledons with a single growing point makes the selection of leaves with the same degree of development easier and the results more reliable. Deficiencies can usually be detected at an early stage and corrected before they have an effect on production.

In general, the sometimes spectacular increase in average yields of tropical tree crops over recent decades has put greater demands on the soil, increased the need for additional nutrients, and shown up deficiencies in soils that formerly appeared adequate at lower yield levels. Increases in yield levels have been partly due to better management and husbandry practices, and to the partial or complete removal of shade for such crops as tea, coffee and cocoa, but are in the main the result of the breeding and selection of higher-yielding varieties and hybrids. The perennial tree crops considered in this chapter are all cash crops, partly or largely grown for export, and often but by no means always produced on fairly large estates. They have benefited from a very great deal of research world wide and this is reflected in, among other things, the very large amount of information available on fertiliser use and responses. It is lamentable that, by comparison, there is so little information available on the many annual subsistence food crops that are grown mainly by the small farmer for his own consumption.

Tea (*Camellia sinensis:* synonym *Thea sinensis*)

Tea originated in South-east Asia and has been used as a beverage for 2–3000 years but its use did not become general until the eighteenth century. Large tea plantations were established in India, Indonesia and Sri Lanka in the second half of the nineteenth century, and in East Africa at the beginning of the twentieth century.

There are two main groups of tea cultivars, the China teas (*Camellia sinensis* var. *sinensis*) and the Assam teas (*Camellia sinensis* var. *assamica*). The first group are slow-growing, dwarf trees that are resistant to cold and adverse conditions but low yielding. The second group are quick-growing, taller trees, well adapted to tropical conditions, and these form the main commercial teas. However, hybrids do occur, and the crop is very heterogenous because in the past seedling tea rather than clonal tea was used. Most new tea is now clonal.

Climatic requirements Tea requires fairly constant temperatures, moderate to high well-distributed rainfall for most of the year, and high humidity. It does not tolerate frost. Mature tea is normally considered to require 100 mm of rainfall a month, though in some areas up to 150 mm a month is considered necessary. Rainfall below 1000–1100 mm is considered marginal, depending on evapotranspiration rates, and rainfall should not be less than 50 mm a month for more than short periods. In long dry spells, flush growth ceases. Mean minimum temperatures should not fall below 31 °C and mean maximum temperatures should not be above 47 °C. Growth is higher at higher temperatures and lower altitudes, but quality is lower. Hail can damage the crop, and in areas subject to hail damage, such as the Kericho area of Kenya, attempts are sometimes made to mitigate its effects by using aeroplanes to seed clouds with silver iodide.

Soil requirements Tea requires a deep, well-drained acid soil. Good soil structure, porosity and aeration are necessary. Soils developed over a wide range of geological formations, including granites, gneisses and volcanic rocks have been successfully planted to tea. A soil depth of 1.5–2.0 m is usually considered minimal, though some tea estates are planted on shallower soils.

Tea has a high aluminium requirement and a high aluminium content in the leaves. It does not obtain sufficient aluminium in neutral or alkaline soils and is therefore one of the crops that thrives best on distinctly acid soils. Soils that are not sufficiently acid are often deliberately acidified. A pH of 4.0–6.0 is suitable, and 5.0–5.5 is considered optimum. Soils with higher reactions than this can be acidified by the use of elemental sulphur during land preparation and planting, and subsequently by the use of acidifying fertiliser such as sulphate of ammonia (Chapter 14). However, in many of the areas that have the heavy rainfalls required by tea, the soils are naturally leached and acid.

Tea is often planted in hilly areas with steep slopes. The soils are vulnerable to erosion during the first two years after planting unless they are protected. They can be protected by a grass or other mulch, or by growing grass or a cover crop between the rows. Once the tea bushes close to form a canopy, erosion virtually ceases and a dense cover of tea bushes provides almost complete protection to the soil since all rain falling is intercepted by the leaves. In addition there is normally a thick mulch of dead leaves and sometimes prunings on the soil which provides additional protection and which adds organic matter and nutrients to the soil. Nutrients are therefore cycled from the whole of the root zone, via the tea bush and leaf fall, to accumulate in the topsoil. There is thus an almost closed cycle of nutrients much as is found under a dense tropical forest. Since only the leaves of the bush are removed, losses through harvest are not great (see next section). A good cover of tea is therefore a very satisfactory form of land use since it protects even steep soils from erosion.

Tea nutrient uptake and deficiency symptoms

Compared with crops such as coffee and cocoa, nutrients removed by the tea crop are very moderate. As far as is known, tea bushes can continue to produce over very long periods provided that they are periodically cut back. Some older estates have tea bushes that are well over 100 years old and still in good health. When tea bushes are uprooted and an estate is replanted, it is usually in order to substitute a better, higher-yielding clone and not because either the tea bushes or the soil are exhausted.

Various estimates based on the analysis of made tea suggest that 100 kg of made tea removes about 5–6 kg of nitrogen, 1–1.5 kg of P_2O_5 and 2.5–3.5 kg of K_2O. Because only the leaves were harvested and nutrient removal was known to be only moderate, it was long considered that fertiliser applications to tea were not needed, or that they would be uneconomic. Nowadays, however, most of the world's tea receives some fertiliser. This is because although nutrient removal is not high, it is continuous, and most tea soils, being leached and acid, are of naturally low to moderate fertility. The main nutrient supplied is nitrogen.

Visual deficiency symptoms of tea can be summarised as follows:

Nitrogen: pale green or yellowish-green leaves of young flushes; decline in shoot formation and yield.
Phosphorus: leaves become dull and matt; excessive phosphorus deficiency leads to die-back of branches after pruning.
Potassium: scorched light grey or light brown leaf edges and silvery-white bark; marked yield decline due to premature defoliation; poor recovery from pruning.
Magnesium: older leaves greenish-yellow and showing a triangular dark green patch along the midrib down to the base.
Sulphur: marked yellowing of the foliage between the veins, often in the first year after planting, with smaller leaves and shorter internodes ('tea yellows' or 'sulphur yellows').
Zinc: very short internodes and dwarfed intervenally chlorotic leaves.
Copper: unsatisfactory tea fermentation after picking.

Tea responses to fertiliser

Nitrogen Since tea is a leaf crop, it is not surprising that nitrogen is the most important fertiliser nutrient supplied. There is often a close correlation between yield and the amount of nitrogen fertiliser used. For unshaded tea, increase in yield is more or less linear up to about 120 kg/ha, but in some areas such as the tea-growing areas of Kenya up to twice or more of this amount is applied and found profitable. Full effects are not obtained in the first year, but responses are to some extent cumulative during the pruning cycle since the frame development of the bush is also affected. Responses may be as much as 4–8 kg of made tea per kg of nitrogen supplied. In East Africa it is considered that if the maximum response to nitrogen is less that 4 kg made tea per kg of nitrogen, then some other factor (such as lack of another nutrient) is limiting yields. In one experiment at Kericho, in the Kenya highlands, for example, the initial nitrogen response was only about 2 kg of made tea per kg of nitrogen, but after a phosphorus deficiency had been corrected it rose to 6 kg of tea per kg of nitrogen. Nitrogen applications have been found to increase not only the plucked tea yield, but also the amounts of leaf fall and pruning litter, thus helping to keep up soil organic matter levels, and to contribute to frame development.

These results apply to unshaded tea. Fertiliser responses on shaded tea are generally much more variable than those for the unshaded crop.

Other nutrients Phosphorus, potassium, magnesium, sulphur and other nutrients (including zinc) are given to tea but generally in smaller amounts

Fig. 16.1 Evaluating tea fertiliser trials in Bangladesh.

and in a less systematic fashion than is nitrogen, depending on the properties of the soils concerned. Some specific cases where nutrients other than nitrogen are recommended are given in the examples that follow.

Some examples of specific fertiliser recommendations for tea

North India In northern India, it is recommended that the soil in the planting hole receives 30 g of single superphosphate, 4–5 kg of cattle manure and 200 g of oil cake. After planting, a compound fertiliser known locally as the YTD mixture is recommended for at least three years. This is a 10:5:10 mixture made by mixing 48.5% ammonium sulphate, 31.5% single superphosphate and 20% KCl (50% K_2O). The mixture is applied only to the area covered by the plants at the rate of 25 g, 45 g and 70 g in the first three years respectively.

For mature tea, current recommendations are of the order of 90–130 kg/ha of nitrogen a year, plus an application once every three years of 20 kg P_2O_5/ha and 40 kg K_2O/ha. Fertiliser for mature tea is thought best applied in a circular band extending from one-third of the spread of the bush to a little outside the spread, or in bands along the row at the same distance from the stems.

The peaty *bheer* soils of Cachar form an exception. Their high organic matter content reduces their nitrogen requirement and in practice they are given no nitrogen for several years until yields begin to fall, after which they are recommended to receive 90 kg/ha.

Sri Lanka In Sri Lanka, recommended fertiliser levels for tea vary with the district and the altitude, but are generally much higher for clonal tea than for seedling tea. For seedling tea they are 90–180 kg/ha of nitrogen, but for clonal tea are 270–405 kg N/ha. The expected yield is taken into consideration. For seedling tea, the nitrogen needed is taken to be 1 kg/ha for every 10 kg/ha of made tea.

The distribution of fertiliser within the four-year pruning cycle is not even, being less in years 1 and 4 than in years 2 and 3.

Phosphorus in Sri Lanka is usually given in the form of small dressings of 'Saphos' rock phosphate. Recommended rates are 22.5 kg P_2O_5/ha for seedling tea and 35 kg P_2O_5/ha for clonal tea.

Recommended potassium applications are 65–70 kg K_2O/ha for seedling tea in up-country and mid-country areas, and 45 kg/ha in the low-country tea districts, with about twice these rates for clonal tea.

Magnesium is frequently given routinely to mature tea in Sri Lanka at the rate of 125 kg/ha of dolomite, but where acute magnesium deficiency is recognised, an additional 125 kg/ha of kieserite (24% MgO) is added into the standard fertiliser or, alternatively, Epsom salts (17% MgO) is given as a leaf spray four times a year to give a total annual application of 56 kg/ha.

Zinc deficiency is thought to be present on many Sri Lankan tea estates, in which case a spray containing 5% zinc sulphate or 1.5% zinc oxide is used.

South India In southern India, tea is grown mainly on the western slopes of the Western Ghats, which receive the full force of the south-western monsoon. The soils are loamy, and derived mainly from gneisses, though heavy clay soils are used in the Nilgiris. Soil pH values range from 4.6 to 5.6. Some of the tea is at unusually high altitudes, up to 2400 m in Kerala, and rainfalls are between 1500 and 5000 mm.

Fertiliser recommendations are generally based on yields of made tea at the rates of 8 kg N/ha and 4 kg K_2O/ha per 100 kg of made tea. Phosphorus recommendations are fixed independently of yield at 33.5 kg P_2O_5/ha. For high-yielding seedling tea, the maximum economic limit was generally considered to be about 135, 33.5 and 67 kg/ha a year of nitrogen, P_2O_5 and K_2O respectively, but some high-yielding estates apply 200 kg N/ha to seedling tea and more than this to clonal tea. Young tea on replanted areas is thought to benefit from additional potassium. Fertilisers are usually either split, or an NPK mixture is given in one season and additional top-dressed nitrogen in another.

Indonesia and peninsular Malaysia In West Java most of the tea is on volcanic soils. Nitrogen dressings at the rate of 120–150 kg/ha a year are recommended for mature seedling tea, with about 30 kg P_2O_5/ha and up to 60 kg K_2O/ha. In peninsular Malaysia, the Cameron Highlands Federal Experiment Station obtained a near linear response to nitrogen up to 134 kg/ha, with an increase of about 15 kg of made tea for each kg of nitrogen applied.

East Africa Fertiliser recommendations for tea in East Africa are given by the Tea Research Institute with headquarters at Kericho, the centre of the Kenya tea-growing industry where tea is grown on deep, friable, red volcanic soils. Fertiliser recommendations are based on yield at the rate of 10–12 kg/ha of nitrogen for each 100 kg of made tea, with up to 380 kg N/ha for a yield of 3000 kg/ha of made tea. The nitrogen is often given in a 25:5:5:5 NPKS mixture. Where application rates are currently well below recommended levels, only a gradual increase is recommended on the grounds that tea takes time to adjust to higher levels of nutrition.

Coffee (*Coffea* spp.)

Coffea arabica is African in origin, its original home being the Ethiopian highlands, but its use as a beverage appears to have begun in Arabia in the fifteenth century, hence the name 'arabica'. *Coffea robusta (Coffea canephora)*, is also African in origin, and occurs wild in African tropical forests, from the west coast to Uganda. *Coffea liberica*, as its name suggests, is indigenous to Liberia, whence it spread to other parts of West Africa.

The greater part of the coffee of commerce is arabica coffee, which is considered to have the best flavour and aroma, but robusta coffee is now produced in large quantities, particularly in West Africa, and is considered particularly suitable for instant coffees as well as for blending with arabica.

Since 1840 the world coffee trade has been dominated by Brazil, and coffee is also the most important export crop in Colombia and other South American countries. Almost all South American coffee is arabica. Since the 1940s increasing quantities of coffee have come from Africa, 80% of it robusta, with the Ivory Coast as the leading producer. Since 1952 coffee has been the second most important commodity in international trade, surpassed only by petroleum and its products.

Climatic requirements

Arabica coffee Arabica coffee is an upland species. Ideal conditions for its commercial growth are found at about 1500–1800 m at the equator and at somewhat lower altitudes away from it where temperatures are about 15–25 °C. Frosts and cold winds are damaging. A rainfall of about 1750–200 mm is considered best, preferably with a drier period of two to three months for initiation of the flower buds, since some water stress seems necessary to break their dormancy. With uniform rainfall distribution, flowering and fruiting occur throughout the year. Climates with a dry season followed by a long wet season, as in Brazil, are considered more suitable than uniformly wet ones since they produce a cycle of vegetative growth, flowering, maturing and harvest, and a rest period for the tree. In São Paulo, Brazil, the ideal rainfall distribution for coffee is considered to be 200–350 mm per month for six months following the first flowering, 50–100 mm during the next three months, and little or none during the remaining three months. In Kenya, two wet seasons, a major and a minor one, are related to one major harvest and one minor one a year. Successful coffee growing is found with both higher and lower figures than the 1750–2000 mm rainfall that is considered ideal; in São Paulo the long-term average is 1400 mm and in Kenya good coffee is grown with only 900–1000 mm, albeit on deep soils with a very good moisture storage capacity. Supplementary irrigation is sometimes economic.

Robusta coffee Robusta coffee is a lowland forest species and differs from arabica mainly in that it prefers higher temperatures that are unsuited to arabica production. A dry period is necessary to encourage heavy and uniform flowering at the beginning of the following rains. In West Africa it is grown with rainfalls mainly in the range 1000–1500 mm with a main dry season of two to four months. In Uganda robusta coffee is grown at altitudes of 600–1300 m, but most West African robusta coffee is grown at altitudes below 500 m.

Soil requirements

Arabica coffee Soils for arabica coffee should be deep, not too heavy in texture, and permeable. *Terra roxa*, a term applied to an important Brazilian coffee soil, is often 20 m deep. Good physical properties are essential, i.e. good aeration, structure, permeability, porosity and water holding capacity. Soils with these qualities are found over a range of parent materials – over basalt, gneiss or sandstone in Brazil, over volcanic ash in Colombia, Java and Central America, or over volcanic lavas in Kenya, Tanzania, the Kivu area in central Africa, and Hawaii. An example of an excellent coffee soil is provided by the Kikuyu friable clays of the Kenya highlands. These deep soils are developed over trachytic lavas and ashes. They have a high clay content of 70–80%, but owing to stable micro-aggregation of the clay particles to silt and fine sand-sized aggregates, they feel and handle more like a loam. They are very porous, very permeable, and have a high available water storage capacity. Many deep Brazilian soils have a similar micro-aggregation, favourable stable structure, high permeability, porosity and water storage capacity.

The need for a porous, well-drained soil is related to the high oxygen requirement of coffee roots. Light sandy soils, while providing the necessary good aeration, are normally little used because of their lower moisture storage capacity and fertility. High water tables, very heavy clay soils and indurated layers all limit root penetration.

A soil reaction of pH 6.0–6.5 is considered best for coffee, though good results are often obtained with somewhat more acid soils.

Robusta coffee Robusta coffee is a very vigorous crop which grows well in a wide variety of soils, provided that they are well drained and well aerated. It thrives sometimes even on shallow soils if the rainfall is well distributed and, as is often the case in West Africa, on soils that contain considerable gravel in the subsoil. Perhaps for these reasons, and because robusta is more often a smallholder crop than is arabica, robusta soils have received less attention from agronomists than have arabica soils.

Nutrient uptake by coffee and deficiency symptoms Coffee puts moderate demands on soil nutrients. It has been estimated that 1000 kg of marketed coffee involves the removal from the soil of about 35 kg of nitrogen, 7 kg of P_2O_5 and 50 kg of K_2O. Additional nutrients are needed for the vegetative development of the plant, particularly if growth is not limited by shade. According to one estimate, a hectare of high-yielding coffee might take up at least 135 kg of nitrogen, 34 kg of P_2O_5 and 145 kg of K_2O. The relatively high requirements of nitrogen and potash are due to the need for the plant to build up reserves of materials that help in maturing a big yield. Too rapid and severe carbohydrate depletion during fruit growth reduces vegetative growth and severe cases of depletion can induce die-back of twig tips and even of primaries. In some cases a very heavy yield is followed by a much lower one the following year. Nitrogen and potassium have been found to be the most effective nutrients in controlling die-back. Phosphorus is particularly important in the younger growth stages of the tree, stimulating vegetative growth and root production. Fertiliser programmes thus often stress nitrogen and phosphorus for young coffee, and nitrogen with or without potassium, depending on the soil, for mature, bearing coffee.

Visual deficiency symptoms of coffee include the following:

Nitrogen: stunted and asymmetrical leaf growth, with leaves turning yellow, starting at the lower branches; leaves may be shed prematurely, causing defoliation.

Phosphorus: stunted root growth and little wood formation.

Potassium: necrotic strips on leaf margins, starting from the leaf tip, reddish-yellow at first turning later to dark brown and black, followed by shedding of older leaves.
Magnesium: dull olive-green leaf discoloration between main leaf veins, spreading out from the leaf centre and later developing to a green herringbone vein pattern on a yellow, orange or bronze background.
Zinc: leaf chlorosis with veins standing out against a pale green to yellow background; leaves small and in terminal bunches because of unusually small internodes.
Iron: similar chlorosis to zinc.
Boron: death of terminal shoots, often with long, distorted, leathery leaves.
Manganese deficiency: discoloration of terminal leaves which turn to bright yellow with white spots, particularly at beginning of rains.
Manganese toxicity: lowered flower production, mottled leaf chlorosis, unbalanced growth and size of leaf pairs; necrosis of apical stem and branch buds.

Leaf analysis is now widely used as a basis for diagnosing nutrient deficiencies and as a guide to fertiliser requirements. Leaf nutrient levels vary widely, however, according to the coffee species, the location, the season, the position of the leaf on the branch and whether or not the plants are shaded. For a reliable interpretation of leaf analyses, all these variables have to be known and taken into consideration.

Mulches Numerous experiments in many countries, including Brazil, Colombia, Tanzania and Kenya, have shown very marked yield improvements due to the use of mulches. Mulches act in a variety of ways so that it is not always easy to determine which factors are responsible. Available soil moisture is usually increased, and the physical and chemical soil conditions improved by the increased supply of humus and of nutrients, the latter sometimes being reflected in improved potassium and phosphorus levels in the leaves. In many cases the normal levels of soil erosion were greatly reduced by mulching and yields improved for this reason. Not all the effects of mulches are advantageous, however. Some farmers feel they encourage termite activity which might then spread to their crops. Too much green material can lead to temporary nitrogen tie-ups and shortages (Chapter 9). In Kenya, grass mulches, usually Napier grass (*Pennisetum purpureum*) were once widely used and recommended for coffee to increase the supply of both nutrients and moisture, but they supplied large amounts of potassium to soils already well furnished with that element and by lowering the Mg/K ratio gave rise to potassium-induced magnesium deficiency. Their use has now virtually disappeared on Kenya coffee, to be replaced by supplementary irrigation by those who can afford it. Coffee pulp forms a good mulch, but surprisingly few farmers feel that the cost of spreading it is worth while.

Shade Although much coffee was grown under shade in the past, the use of shade trees has tended to decline. In Brazil and Kenya, coffee is grown without shade because rainfalls are too low to provide enough moisture for both the coffee and shade trees, thus illustrating the fact that the two inevitably compete for water and nutrients. Despite this, shade trees are thought to be justified by their advocates because of their effects on the microclimate and the soil. Microclimatic effects of shade trees include protection of the coffee from possible excessive sunshine and from temperature fluctuations, frosts and cold winds. As regards the soil, it is argued that shade trees control soil erosion, produce organic matter, bring up nutrients from the lower soil layers and make them available in the topsoil through leaf fall, improve the nitrogen supply (particularly if the shade trees are legumes) and suppress weed growth. It would be difficult to quantify these effects, but one important fact is generally accepted: growth under shade is generally slower and yields lower because of the brake put on plant metabolism when there is less sunlight and because lower light levels also reduce flower initiation. In consequence fertiliser needs and recommendations are usually lower for coffee under shade than for unshaded coffee, and for maximum yields shade should be much reduced or completely eliminated.

Coffee responses to fertiliser

Nitrogen Adequate nitrogen is essential for vegetative growth and for good yields of coffee. Nitrogen requirements are higher for unshaded coffee than for shaded coffee. Fertiliser needs depend on the natural nitrogen supply which in

Fig. 16.2 Fertiliser trials on bearing coffee, Costa Rica.

turn is related to the soil organic matter content and rate of mineralisation (Chapter 9). Typical nitrogen application rates are of the order of 60–120 kg/ha for unshaded trees, and somewhat more for unshaded ones depending on yield levels. The need for nitrogen is highest during the rainy season and nitrogen applications are most needed at the beginning of the rains, though in many areas additional split applications are also recommended. In some cases conventional fertiliser applications have been supplemented by foliar sprays of urea, but since coffee is sensitive to biuret, only low-biuret urea should be used.

Phosphorus Phosphorus is particularly needed by young trees and the best way of supplying it is to put a phosphatic fertiliser, preferably mixed with compost or farmyard manure, in the planting hole. This encourages a good root system which, once developed, often enables the older tree to exploit a sufficient volume of soil for it to be able to extract all or most of its phosphorus needs. For this reason, mature coffee seldom shows a very clear or marked response to phosphorus. Applications to mature trees, particularly those on soils such as *terra roxa* with a high phosphorus fixation capacity, often result in only a small percentage of the added nutrient finding its way into the plant if fertiliser is applied as a top-dressing or a in a semi-circular furrow round the tree; leaf sprays have proved more efficient.

Potassium Potassium needs are high, particularly during berry formation, during which time the potassium content of leaves may decrease considerably. Potassium applications are usually made at the beginning of the rains, with nitrogen, but may be split into two.

Calcium, magnesium and sulphur Calcium requirements are fairly high. Although coffee can grow on soils as acid as pH 5.0, periodic moderate liming is recommended in such cases. Magnesium availability is influenced by Mg/K ratios. When these fall owing to high potassium applications, potassium-induced magnesium deficiency may result. In Kenya when Napier grass mulches added potassium to the soil, magnesium deficiencies were found to occur when the exchangeable K/ exchangeable Mg ratio fell to 1.5:2.0. Magnesium deficiencies can be cured by using dolomite ($MgCO_3$), dolomitic limestone ($CaCO_3$ and $MgCO_3$) or kieserite ($MgSO_4$), but the application of these amendments may need 12 to 18 months to take full effect. Quicker results can be achieved with fortnightly sprays of a 1–2% solution of Epsom salts.

Sulphur deficiency in coffee has been reported from New Guinea.

Trace elements Zinc deficiencies have been reported in many coffee producing areas, including areas of *terra roxa* soils in Brazil derived from basalt. Zinc sulphate is applied to the soil or as neutralised 0.4% solutions to the leaves.

Boron deficiency seriously affects coffee production in the Meseta Central areas of Costa Rica, where applications of 30–60 g of borax per tree have given very large yield increases.

Iron deficiencies may occur on soils with a high pH, or be induced by high phosphorus levels, and are best alleviated by the use of iron chelates at rates of up to 15 g per tree.

Manganese deficiency occurs in Costa Rica, the symptoms being particularly marked at the beginning of the rains. It is most likely to occur on soils with a high pH. A spray containing 0.4% manganese sulphate is used.

Manganese toxicities occur in Costa Rica and in Puerto Rico on soils of low pH that are high in reducible manganese, and can be controlled by careful liming, bearing in mind that overliming can lead to deficiencies of trace elements such as zinc and boron.

Some examples of specific fertiliser recommendations for coffee

Brazil In the most important coffee growing country, Brazil, coffee growing in the early days exhibited a certain 'nomadic' character, with older, exhausted coffee farms being abandoned in search of virgin soils on which new farms could be established. To illustrate the nature of this exhaustion, an investigation published in 1935 compared a virgin *terra roxa* soil with one that had been under coffee for 22 years. The worn-out coffee soil appeared to have lost about 39% of its organic matter, 50% of its nitrogen, 37% of its phosphorus, 93% of its potassium, 85% of its magnesium, and 80% of its calcium. The pH appeared to have dropped from 6.7 to 5.7. Since it is now becoming increasingly difficult to find soils that are virgin or which have been under a long natural fallow, much more attention is being given to improved cultural practices and better soil management, including fertiliser use.

In the 1960s a survey of Brazilian coffee-growing areas led to the estimate that 90% of the land needed nitrogen and potassium, and 10% needed phosphorus. In 1970, one investigator gave the following tentative fertiliser recommendations in relation to soil type in terms of grams of nutrient needed per tree to obtain maximum yields:

	N	P_2O_5	K_2O
Coffee on *terra roxa*	200	25	150
Coffee on sandy Latosols	100	25	75
Coffee on Quaternary sands	400	100	300

These figures show that fertiliser needs were considered higher on the poor Quaternary sands than on the *terra roxa* soils or the sandy Latosols, but that on all three soils nitrogen and potassium recommendations are higher than those for phosphorus.

Fertiliser programmes should take into account the nature of the soil and the age of the trees, distinguishing between fertiliser needed at planting, for young trees, and for mature trees in full production.

In Brazil the recommended practice at planting is that each planting hole should receive 30 kg of farmyard manure mixed with 750 g of ammonium sulphate, 500 g of single superphosphate and 400 g potassium chloride. In addition, 5 g of borax and 10 g of zinc sulphate are recommended to prevent boron and zinc deficiencies, while for acid soils an additional 300 g of limestone is advised.

For young trees not yet producing, the Brazilian recommendation is for 5–10 g each of nitrogen and K_2O to be applied to each tree four times a year, at the beginning of the rains (September–October), and again in November–December, January–February and March–April. The first application should also include 20–40 g P_2O_5 and 5 g of borax, and the trees given a 0.5% zinc sulphate spray.

For trees in full production the Brazilian recommendations also involve four applications a year, each of 25–100 g of nitrogen and 37–80 g of K_2O per tree, depending on the soil, and one application of 25–1000 g of P_2O_5, with additional boron and zinc, farmyard manure if available, and lime if the soil remains acid.

The above recommendations are all for unshaded coffee exposed to full sunshine. Recommendations for shaded coffee are often considered to be 30–50% less than those given for unshaded coffee.

Colombia and Central America In Colombia, fertiliser applications are recommended, as in Brazil, four times a year, with rates of total fertiliser per tree for the whole year of between 100 g and 1 kg of á NPK mixture. Common mixtures are 12–6–22, 12–12–17 and 12–6–24, which mainly supply more nitrogen and potassium than phosphorus. Magnesium and boron are also given.

In Costa Rica and Guatemala, sulphur at the rate of 15–20 g per year is recommended in addition to nitrogen, phosphorus and potassium.

Kenya In Kenya, where the soils are naturally well supplied with potassium, only nitrogen is normally given to coffee. Recommendations vary according to whether estimated yields are moderate (less than 6 kg of cherries per tree), good (6–9 kg) or heavy (over 9 kg). For moderately yielding coffee, 90 g per tree per year of nitrogen is recommended, for good-yielding coffee, 110 g, and for heavy crops, 160 g of nitrogen is recommended, to be applied in three split applications in each case. The form of nitrogen fertiliser recommended depends on the soil pH. For neutral to alkaline soils, sulphate of ammonia is considered satisfactory, but for moderately acid soils, ammonium sulphate nitrate (ASN) or calcium ammonium nitrate (CAN) are preferred. The more acid the soil the more should CAN be used in preference to other forms since it has no net acidifying effect on the soil (Chapter 14).

Robusta coffee in the Ivory Coast In all the countries listed above, coffee is of the arabica type. Fertiliser requirements of robusta coffee, often considered a more vigorous tree, are thought to be comparable to or higher than those of arabica, but in practice fertiliser use on robusta coffee, particularly in Africa where it is often a smallholder crop, is much less widespread than it is on arabica.

The main producer of robusta coffee is the Ivory Coast. In that country fertiliser recommendations, often assisted by leaf analyses, distinguish between coffee on Tertiary sand formations and coffee on the more extensive areas of pre-Cambrian rocks of the basement complex – mainly schists, phyllites, gneisses and granitised rocks. On Tertiary sands nitrogen is recommended at 30–50 g per tree per year and K_2O at 60–100 g, it being thought that on these soils the K_2O/N ratio should be between 2.0 and 2.5, and the N/P_2O_5 ratio from 1.0 to 1.5, and that these dressings be supplemented with 10 kg per tree per year of compost or animal manure. The pre-Cambrian soils are more varied than those over Tertiary sands, and are sometimes better supplied with potassium and less well supplied with phosphorus, so that fertiliser recommendations are modified accordingly. The recommended K_2O/N ratio is lower, at 1.0–1.5. Farmers are told to apply all the phosphorus in April, at the beginning of the main rains, and the nitrogen in three split applications in April, July and October.

In practice these recommendations are largely ignored by most small farmers, and the amount of fertiliser used, if any, varies considerably from area to area and from year to year. Since the most easily available fertiliser is a 10–18–18 NPK mixture, more sophisticated recommendations based on soil differences and involving single fertilisers are seldom followed. Fertiliser applications are of relatively little use unless the farm is also well weeded and maintained, and many small farmers prefer low yields with little effort, to the higher yields that result from better management.

A recent development in the Ivory Coast has been the development of an arabica–robusta hybrid, known as 'arabusta', now grown commercially but in small quantities.

Cocoa (*Theobroma cacao*)

Cocoa originated in the forests of South America. It is a forest understorey tree adapted to shade conditions, though in some cases yielding better when cultivated without shade. Cocoa was cultivated from an early period in Central America. In the seventeenth century it was taken to the island of Fernando Po, in the Gulf of Guinea, and in 1879 was taken from there to what is now Ghana. Cocoa-growing spread from small beginnings in the south-east of the country to occupy most of the forest zone except the very wet south-west, and for a long period Ghana was the leading producer, though it is now severely challenged by its neighbour, the Ivory Coast, and by Nigeria. West Africa is the main area of production, but lesser amounts are grown in Brazil, Ecuador, the West Indies, New Guinea and elsewhere.

Climatic requirements Cocoa has relatively narrow temperature tolerances, a mean annual temperature of 25–26 °C and a range of 21–30 °C being best. For this reason cocoa is grown mainly at low altitudes between 15° N and 15° S.

For good yields rainfall should normally be about 1250 mm or more a year, with a dry season less than three to four months long. A dry period at harvest helps in drying the beans.

Soil requirements Cocoa requires a relatively rich soil, preferably well supplied with calcium and magnesium, with a good organic matter content and with a pH between 6.0 and 7.0. A deep well-drained soil is usually recommended. Cocoa can stand occasional flooding if the water is moving and well oxygenated, and is sometimes grown on river bank alluvium. Traditionally it was planted on soils newly cleared from high forest with a high topsoil organic matter content.

In West Africa, the main cocoa belt is found in those areas of the semi-deciduous forest zone with 1250–1750 mm of rain and a topsoil pH in the range 5.5–7.0. In the wetter evergreen rainforest areas to the south of this belt, rainfalls rise to over 2000 mm and the soils are generally too leached and acid to support cocoa, with topsoil pH values below 5.0. In the drier semi-deciduous forest areas to the north of the cocoa belt, soils are often fertile enough to support cocoa but the crop suffers from dry season water shortages. Within the cocoa belt itself cocoa is grown mainly on deep Alfisols and Ultisols developed over Pre-Cambrian schists, phyllites and granitised rocks of the basement complex, but the highest-yielding cocoa is on soils developed over generally small patches of more basic rocks.

Cocoa farms were traditionally established by clearing virgin high forest, and planting a mixture of plantains and cocoyam in the shade of which the cocoa seedlings grown in a nursery were established. After three or four years food crop production ceased and the cocoa took over, the traditional Amelonado varieties beginning to yield in about the sixth year. Now, lack of new areas for clearing and the need to rehabilitate old ones have led to cocoa being established on previously cropped soils, often using more vigorous new varieties such as Amazon or Amelonado–Amazon hybrids, and occasionally fertilisers.

Cocoa is a relatively shallow feeder, with about 80% of its roots in about the top 15 cm of soil. In a well-established cocoa plantation, little sun filters through to the ground surface and there is often a well-developed cocoa leaf litter which decomposes and recycles nutrients.

Nutrient uptake by cocoa and deficiency symptoms Cocoa makes relatively heavy demands on the soil, depending on yield levels and whether it is grown with or without shade, and does well only on soils that are well supplied with exchangeable calcium and magnesium; hence its intolerance of most acid soils.

Deficiency symptoms are thought to include the following:

Nitrogen: small, yellow leaves, various degrees of necrosis.
Phosphorus: bronze tinge to leaves, especially at edges and tips of younger leaves which develop a marginal necrosis; much leaf fall, leaves turning orange-yellow before falling.
Magnesium: chlorosis between veins, spreading from midrib to leaf edges.
Zinc: distorted leaf shapes, including sickle leaves.
Boron: over-profuse flower and leaf production followed by dying of growing points and of whole tree.

Cocoa responses to fertiliser Most of the cocoa planted before the 1960s was relatively low yielding, including the fairly homogenous Amelonado plantings in much of west Africa, and the crop was grown under shade. Under these conditions the nutrient demand on the soil is not as high as when more vigorous varieties are grown, and when there is no shade to retard plant metabolism and photosynthetic activity. Responses to fertiliser applications to traditional peasant cocoa farms under shade have often been poor or erratic. With shade removal and the use of more vigorous varieties, the demand for soil nutrients is increased, and sooner or later most soils begin to show deficiencies and to require fertiliser applications if the yield potential is to be obtained. In West Africa one of the advantages of shade was that it controlled capsid attack, but shade removal became practicable with the simultaneous use of insecticides to control capsid. It was then found that shade reduction or removal from healthy cocoa on good deep soils where the dry season was not severe, increased yields even without fertiliser, and that there was also more likelihood of a fertiliser response. On poor soils and in areas with a long dry season, shade removal was not beneficial and sometimes had adverse effects.

The relationship between light intensity and response to fertiliser was shown dramatically in experiments at the West African Cocoa Research Institute station at New Tafo, Ghana, in the late 1950s, and also on long-term experiments on cocoa in Trinidad. In New Tafo, plots with no shade and no fertiliser yielded 1000–1500 kg/ha of dry cocoa,

Fig. 16.3 Two typical scenes from small West African cocoa farms. Left: densely planted amelonado type cocoa forming a closed canopy which lets in little light. The soil is covered with a thick layer of cocoa leaves and at left centre the trunk of a shade tree can be seen. Fertilisers are little used on generally low-yielding peasant farms of this type, but responses have been obtained on the better farms when the shade trees are removed. Right: splitting open the cocoa pods prior to fermentation.

but very similar cocoa from which the shade had been removed yielded about twice as much, whereas shade removal plus fertiliser applications caused yields to increase to about three times that of the shaded and unfertilised plots, to about 3500 kg/ha of dry cocoa.

In Ghana, where cocoa is a smallholder crop and where fertilisers are not always freely available, very little fertiliser is used on old-established cocoa, despite the evidence of yield increases at New Tafo. Fertilisers are recommended for cocoa establishment in areas not previously under high forest. Studies of cost/benefit ratios obtained from fertiliser applications to cocoa indicated that applications are in general profitable on farms that are initially medium to high yielding. In Ghana yields are low, averaging only about 300–400 kg/ha, and fertiliser applications are likely to be profitable only on those farms with an above-average yield of 600 kg/ha or more, and then only if accompanied by good maintenance, including pest and disease control and adequate weeding. Most of the existing shaded cocoa seems to obtain enough nitrogen, and recommended fertiliser applications are limited initially to 45–90 kg P_2O_5/ha, applied as triple superphosphate, though if yields are increased by the phosphate, then potassium and possibly magnesium deficiencies are thought likely to occur with time and to require correction.

In Nigeria, where most of the cocoa has little or no shade, fertiliser experiments have suggested that phosphorus is the main limiting nutrient, followed, in unshaded conditions, by nitrogen. Fertiliser recommendations for young cocoa depend on whether the ground has been newly cleared from high forest, or was previously cropped. Where ground has been newly cleared from high forest, only nitrogen applications are recommended until the tree comes into bearing, when phosphorus is also needed. In instances where the ground was cropped, after previous tree or arable crops, both nitrogen and phosphorus are recommended from the beginning, with low applications at planting rising to a maximum in the sixth year of 65–100 kg N/ha and 50 kg P_2O_5/ha. These applications are considered likely to give average yield increases of at least 30% if the soils are physically suitable and pest and disease control adequate.

Potassium applications to cocoa in Nigeria are recommended only for certain soils derived from sediments and for very sandy soils. Potassium at present is not usually needed for cocoa on the main cocoa soils derived from the metmorphic rocks of the basement complex. On these soils a 20–20–0 or similar mixture is considered suitable.

In the Ivory Coast, where cocoa production has increased very substantially in recent years to make the country the leading exporter, recommendations for mature cocoa of the more vigorous new varieties grown under little or no shade are 60 g N, 50–80 g P_2O_5 and 80 g K_2O per tree. Higher potassium applications (up to 120 g per tree) are recommended for the leached soils of the wettest areas of the country with over 1800 mm of rainfall, but these areas are less suited to cocoa than the remaining forest areas of less heavy rainfall.

In the Yaoundé area of Cameroon, trials have indicated a need for nitrogen and phosphorus applications, together with calcium and magnesium.

In Grenada, in the West Indies, the soils are derived from volcanic rocks. Cocoa is planted under bananas and both receive NPK applications. In the fifth year the bananas are removed, and each cocoa tree receives 450 g of a 12–8–24 mixture, rising to double this amount in the sixth and following years.

In Ecuador, cocoa is grown mainly in the riverine belt, and most of these alluvial soils are amply supplied with potassium. The main fertiliser need for cocoa is nitrogen, followed by phosphorus. In some areas magnesium is deficient.

In the state of Bahia, Brazil, fertilisers for mature cocoa are recommended only if trees are healthy and vigorous, if there is little or no shade, and if the soils are deep. For newly planted cocoa, a mixture consisting of 50% ammonium sulphate, 30% triple superphosphate and 20% potassium–magnesium sulphate is recommended, commencing with 65–100 g of the mixture applied in the planting hole. Applications increase progressively to 300–500 g for mature trees. The fertiliser is placed in a circle round the tree when it is young, but for mature trees is broadcast.

In the West Indies and Central America, a general recommendation for the establishment of cocoa has been to use 6–7 kg of well-rotted farmyard manure per planting hole, supplemented by 225 g of an NPKMg mixture and by a trace element dressing. The mixture consists of 30% ammonium sulphate, 30% triple superphosphate, 30% potassium sulphate and 10% magnesium sulphate.

Oilpalm (*Elaeis guineensis*)

The oilpalm originated in the high forest areas of West Africa where it grew mainly along rivers. Semi-wild groves associated with human disturbance and selective weeding are still found and used in West Africa, but the great bulk of palmoil production now comes from large commercial estates using improved varieties in West Africa, central Africa, Malaysia and other areas of South-east Asia. Yields have been greatly increased by breeding and the oilpalm has the highest yield of oil per unit area of any crop.

Climatic requirements The oilpalm needs a high, well-distributed rainfall, high temperatures and high levels of solar radiation.

The best temperature for growth is said to be about 28 °C, and growth at 20 °C to be at only half the rate of that at 25 °C. In most oilpalm-growing areas, average minimum temperatures are over 21 °C in all months. Temperatures below this delay bunch development and give lower yields. Since temperature decreases with altitude, palms in Sumatra above 500 m are said to come into bearing one year later than palms in the lowlands.

In areas of highest production, such as those in West Malaysia and Indonesia, rainfall is about 2000 mm a year with no very marked dry seasons, mean maximum and minimum temperatures are 29–33 °C and 22–24 °C respectively, and sunshine is at least five hours a day. Oilpalm is grown profitably in areas with both higher and lower rainfalls than this (with up to 5000 mm in Cameroon), and the palm is adapted to dry seasons up to three months long, although this affects yields. The effects of a dry season on bunch production are shown by a lowering of yield some 33 months later. The adverse effects of dry spells depend to some extent on the water storage capacity of the soil, and deep, medium-textured soils make the palm less dependent on rainfall.

Soil requirements The oilpalm is grown on a wide range of soils, from loamy sands to heavy clays. The depth of the soil is important, since the palm has a very deep rooting system. The ability of the soil to store water, especially where rainfall is below optimum or not well distributed, may be more important than its chemical fertility. To some extent the oilpalm can compensate for low fertility, if its root system is allowed to develop normally, by exploiting a very large volume of soil.

Well-drained soils are considered desirable, though the palm can tolerate short periods of flooding if the water is not stagnant. Poor drainage affects young palms in particular. Oilpalms are grown on very heavy clay soils in peninsular Malaysia, Brazil and Colombia, but then require closely spaced deep drainage ditches. They cannot be grown where there is a high water table that cannot be lowered.

Soils with a high gravel content are not favourable if the gravel content is enough to obstruct root development. Gravel and stones lower the water holding capacity of the soil by reducing the fine earth content (Chapters 2 and 4). Hard layers of ironstone or plinthite, or rock, may limit root development and thus lead to water shortages even where rainfall is adequate for palms growing on deeper soils. In West Malaysia it has been found that peat soils are not suitable if the peat is deep (over 2.5 m) since the palms are not able to develop a root system to provide sufficient anchorage, but if the peat is shallow (less than 1.2 m) then the palm may be able to anchor itself in a suitable soil layer below the peat.

In peninsular Malaysia the most extensive oilpalm soils are derived from inland sedimentary and metamorphic rocks and are of very variable texture and productivity. Oilpalm in Malaysia is also grown successfully on heavy but chemically rich coastal clays developed from marine alluvium.

In Sumatra and Sabah, oilpalm is grown on riverine and coastal alluvium and on volcanic soils.

In West Africa, particularly in the Ivory Coast and Nigeria, oilpalm is grown on a range of soils developed over metamorphic rocks of the basement complex which range in texture from loams to sandy clays, and also in sandy loam to sandy clay loam soils developed in Tertiary sand deposits. These Tertiary sand soils are often deep and physically suitable, but leached and low in nutrients.

Tropical soils developed under the high rainfalls preferred by the oilpalm are normally subject to considerable through-leaching and are therefore typically acid in reaction. The palm can grow in soils as acid as pH 4.0. The normal pH is considered to be 4.0–6.5, with 5.0–5.5 probably being optimal.

Nutrient uptake by the oilpalm and deficiency symptoms The average quantities of nutrients removed in 10 t of oilpalm bunches (see Table 16.1) indicate that the demand for nitrogen and potassium is moderately high, but that for phosphorus is low:

Table 16.1 Nutrients removed in 10 tonnes of oilpalm bunches

	Nutrients										
	N	P	K	Mg	Ca	Mn	Fe	B	Cu	Zn	Mo
Amount (kg)	29	4	37	8	8						
Amount (g)						15	25	21	48	49	0.08

Deficiency symptoms include the following:

Nitrogen: leaves of young palms and seedling turn pale green, then yellow.
Potassium: orange spotting and mid-crown yellowing.
Magnesium: pale yellow young leaves, orange older leaves (this condition was formerly known as 'orange frond disease').

Much research has been devoted to oilpalm leaf analysis, but critical levels vary with local soil and climatic conditions.

Oilpalm responses to fertiliser For seedlings in the nursery, the West African Institute for Oilpalm Research (WAIFOR) recommends a mixture containing equal parts of ammonium sulphate, single superphosphate, muriate of potash and magnesium sulphate at the rate of 120 g per seedling split into three applications.

Fig. 16.4 Panoramic view of a commercial oilpalm plantation in Ivory Coast where oilpalms have been extensively planted in deep Tertiary sands in areas of heavy rainfall. The photograph shows blocks at different stages of growth. Fertiliser applications are often guided by routine foliar analyses and may include nitrogen for young palms and potassium and/or other elements for mature palms, depending on soil conditions.

Nitrogen is sometimes applied to young palms but not usually to mature ones. Nitrogen deficiencies may result from low organic matter levels in the soil, or from waterlogging, or from weed competition. The use of a leguminous cover crop, such as a mixture of *Calopogonium mucunoides*, *Pueraria phaseoloides* or *P. javanica*, and *Centrosema pubescens* is strongly recommended. Nitrogen deficiencies may develop after forest clearing if large quantities of wood and debris are left to rot. In South-east Asia it is thought that the noxious weed *Mikania scandens* can inhibit nitrification and thus lower the natural nitrogen supply.

Typical nitrogen application rates for young palms are 0.22 kg of nitrogen per palm in the planting year, and 0.45 kg per palm per year of age for 3–4 years after planting. If the nitrogen is given as sulphate of ammonia it should not be put in the planting hole because of the possibility of scorching new roots.

Potassium fertilisers are recommended by WAIFOR throughout the life of the palm at rates of 225 g KCl (60% K_2O) per palm six weeks after planting, 450 g KCl per palm in the first year at the beginning of the rains, and 675 g KCl per palm at the beginning of the rains in subsequent years. Potassium sulphate can be used instead of muriate of potash.

For the Ivory Coast and Togoland, the French Oilpalm Research Institute, IRHO, recommends 1–1.5 kg KCl per palm every year or every two years, depending on soil analyses. For Dahomey the IRHO recommendation for potassium is 1.25–1.5 kg KCl per palm per year depending on yield levels.

In Malaysia, fertiliser recommendations are based mainly on leaf analysis, and differ markedly according to the soil. On the relatively rich coastal soils developed in marine alluvium it is thought that only an initial nitrogen dressing is needed after planting. Of the inland soils, for those developed over granites the primary need is for potassium or for potassium and magnesium, and heavy applications may be needed to maintain yields. Magnesium deficiencies have been induced in the past by heavy potassium applications, and trace element deficiencies, particularly of manganese and boron, are likely to develop. The inland soils over sedimentary rocks, in contrast to the granite-derived soils, are thought to be well supplied with potassium but to need nitrogen and phosphorus. Where phosphorus is recommended it is given as rock phosphate at rates of 4 kg or more per palm per year over a prolonged period, and nitrogen applications have been found to improve the phosphorus uptake. The peat soils need potassium, magnesium and copper.

In Sumatra the best oilpalm soils are developed in volcanic deposits. On clearing from forest they need phosphorus, and a basal dressing of 0.5–1.0 kg of rock phosphate in the planting hole was recommended, plus an additional 150–250 kg/ha of rock phosphate broadcast on the soil to stimulate the growth of a leguminous cover crop. Additional nitrogen to that provided by the cover crop was given on sandy soils and those low in organic matter. Potassium and magnesium deficiencies are widespread both in volcanic soils and in those over Tertiary formations. Fertiliser recommendations are based on leaf analysis and are generally highest for 10–20 year-old palms. Chlorine leaf levels are often low, so that potassium is best given as KCl, provided that sulphur is also given as ammonium sulphate to prevent yellowing due to sulphur deficiency.

In Para state, Brazil, nitrogen is recommended during the first two years at 250 g per palm for the first year and twice this amount for the second year. Triple superphosphate is recommended for years 1 to 4 at 250 g per palm in the first year, rising by equal steps to 1 kg per palm in the fourth year.

Rubber (*Hevea brasiliensis*)

Natural rubber can be obtained from hundreds of species of plants, including trees, shrubs and vines, but the term 'rubber' (or Para rubber) now refers almost entirely to the product obtained from the latex of a single species, the tree *Hevea brasiliensis*. The tree is native to the forests of South America, and originally most rubber was obtained by tapping wild trees in the Amazon jungle. In 1876 both seeds and seedlings were taken from Brazil to England and thence to Sri Lanka, and there was a marked expansion of planting in Sri Lanka and Malaysia in 1900–10. The bulk of the world's rubber is now grown in South-east Asia, with lesser amounts in West Africa and elsewhere. Most rubber is grown on estates, but it is also successfully grown in smallholdings.

The growth of the rubber industry was revolutionised in 1839 by the invention of vulcanisation, a process involving the application of heat to a mixture of rubber and sulphur. Synthetic rubber manufacture was stimulated by shortages of natural rubber in the Second World War, but despite the growth of synthetic rubber manufacture, some natural rubber is essential or desirable for the manufacture of a range of rubber products, including tyres. Recent advances in rubber-growing technology and efficiency have improved the competitiveness of the rubber-growing industry. These advances include increasing yields by the selection of improved clones.

Climatic requirements Para rubber is indigenous to the tropical rainforest of the Amazon basin, and most planted rubber is grown under similar conditions between 15° N and 10° S with average temperatures in the range 23–35° C. The most productive areas are generally within 6° of the equator, but rubber grows well at latitudes above 20° S in the state of São Paulo in Brazil, where it is said to resist occasional frost better than coffee. Decreasing temperatures with altitude retard the growth of the rubber tree, and trees need 3–6 months longer to reach tappable size for each additional 200 m above sea level, though latex production may not be affected. An altitude of 600–700 m above sea level is usually taken as the upper desirable limit for rubber growing unless conditions are particularly favourable, as in the Mount Cameroon area of West Africa. Studies in Brazil have suggested that an annual mean temperature of 20 °C is the lower limit of temperature adaptation of rubber.

Most rubber is produced in areas with rainfalls of 1900–2500 mm, but the range of rianfalls in which rubber is grown is 1500–6000 mm. Marked dry seasons are not ideally suited to the rubber tree, but are tolerated if the soil allows deep root penetration. About 100–150 rain days a year are considered desirable. More rain days than this may interfere with tapping, particularly if rain falls in the early morning.

Soil requirements It appears that the wild *Hevea brasiliensis* of the Amazon is of at least two types, one adapted to well-drained soils and one to swampy conditions. The type collected in 1876 that was sent to Sri Lanka and peninsular Malaysia was of the type needing good drainage, though an erroneous idea persisted for some time that rubber grew best in swamps. Subsequently, poorly drained soils were found very unsuitable because of the restricted root system caused and because of root attack by the fungus *Fomes lignosus* which causes white root rot. By grafting selected clones to *Fomes*-resistant root stocks, a rubber tree can be produced that is resistant to *Fomes* and therefore more suited than the traditional rubber tree to poorly drained soils formerly considered unusable for rubber.

The rooting habits of the rubber tree vary according to soil conditions. Research at the Rubber Research Institute of Malaysia has shown that taproots in light, sandy, well-drained soils reached 1.5 m at three years and 2.4 m at seven to eight years. Taproots up to 10 m deep have been excavated in mature rubber on deep fertile basalt-derived soils in Vietnam and Cambodia, and there was a profuse development of laterals all along the tap roots. This finding is not in accord with the general view that the rubber tree has only shallow lateral feeder roots to 20 cm.

In the light-textured peninsular Malaysian soils referred to, the lateral roots were 6–9 m long at three years. At about four years the feeder roots from adjacent rows had met in the centre of the inter-row area. In 7–8 year-old trees the greatest density of feeder roots was in the centre between the rows because of intermingling of roots from each row. At 15 years and over, root density between rows had become more even. The root system thus changes with age.

Root growth may be impeded by poor aeration, and also by a compact or indurated soil layer. In shallow soils the taproot may be stunted, reaching only 30–40 cm, but there may be a more vigorous growth of laterals.

The Selangor series in West Malaysia is an example of a soil with a hard pan within rooting depth. Restricted root development has been held responsible for the low rubber yields usually obtained on this series. Where a soil pan results in shallow root development, rubber trees may be uprooted by strong winds. In addition to compact and indurated soil horizons, root development may be restricted by large amounts of gravel in the subsoil. Horizons containing large amounts of pea gravel are found in the soils of the Firestone rubber plantations in Liberia and are widespread in other areas of West Africa, though the gravel content has to be very high, over 60–70% by weight, before root penetration is markedly affected.

Since *Hevea* rubber grows best in high rainfall areas, and high rainfalls are often associated with leached and acid soils, it is to be expected that the rubber tree tolerates soils with an acid reaction. Rubber is grown on soils with a pH range of 3.5–7.5, with a pH of 5.0–6.0 probably being optimum. Rubber is known for being able to grow well on a wide range of soils, including chemically very poor ones, provided physical soil conditions are satisfactory. It has been claimed that 'there is scarcely another plantation crop that makes as few specific demands on the soil as the rubber tree'. Much of the early planting of rubber in South-east Asia was on soils exhausted by the production of other crops. The nutrient needs of the tree are lower than for most other tropical tree crops, including cocoa and oilpalm. The amounts of nutrients exported in the latex are very low (Table 16.2). Yields are nevertheless higher on richer soils than on very poor ones. Wild trees tapped in the Amazon were said to yield best on the more fertile basalt soils. The highest commercial rubber yields recorded have been on fertile basalt-derived soils in Vietnam and Cambodia.

Table 16.2 Nutrients (in kg) contained in 1500 and 3000 kg/ha of latex, and in a plantation of 33-year-old rubber

	Removed in latex annually		Contained in 1 ha of
	1500 kg/ha	**3000 kg/ha**	**33-year-old trees**
N	9.5	18.9	1389
P	1.9	3.8	218
K	6.5	12.9	989
Mg	–	–	328

Rubber responses to fertiliser The early view was that since rubber grew well on very poor soils, little attention need be given to the use of fertilisers. Recently there has been greatly increased interest in fertiliser use on rubber due to the realisation that:

1 nutrients removed in the latex, though small in amount, are only a tiny fraction of the total in the tree;
2 it is profitable to use fertilisers on immature trees to bring them to tapping girth earlier;
3 fertilisers correctly used can maintain the vigour and yield of the new higher-yielding clones; and
4 much rubber is planted on impoverished soils that need careful husbandry, including appropriate fertiliser use.

Fertiliser use on rubber can usefully be divided into three stages: an initial stage in the nursery, a second stage involving the fertilisation of immature trees up to the time they reach tapping girth, and the third stage of fertiliser application to trees in full production.

Although nurseries are usually established on soil with a good organic matter content, additional fertilisers, sometimes mixed with compost, have been found to improve seedling vigour.

It is often profitable to apply fertilisers to immature trees between planting and the onset of tapping in order to reach tapping size earlier. Tapping begins when the tree has reached a certain girth, not a certain age. In some instances, fertilisers have reduced the tapping age from eight years to less than five. Fertiliser applications of this stage have been found to improve later bark renewal and thus to have a long-term effect on yields.

Phosphorus is important at establishment to stimulate rooting. Large responses to phosphorus applied to immature rubber have been recorded on inland soils in peninsular Malaysia. In wet areas of acid soils, including most of the Malaysian rubber-growing areas, the basal phosphorus dressing can be in the form of rock phosphate, and in peninsular Malaysia an initial dressing of 0.5–1.0 kg rock phosphate per tree is recommended, with additional annual applications to bring the total to 2.0 kg rock phosphate during the period of tapping. In one experiment, trees that had received this amount over the first six years were producing twenty years later twice as much latex as the control, and had better height, girth and bark renewal. On average they had reached tapping size 22 months earlier than trees that received no phosphate after the first year.

Immature trees also need nitrogen to encourage shoot and leaf growth. Applications of nitrogen and phosphorus together, sometimes as ammonium phosphate, are often recommended at six-monthly intervals during the first five years, the amounts of nitrogen recommended being influenced by possible nitrogen contributions from a leguminous cover crop, and the balance between crown and trunk. The progressive shading out and dying of a leguminous cover crop, usually from the third to the fifth year, can supply substantial amounts of nitrogen. Fertiliser nitrogen may need to be reduced during this period since too much nitrogen causes the crown to develop faster than the trunk, leading to bending in young trees and trunk snap in older ones.

Some rubber clones are more sensitive to potassium deficiencies than others, but the need for potassium fertilisation depends mainly on the soil. Potassium deficiencies have been reported on sandy soils, and from impoverished and highly weathered soils in Malaysia, Sri Lanka, Indonesia, Vietnam and Cambodia, but in general the need is patchy and the application of potassium fertilisers to rubber is less widespread and general than is the use of fertiliser nitrogen and phosphorus.

Fig. 16.5 Rubber trials.

Magnesium deficiency is found in both old and young rubber in western Malaysia and Sri Lanka, and in young rubber in Nigeria. It has occasionally been induced by potassium applications. Magnesium has been corrected in some instances by applying 125–250 kg/ha of dolomitic limestone to the cover crop, which later releases magnesium as it decomposes, or by giving 1.0–1.5 kg of dolomitic limestone per tree. Too much magnesium adversely affects latex composition, leading to precoagulation, and this is corrected by applying potassium.

Manganese deficiencies have been reported in peninsular Malaysia on highly weathered alluvial and granite soils. Manganese sulphate can be used to correct the deficiency, but with care, since repeated applications can lead to a toxicity of the element. The manganese content of the bark scrap should remain below 10 ppm.

Higher-yielding clones and increased fertiliser use have led to more attention being given to trace elements, particularly to copper, boron, molybdenum and zinc.

Boron toxicity has been reported in rubber in acid and often somewhat poorly drained coastal alluvial soils in Malaysia. In another Malaysian instance, boron toxicity was ascribed to the presence of tourmaline, a boron-containing mineral, at shallow depth in the soil.

Copper has given yield responses on copper-deficient soils in Vietnam and Cambodia. Zinc deficiency has been reported in Sri Lanka, and very temporarily in peninsular Malaysia in the immediate area of phosphate application due to a phosphate–zinc interaction.

The above discussion applies primarily to fertiliser applied to young rubber before it reaches tapping size. Fertiliser application to mature rubber has given variable and generally less encouraging results. In some cases of gross deficiencies, yield improvements have been obtained in a year or less, but normally the effects, if any, of fertiliser applications to mature rubber take three or more years to show themselves. Lack of genetic potential of the unselected seedlings that were often used in the past to respond to an increased nutrient supply, has also contributed to the fact that the effects of fertilisers applied to mature rubber have often been absent, or slow and uncertain.

Western (peninsular) Malaysia Recommended fertiliser dressings for immature rubber are summarised in Table 16.3. These are in grams per plant, to be given at various intervals after the initial dressing in the planting hole of 120 g rock phosphate. Recommended rates increase substantially with the age of the tree. On very sandy soils, more potassium is recommended than on loam and clay loam soils. The schedule should be modified according to the observed growth of the tree. If crown development is too fast, particularly in years 3 to 5, applications should be reduced or stopped. For soils known to be magnesium deficient, additional magnesium is recommended in the form of kieserite $MgSO_4$ or the slower acting but cheaper dolomite or dolomitic limestone.

The West Malaysian recommendations for mature rubber on most inland soils during the first 15 years of tapping are for annual applications of 325 kg/ha (equivalent to 1.125 kg per tree) of an NPKMg mixture known locally as mature mixture 1. This mixture contains 10.5% N, 6.8% P_2O_5. 9.4% K_2O and 4.1% MgO. These recommendations for mature rubber should also be modified as necessary. Phosphorus may not be needed if adequate amounts have been applied in earlier years. Potassium applications should be increased for Malaysian soils derived from intermediate and basic igneous rocks that are low in potassium, and reduced for those over sediments that are better supplied with this element.

Sri Lanka The more important rubber soils in Sri Lanka are similar to some of the inland rubber soils of peninsular Malaysia. There is a general need for NPK fertilisers supplemented with magnesium, except for soils over limestone. Standard recommended fertiliser mixtures contain 7–9.1% N, 9.5–12.4% P_2O_5 in the form of rock phosphate,

Table 16.3 Recommended fertiliser application rates for immature replanted rubber on inland soils in peninsular Malaysia after a basal dressing of 120 g rock phosphate in the planting hole. Nitrogen rates assume a leguminous cover, and should be substantially increased if this is absent.

Months after planting	**Fertiliser application in g per tree**							
	Loams or clay loams				**Very sandy soils**			
	N	**P_2O_5**	**K_2O**	**MgO**	**N**	**P_2O_5**	**K_2O**	**MgO**
1,3 and 5	10	18.4	3.4	2.4	9.6	16.4	8.2	2.4
8	15	27.6	5.1	3.6	14.4	24.6	12.3	3.6
12, 15 and 18	20 +	36.8	6.8	4.8	19.2	32.8	16.3	4.8
24 and 30	30 +	55.2	10.2	7.2	28.8	49.2	24.5	7.2
36	60 +	110.4	20.4	14.4	57.6	98.4	49.0	14.4
42, 54 and 66	180 +	78.2	32.7	19.1	97.2	94.5	65.4	19.1
Total: 13 applications								
5½ years	765 +	649	175	111	508	652	380	111

4.2–8.3% K_2O and 0–2.6% MgO. Recommended applications vary between soils but are generally similar to or somewhat less than the standard Malaysian recommendations for immature rubber on inland soils given above, except that applications of dolomite are recommended at the end of the first, third and fifth years for most soils except those over limestone.

Indonesia The soils of Java differ from those of Sri Lanka and Malaysia in that they are developed over volcanic materials. The young volcanic soils of eastern Java are relatively well supplied with potassium, and dressings of only nitrogen and phosphorus are recommended. The older volcanic soils of western Java are more weathered and contain less potassium, so a complete NPK fertiliser is advised. A complete application, possibly with magnesium, is recommended in Sumatra.

Vietnam and Cambodia Rubber soils in Vietnam and Cambodia are of two contrasting types. Red soils in both countries are derived from basalt, and are deep, with good physical properties. These soils are well suited to rubber. Some of them respond to potassium applications. In Vietnam, poorer sandy clay soils developed in old, weathered alluvium are also used for rubber. These require phosphorus in particular, but also nitrogen and potassium, and good yield responses have been obtained to fertilisers with a 1:2:1 $N:P_2O_5:K_2O$ ratio.

India Rubber in India is confined to the southwest, mainly in Kerala state. Much of the rubber is grown on highly weathered lateritic soils, free draining but low in nutrients. On these soils balanced NPK applications are usually profitable provided that the rubber is yielding at least 560 kg/ha of dry rubber. Yield increases vary according to the type of rubber tree, its age and the initial soil fertility status.

West Africa Rubber is grown in Eastern Nigeria on deep sandy soils. In Ghana and the Ivory Coast, rubber is grown in the wetter parts of the forest zone with rainfalls over about 1700 mm on clay soils developed over phyllites and schists, on sandy clay soils over granites, and on deep sandy soils over Tertiary sands. In all three countries the major nutrient element needed appears to be phosphorus.

In Liberia rubber is grown over gravelly sandy clay soils developed over metamorphic rocks of the basement complex. The main responses are to phosphorus applied at planting and during the early years. Potassium, or potassium and magnesium with a K_2O/MgO ratio of 2:3, may be necessary in later years. A good leguminous cover crop, normally *Pueraria phaseoloides*, supplies most of the nitrogen needed in early years, but nitrogen responses may be obtained after the cover has been shaded out. Fertilisers, particularly magnesium, are more likely to be needed when rubber is planted on previously cropped soils than when virgin forest soils are used.

Coconut palm (*Cocos nucifera*)

Some investigators maintain that the coconut palm originated in South America, others that it originated in Melanesia. Whatever its origin, the coconut palm is now very widely distributed in coastal areas of the tropics between latitudes 23° N and 23° S. Nuts carried by the sea occasionally germinate, but the present wide distribution is due mainly to human action.

The leading coconut-growing countries are the Philippines, India, Indonesia, Sri Lanka, Malaysia, the South Pacific islands, Mexico, the West Indies and Brazil.

Coconut growing and its problems have been intensively studied by research organisations in the Philippines, Trinidad, India, Sri Lanka, Malaysia and francophone West Africa.

Climatic requirements The optimum annual temperature for coconuts is about 27 °C, with only a small diurnal range of 6–7°. The palm grows best where temperature is uniform throughout the year. Since the sea helps to reduce temperature changes, these conditions are usually found near tropical coasts, though nearness to the sea is not in itself essential.

Since temperature falls with altitude, it is the temperature requirement of coconut that excludes it from the higher tropical areas. In East Africa, coconuts grow near the equator at altitudes up to 1300 m, but in India, Sri Lanka and the Philippines, commercial plantations do not often occur much above 600 m above sea level, while in Jamaica commercial coconuts are found mainly below 150 m.

Optimum rainfalls for coconut are between 1300 and 2300 mm a year. Much higher rainfalls are tolerated if the soil is well drained. Ideally, the rain should be evenly distributed through the year. If rainfall falls below 100 mm a month, and the palm has no access to a water table, yields are depressed. Nuts are produced continuously, taking about 12 months to mature, and production is adversely affected by water shortages at any time of the year. The adverse effect of a dry period on nut numbers is felt 28–30 months later, but the effect on nut size and copra content is felt earlier, about one year after the water shortage.

Coconuts grown on and near coastal beaches often benefit from water seeping towards the sea from swamps and lagoons behind the coast. In such cases the palm has an additional source of water which is of particular value when there is not enough rain. The fresh water seeping seawards may contain dissolved nutrients and thus allow the palm to grow on very poor sandy soils that would otherwise be too infertile.

The palm wilts when the soil moisture content is higher than that at which wilting occurs in many other tree crops, thus indicating that the palm is not as efficient as most plants at extracting water at the higher soil moisture tensions (Chapter 4). The older the palm, the more susceptible it is to water shortages. Palms are also more susceptible if their root development has been restricted by soil conditions or by a fluctuating water table.

The coconut palm is a sun-loving plant. It does not grow well in shade or in cloudy areas. Yields are reduced where total sunshine is less than 2000 hours a year. The palm tolerates strong winds quite well, but may be destroyed by hurricanes or cyclones which kill the palm by blowing it over or breaking off the crown. In many tropical countries, particularly Malaysia and Sri Lanka, palms are killed by lightning. Intense heat from the strike bursts the vascular system, and the palm dies either at once or in a period of months.

Soil requirements The coconut palm is grown on a wide range of soils, including sands, clays and even peats. It will tolerate a wide range of soil pH, from pH 8.0 in coastal sands to pH 5.0 or a little below in acid clays.

The soil requirements of the coconut palm are related to its root system, which is that of a typical monocotyledon. There is no tap root, or main root stock with dividing branches, but instead there are between 2000 and 10 000 relatively uniform roots of about the same diameter throughout their length. Since these roots are relatively thick, the major ones being up to 4 mm in diameter, they cannot easily penetrate compacted soils, nor do they tolerate waterlogging.

The palm grows best on well-drained soils, and soils providing adequate rooting depth, i.e. on what some agronomists term 'deep, open soils'. Features that may limit rooting depth include indurated soil layers or pans, and water tables and perched water tables near the surface.

The coconut palm is not demanding as regards soil fertility. It often grows on soils that are too poor for any other crop, though it yields better on richer soils and responds to fertiliser applications when grown on poor soils (see below).

In many tropical areas coconuts are found on beach sands. They may be almost sterile but the palm can flourish if water seeping to the sea brings additional nutrients dissolved in it.

Alluvial deposits, if they are well drained, often form very good coconut soils, as in India and Sri Lanka, particularly if the soils are sandy or loamy in texture. In the Philippines coconuts grow well in alluvial soils derived from limestone. Other extensive coconut areas in India and Sri Lanka are on soils loosely described as 'lateritic'. These are soils that are high in iron, either with an iron-rich layer in the subsoil, usually hardened, or containing much ironstone gravel. Highly leached ironpan soils in high rainfall areas of Sri Lanka are used for coconut palms but growth is poor unless the soil is relatively deep and supplied with potassium and phosphorus.

Soils derived from volcanic ash or tuff support good coconut plantations in Indonesia, the Philippines and New Guinea. These volcanic rocks weather quickly to give nutrient-rich, well-drained soils.

The most difficult soils for coconut production are heavy clay soils, particularly those that are waterlogged during rains but which dry out and crack in the dry season. In peninsular Malaysia, most coconuts are grown on relatively heavy soils. In some cases these soils contain peat which makes them less heavy to work and improves drainage.

Nutrient uptake by the coconut plam Coconut roots do not have root hairs, but have small pneumatophores which exchange gases with the soil air. Absorption of water and of nutrients takes place from that part of the main roots just behind the well-developed root cap, and from shorter-lived secondary and tertiary roots which arise, particularly in the upper soil layers, from the main ones. The stem of the palm does not become thicker with age, since there is no cambium and no secondary thickening. A narrower stem than usual is thought to be due to water and nutrient deficiencies, and may result in 'tapering palm' and 'pencil point' symptoms.

Much of the nutrient uptake of the coconut palm goes into nut production. The analysis of nuts from different areas shows a range of nitrogen and phosphorus levels, but all show relatively high potassium levels (Table 16.4). Nearly half the potassium in the nut is in the husk, and if left in the field to decompose is a useful source of potassium. The relative order of importance of nutrients for an adult bearing palm is potassium, nitrogen, phosphorus, calcium and magnesium.

Leaf analysis has proved useful in diagnosing nutrient deficiencies. In West Africa, using the 14th leaf from the youngest, a good correlation has been obtained between leaf potassium and nut numbers and weight.

A useful method of measuring potassium levels in the palm is to use the nut milk, which is a ready-made solution for analysis. Its composition has been found to reflect potassium, calcium and magnesium levels in the palm and in the soil.

Coconut responses to fertiliser Fertiliser responses with coconuts have been varied, and depend on the soil and the climate and on genetic differences between palms. There is usually a long delay before the effects of the fertiliser are seen. In some fertiliser trials where the applied nutrients were leached down to the water table, there was a response both from palms receiving and palms not receiving fertilisers, because both groups benefited from the enrichment of the groundwater.

In general, young palms respond to applications of N, NP or NPK. The main responses from mature palms have been to potassium.

Nitrogen promotes leaf development and early flowering. Nitrogen given to young palms, often with phosphorus and occasionally with potassium, causes them to come into bearing earlier that they would otherwise do.

Nitrogen deficiency results in a general yellowing of the leaves, particularly the older ones. Nitrogen deficiencies may be associated with soils that are very low in organic matter, or soils with very low soil nitrification in the dry season, or, as in Polynesia, with very calcareous soils. Grass or weed competition may also lead to nitrogen shortages.

Table 16.4 Nutrients (kg/ha) removed by 1 ha of high-yielding hybrid coconuts (138 palms) in the Ivory Coast, yielding 6700 kg/copra per ha. This appears to be the first time that chlorine has been included in coconut analyses. A feature of the results is the high total annual chlorine uptake, equal to that of potassium and higher than nitrogen. It has been suggested that in some cases responses of coconuts in the past to potassium fertilisation with KCl may have been due to a chlorine response.

Component	Dry weight	N	P	K	Ca	Mg	Cl	S
Spikelet	492	3	0.14	14.0	2.0	2.0	11	1.0
Stalk	349	1	0.12	7.0	0.3	1.0	6	trace
Husk	7,843	19	1.0	116.1	5.0	4.0	92	1.0
Shell	3,849	5	0.13	9.0	1.0	0.4	4	0.5
Albumen	6,375	80	13.0	47.0	1.0	8.0	12	6.0
Total	18.908	108	14.7	193.1	9.3	15.4	125	8.5
Annual uptake whole stand, including trunks and fronds		174	20	249	70	39	249	30

Source: **Ouvrier, M. and Ochs, R.** (1978) 'Nutrient removal by hybrid coconut Port-Bouet 121 (MAWA)', *Proceedings, International Conference on Cocoa and Coconuts, Kuala Lumpur, 1978*. Malaysia Agricultural Research and Development Institute and the Incorporated Society of Planters: Kuala Lumpur.

Fig. 16.6 Fertiliser trials on hybrid coconuts in the Solomon Islands planted on clay loam soils (Alokan series). One plot received no fertiliser. The other received 2.6 kg ammonium sulphate, 3 kg triple super and 7.8 kg KCl per palm over the previous four years and in the 12 months after the photographs were taken yielded 2.44 tons/ha dry copra as against only 0.82 tons for the unfertilised plot.

In Trinidad, young palms on beach sands have responded to nitrogen but not to phosphorus or potassium. In Sri Lanka it is recommended that 2-year-old palms receive 680 g of a mixture consisting of two parts sulphate of ammonia, two parts single superphosphate and three parts potassium chloride, the dressing being increased progressively to twice this amount at six years.

Phosphorus deficiencies in coconut palms are relatively uncommon.

Potassium is the deficiency that is most frequently encountered. This is related to the heavy demand for this nutrient when nut yields are good. Potassium shortages are shown by an uneven yellowing of the leaves accompanied by numerous irregular brown blotches. Yellowing is more in middle leaves than in higher ones, and more round the edge of each whole leaf than in the centre. Severe potassium deficiency gives a yellow, sickly palm with few, short leaves. A general recovery begins about six months after potassium fertilisers are applied.

Responses to potassium in Sri Lanka, West Africa and many other areas have often been spectacular and very profitable, typically increasing both nut numbers and copra contents. In the Ivory Coast, applications of 1.5 kg potassium chloride per palm increased copra yields by over 1000 kg/ha, and rates up to 2.5 kg per tree per year have been recommended. Recently, responses in the Ivory Coast to chlorine (Table 6.4) have suggested that some earlier responses attributed to potassium applied as KCl may have been at least partly due to the chlorine applied.

There is no record of palms suffering from calcium deficiencies, but magnesium deficiencies are often reported. Magnesium is most likely to be limiting on very acid soils and in soils in which the Mg/K ratio is low and magnesium uptake is inhibited. Magnesium shortages are most frequently induced by heavy potassium applications. Potassium dressings lower the leaf levels of both calcium and magnesium, and magnesium is sometimes included in the mixture given to young palms.

Iron and manganese deficiencies may occur on very calcareous soils, such as are found on coral islands, because of the high levels of soil calcium. Special methods evolved in these cases include the application of 10 g of iron sulphate and 5 g of manganese sulphate to the nut husk at planting. In the case of adult palms, iron and manganese compounds can be injected into the palm but it is better to place iron and manganese sulphate in a small hole in the ground near the palm base, thus supplying the palm over a period with the small amounts needed.

Applying fertilisers and manures to coconut palms

Organic manures are beneficial, particularly at establishment, when the placing of 50 kg of husks (a good source of potassium) and 50 kg of well-decomposed farmyard manure in the planting hole is recommended. Nitrogenous fertilisers, or NP or NPK mixtures are also commonly given at planting, and are best mixed with the material in the planting hole rather than applied as a top-dressing. Ammonium-containing fertilisers placed too near the tender neck of the young seedling can cause scorch.

Fertiliser applications to older palms may be given at any time of year, but in areas of heavy seasonal rain it may be best to apply them at the end of the main rains when roots are still active but leaching losses are lower. Where rainfall is marginal, then fertilisers are best applied just before the main rains. Fertilisers are mainly applied by hand to a circular area round the palm which increases in radius with the age of the palm from 50–100 cm for young palms to 200 cm for adult palms. The fertiliser is then lightly forked in. In steeply sloping areas that are subject to runoff, fertilisers are best placed in a circular trench 15 cm deep and 90 cm away from the palm.

Recommended fertilisers and fertiliser application rates are broadly related to soil differences and to the age of the palm, but vary considerably between different areas and countries. The selection of recommendations from eight countries given in Table 16.5 shows considerable variation both in fertiliser amounts and the ratios between nutrients. In general, amounts are small for young palms and then reach a maximum for adult bearing palms. In the examples listed, applications vary between 2.0 and 4.4 kg of fertiliser of different compositions per palm per year.

Sisal (*Agave sisalana*)

Sisal is the main commercial hard fibre, used for ropes and twines, the others being henequen (*Agave fourcroydes*) and abaca or manila hemp (*Musa textilis*). All the species of the *Agave* genus are natives of tropical North and South America. In the late nineteenth century sisal was introduced to East Africa, and Tanzania is now a large producer,

Table 16.5 A selection of fertiliser recommendations for coconut palms

Country	Age of palm	Recommended application per palm per year
Jamaica[1]	1 year	450 g SA
	2–4 years	450–900 g SA, plus same amounts NPK mixture
	5 years and over until bearing	1.125 kg SA, 1.125 kg NPK
	Bearing palms, few female flowers	As above
	Bearing palms, many female flowers	1.125 kg NPK, 1.125 kg KCl
Trinidad	Adult palms	2.25–4.5 kg 10–5–10 mixture
Fiji	1 year	225 g 9–10–15 mixture
Brazil	1–5 years	600 g rising to 2.0 kg 5–15–15 mixture
	Bearing palms	3 kg 5–15–15 mixture plus organic manures
Dahomey	Up to 3 years	300 g SA, 400 g DP, 250 g KCl
	3 years to bearing	500 g SA, 500 g KCl
	Bearing	1–1.5 kg KCl
India (2)	Adult palms	500 g N, 300 g P_2O_5, 1.2 kg K_2O
Sri Lanka	Adult palms	3.6–4.5 kg of mixtures containing 10.3% N, 5.5–6.9% P_2O_5 and 15–18% K_2O, depending on soil type
Philippines	Adult palms	2.5 kg 12–14–12 mixture

SA = sulphate of ammonia
DP = dicalcium phosphate
KCl = muriate of potash

[1] Coconut Industry Board, Jamaica
[2] Central Coconut Research Station, Kasaragod

Source: **de Geus, Jan G.** (1973) *Fertilizer Guide for the Tropics and Subtropics*. Centre d'Etude de l'Azote: Zürich.

though recently production has been much affected by price declines and fluctuations.

Climatic and soil requirements Although sisal survives drought conditions, for commercial production a well-distributed rainfall of 1250–1500 mm is considered desirable. Under drier conditions the production of new leaves is retarded and irregular, and the life cycle prolonged, but deep soils with a good moisture-holding capacity offset rainfall irregularities.

Sisal needs a well-drained soil. Poor drainage results in pale and stunted plants and low yields. On heavy clay soils in East Africa, sisal may be grown on cambered beds, or the soil drained with deep ditches.

Sisal will grow on relatively infertile soils, but successful commercial production requires a soil that is fertile enough to produce high fibre yields in a relatively short cycle, usually 7–11 years but longer for some of the vigorous new hybrids. Continued growth of sisal cycle after cycle can substantially deplete soil nutrient reserves. This can be offset by returning sisal waste to the fields and supplying fertiliser nutrients. As the average fibre content of the leaves is only about 5%, returning sisal waste means that most of the nutrients removed are returned, and the waste has beneficial effects on soil physical conditions and moisture retention. However, sisal waste is rarely used in practice because of the expense of transport. Well-rotted waste at 50 t/ha is recommended for the bulbil nurseries.

Agaves are calcium-loving plants and have high calcium requirements. Acid soils are sometimes limed, preferably with dolomitic limestone containing magnesium as well as calcium. In very acid soils sisal may take up toxic amounts of trace elements, particularly manganese and cobalt.

Fertiliser use on sisal In East Africa, the major area of sisal production, sisal is fertilised only rarely. Nitrogen is given on soils with a marked nitrogen deficiency at rates of 50 or 100 kg/ha. Nitrogen is sometimes supplied to some extent by a leguminous cover crop of tropical kudzu (*Pueraria phaseoloides*) sometimes mixed with centrosema (*Centrosema pubescens*) or other species.

Phosphorus is taken up by the sisal plant only in small quantities but deficiencies have been reported on some Alfisols in Tanzania over calcareous parent materials and metamorphic rocks of the basement complex. On low-phosphate soils, dressings of 62–124 kg P_2O_5/ha have been recommended, placed in the soil near the roots or in the planting hole.

Fig. 16.7 Sisal plantation near Tanga, Tanzania.

Potassium deficiency results in a condition known as banding disease in East Africa, leaf foot disease in Indonesia, and leaf base disease in Brazil. This begins as yellow pin-head dots on the underside of the leaves at the neck and then develops to form a weak band of dark, necrotic sunken tissue which, in advanced stages, causes the leaf to collapse at the base. New leaves become shorter and the whole plant is stunted. On soils that are marginal in potassium, deficiencies may be induced by calcium or nitrogen applications which reduce the potassium content of the leaf. Recommended potassium application rates in such conditions are 250–500 kg/ha of 60% KCl per cycle, split into five dressings placed along the sisal lines.

Sisal appears to have a relatively high boron requirement, and molybdenum responses have been reported in East Africa.

17 Soil management in shifting cultivation systems

Types and extent of shifting cultivation

Shifting cultivation, also known by various other names, including land rotation cultivation, and slash and burn agriculture, is the most widespread form of agriculture in the tropics. It is the dominant form in much of Africa and South America but is also important and widespread in Asia outside the main areas of dense population and intensive agriculture.

Shifting cultivation, as would be expected from a practice that is so widespread, takes many forms. In most areas these various forms are undergoing changes and modifications in response to increasing population densities and other factors.

Shifting cultivation involves three essential features:

1 The clearing of natural vegetation.
2 A cropping period, usually one to three or more years.
3 A fallow period during which the land is abandoned to natural regrowth vegetation.

By definition, the fallow period must be longer than the cropping period. Traditionally in many areas it was very much longer, perhaps 10–20 years or more, but with increasing population densities it has generally become progressively shorter. The ratio of the cropping period to the fallow period can be expressed in various ways, such as the 'land use factor', obtained by the formula:

$$\frac{\text{years of cultivation} + \text{years of fallow}}{\text{years of cultivation}}$$

Thus 2 years of cultivation followed by 10 years of fallow would given a land use factor of 12/2, or 6. This means that one-sixth of the land is cultivated at any one time. Similarly 3 years of cropping followed by 6 years of fallow gives 9/3, or 3, indicating that one-third of the land is cropped and two-thirds is in fallow.

The land is cleared from original, little-disturbed vegetation, either forest or savanna, or from regrowth vegetation of different ages. Methods of land clearing vary with the groups of people involved, the climate and the vegetation, but normally involve cutting all or most of the vegetation, allowing it to dry, and burning it. The effects of the ash produced on burning are often important in supplying nutrients and raising soil pH values. The cropping period usually involves a mixture of crops, and/or different crops at different stages of the cropping period. The cultivated area is abandoned when the farmer decides that it is easier or more profitable to farm a new area instead. The decision to abandon is influenced by declining productivity, which may be due to a lowering of soil fertility or to increasing difficulty of coping with weeds, and by the availability or otherwise of alternative land to clear. The availability of alternative land for new clearing will also influence the length of the fallow period. Sometimes the abandonment is progressive over a period during which the farmer no longer weeds but continues to harvest remaining crops such as cassava or pineapples among the weeds.

During the fallow period a succession of types of regrowth vegetation progressively restores the productivity of the land. It does this by improving the chemical fertility and physical properties of the soil, largely by increasing its organic matter content, and by suppressing the weeds of cultivation. The speed and efficiency with which the fallow brings about these changes is related to the type of regrowth vegetation and the species involved. Generally, soil productivity is restored more quickly under forest vegetation than under grass.

To the soil scientist, the study of soil changes under shifting cultivation can conveniently be divided into the three phases of initial clearing and burning, the cropping period, and finally the fallow period which gradually restores soil productivity. An examination of the processes and changes involved in each of these three stages is fundamental to the understanding of how different systems of shifting cultivation function in different environments, and what measures might be undertaken to improve their efficiency and productivity.

Soil changes during clearing and cropping

The nutrients in the soil–plant system are stored partly in the soil and partly in the vegetation. These nutrients move in a circle (or cycle) from soil to plant and then back to the soil as leaf litter and other parts of the vegetation decay. Under high forest, losses from such a closed cycle might be very small indeed, and the pool of stored nutrients may be gradually increased through additions from decomposing rocks and minerals and from the atmosphere.

In some cases, as when a long-established high forest grows on relatively poor soils, there may be more nutrients stored in the vegetation than in the soil. It is of obvious importance to the soil scientist and agronomist to know what happens to this store of accumulated nutrients when the vegetation is cleared and burnt.

When vegetation is burnt, the nitrogen and sulphur, depending on the heat of the burn, are largely lost as gases to the atmosphere, but the other nutrients generally remain in the ash. The ash acts in some ways much as an application of fertiliser, providing a range of nutrients, other than nitrogen and sulphur, and often also affects the pH of the soil. This may be particularly important in the case of acid soils initially below pH 5.0 containing free aluminium. The raising of the soil pH, commonly to pH 6.0 or more, increases soil productivity for most crops by eliminating free aluminium (Chapter 11). It also affects microbial activity. The effects of the ash commonly last for a season or more before being progressively lost.

An alternative method of clearing would be to kill the standing vegetation with hormones such as 2–4D, and let it decay over a period, or to fell the vegetation but to let it decompose slowly without

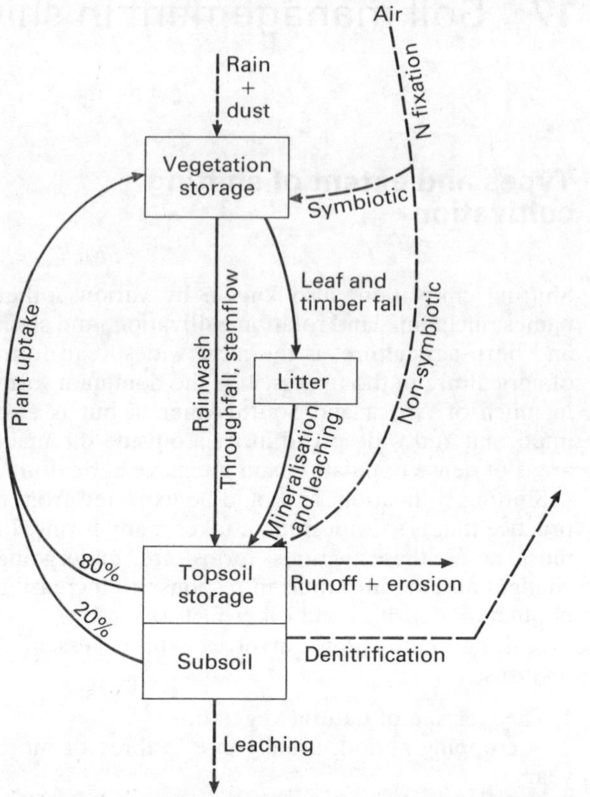

Fig. 17.1 A simplified forest soil nutrient cycle. Broken lines represent additions to or removals from the system. Source: Sanchez, 1976.

burning. These methods might serve to conserve organic matter, but have practical disadvantages, for fire is the cheapest and most convenient tool of the unmechanised cultivator, and the fertiliser and pH-raising effects of the ash which often contribute substantially to raising yields in the first year or two of cropping would be lost. In experiments carried out on acid soils at Kade, in the wet forest zone of Ghana, clearing by the traditional method – felling most of the forest vegetation, except for a few shade trees, letting the dead material dry out, putting it into heaps and burning it – was compared with the cheaper method of killing the forest vegetation by painting on a hormone. Yields of plantains, and cocoyam (*Xanthosoma sagittifolium*), were considerably higher in the traditionally cleared

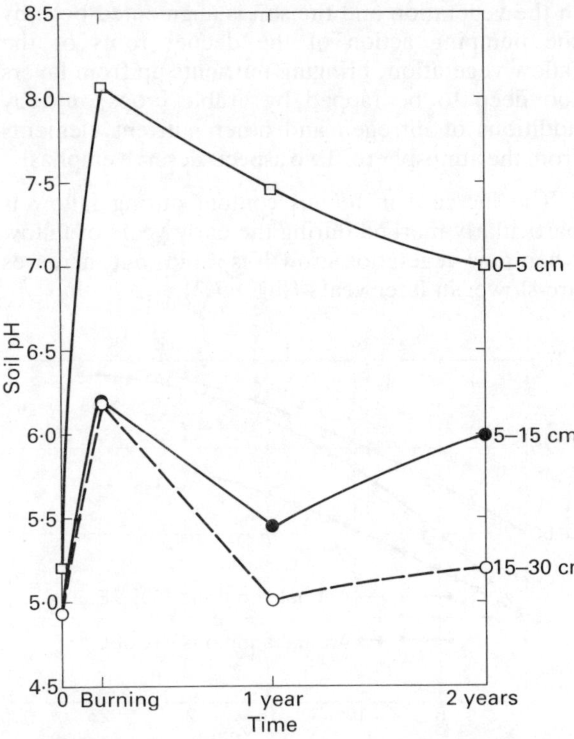

Fig. 17.2 Changes in soil pH after burning original forest at Kade, Ghana. Source: **Nye, P. H. and Greenland, D. J.** (1964) 'Changes in the soil after clearing a tropical forest', *Plant and Soil* **21**, 101–12.

plots than in those that had been cleared by poisoning, with the added inconvenience in the latter case that dead trees and branches fell progressively over a period.

The nature of the original clearing affects both what happens to the nutrients stored in the vegetation and the subsequent soil changes during cropping. Methods of clearing differ considerably. In some areas all the standing vegetation is felled and burnt, in others some trees are left as shade. Putting material in heaps prior to burning concentrates both the effect of the heat on the soil, and the distribution of the ash produced, and may thus contribute to the natural heterogeneity of most tropical soils. Ash may be washed into the soil, or lost by wind or water.

Soil changes during cropping might be abrupt at first since a soil formerly protected and shaded by natural vegetation is now exposed to sun and rain. Erosion might be particularly severe in the period during which the soil is unprotected between clearing of the vegetation and the development of a crop canopy. The organic matter content of the soil declines during cropping owing to the removal of the shade and the rise in soil surface temperatures (Table 17.1), and to the fact that the soil no longer receives the organic material supplied by the vegetation that provided the raw material for humus formation.

Table 17.1 Relative humidity and soil temperature differences between adjacent high forest and cleared patches at Kade, southern Ghana, as recorded by University of Ghana students on a field exercise. The cleared plot has a lower relative humidity and much higher soil surface temperatures than the adjacent plot under forest.

		High forest	Adjacent cleared plot
Relative humidity	2.30 pm	91%	62%
	5.30 pm	100%	89%
Soil temperature at 3 pm (°C)			
	Surface litter	28.0	40.3
	at 7.5 cm	25.0	30.0
	at 36.0 cm	24.8	28.0

With the decline in organic matter levels during cropping comes a deterioration of related soil properties such as structure, porosity and rainfall acceptance. These changes are independent of any additional physical loss of topsoil due to removal by accelerated erosion.

Yield declines during cropping are often more marked for soils cleared from forest or forest regrowth vegetation than they are for savanna soils. The latter are normally initially poorer in organic matter and nutrients than forest soils but show slower yield declines.

Declines in crop yields during the cropping period are associated with the following:

1 Multiplication of pests and diseases.
2 Increase of weeds.
3 Erosion of the topsoil.

4 Deterioration in the soil physical condition.
5 Deterioration in the soil nutrient status.
6 Changes in the numbers and composition of soil fauna and flora.

Of the above, (4) to (6) are much influenced by declines in soil organic matter. The progressive lowering of the soil organic matter content is normally associated with the following:
1 Loss of crumb structure in the topsoil.
2 Lower total porosity and lower macroporosity.
3 Poorer aeration.
4 Poorer rainfall acceptance.
5 Increased likelihood of runoff and surface erosion.
6 Lower cation exchange capacity.
7 Lower release of nutrients due to the smaller amounts of humus mineralised.

Changes in soil productivity during cropping can be assessed and monitored through the evidence of (a) crop yields, (b) soil analyses, and (c) changing responses to fertiliser applications. The evidence of crop yields is complicated by the fact that crops are often grown together, in mixtures, and that the cropping period may involve a succession of different crops, the yields of which cannot be directly compared. It often appears that a farm patch is abandoned more because of weed problems and the increasing amount of work needed to weed it than because of changes in soil fertility. It may need less work to clear a new area than to continue to weed the old, and pest and disease problems may reinforce the desire to abandon a plot even if it is still fairly fertile. Clearing is done in the dry season when other farm work may be at a minimum, but weeding comes during the rains when there may be too much to do. Although in some cases soil exhaustion is the main cause for abandoning a plot, in others it clearly is not.

Soil changes during fallow

The soil changes during the fallow period are comparable to those that take place during cropping, but are in the reverse direction. As in the case of changes during cropping they are very much associated with changes in soil organic matter levels which rise again during the fallow period and bring associated improvements in the physical and chemical properties of the soil. The nutrient pool stored in the vegetation and the soil is augmented both by the pumping action of the deeper roots of the fallow vegetation, bringing nutrients up from layers too deep to be tapped by arable crops, and by additions of nitrogen and other nutrient elements from the atmosphere. Two aspects deserve emphasis:

1 The increase in humus content during fallow is particularly marked during the early years of fallow when new vegetation growth is rapid, but increases are slower in later years (Fig. 17.3).

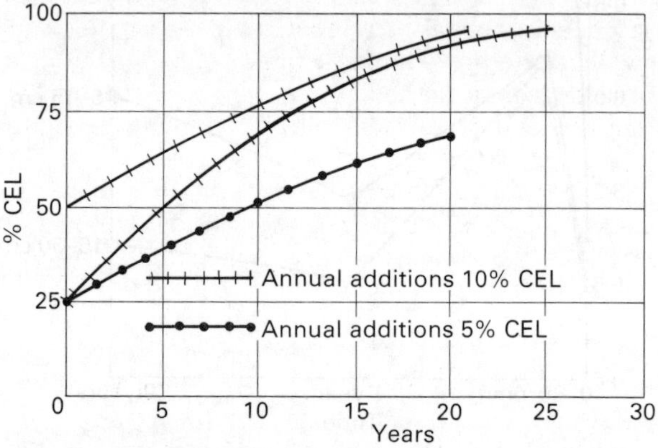

Fig. 17.3 Graph showing the calculated increase in humus content during 20–25 years of forest fallow when the fallow vegetation adds humus carbon at annual rates equivalent to 5% (lower line) and 10% (upper lines) of the carbon contained in the soil at equilibrium level. The carbon equilibrium level (CEL) is reached when gains balance losses and there is no further net change. Assumed initial humus levels are 25% and 50% of the carbon equilibrium level.

The curve showing humus increases is steepest with the lower initial carbon level and the higher rate of annual additions. In all cases, annual carbon increases are highest in the first years, then decrease progressively as the curve flattens. The optimal length of fallow should include the steep initial part of the curve when annual additions are highest. Later annual humus increments become smaller and smaller so that there is less and less gain from keeping the land out of production, but other considerations, such as the need to smother noxious weeds, may also be locally important. Source: **Ahn, P. M.** (1979) 'The optimum length of planned fallows' in **Mongi, H. O. and Huxley, P. E.** (eds), *Soils Research in Agroforestry*, 539–68. ICRAF: Nairobi.

2 Forest regrowth vegetation usually grows faster and is more effective at restoring humus levels than is savanna regrowth.

A knowledge of soil changes during the fallow period is helpful if we wish to consider ways of inducing the same changes more quickly or by alternative methods, such as by planting a quick-growing leguminous fallow rather than relying on spontaneous natural regrowth vegetation. The rates at which changes take place and organic matter levels increase is also important. If the build-up of organic matter is greatest in the first few years, and subsequent annual increments become smaller and smaller as a new humus equilibrium level is approached, then the benefit of each succeeding year that the land is unproductive also becomes less and less.

A desirable length of fallow would be one that continues as long as the net gain curve is still steep, but is stopped when the curve begins to flatten and annual humus increments become very small. However, in practice fallow lengths depend on the availability of land. Other functions of the fallow, including that of suppressing obnoxious weeds of cultivation, such as *Imperata cylindrica*, and giving a disease break, have also to be considered.

Advantages and disadvantages of shifting cultivation

Although initially criticised as primitive and wasteful, most agronomists and soil scientists now recognise that shifting cultivation works well, is labour efficient, and is adapted to areas of the tropics with fairly low population densities. In many areas of the tropics it has proved difficult to introduce acceptable, alternative systems of annual crop production that do not require a fallow period. The spectacular failure of some attempts to introduce sudden changes suggests that, in order to succeed, innovations may best be in the form of progressive improvements to existing systems which aim at retaining their advantages while minimising or eliminating their disadvantages.

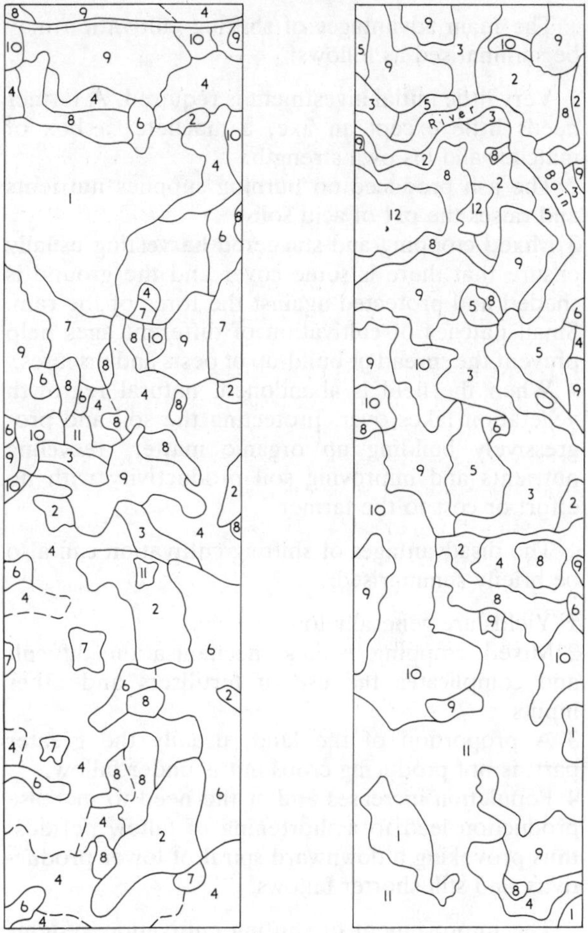

Fig. 17.4 The pattern of shifting cultivation: vegetation mosaics typical of moderately closely farmed forest areas. The vegetation and land use were recorded on two sample strips, each 1.6 km long and 0.4 km wide, in the western forest zone of Ghana.

1 Original forest.
2 Secondary forest.
3 Secondary swamp forest.
4 Thicket regrowth.
5 Swamp thicket.
6 Forb regrowth.
7 Newly cleared land.
8 Land under current food crop cultivation.
9 Cocoa.
10 Abandoned cocoa amid forb or thicket regrowth.
11 Abandoned cocoa or coffee amid secondary forest.
12 Built-over land (village site).

Source: Ahn, 1970.

The main advantages of shifting cultivation may be summarised as follows:

1 Very little initial investment is required. A farmer needs little except an axe, a machete, a box of matches and his own strength.
2 The ash produced on burning supplies nutrients and raises the pH of acid soils.
3 Mixed cropping and staggered harvesting usually ensure that there is some cover and the ground is shaded and protected against the force of the rain. Small patches of cultivation of different ages help prevent the spread or build-up of pests and diseases.
4 When the field is abandoned, natural regrowth vegetation takes over, protecting the soil and progressively building up organic matter, recycling nutrients and improving soil productivity with no effort or cost to the farmer.

The disadvantages of shifting cultivation can also be briefly summarised:

1 Yields are generally low.
2 Mixed cropping makes mechanisation difficult and complicates the use of fertilisers and other inputs.
3 A proportion of the land, usually the greater part, is not producing crops but is under fallow.
4 Population increases and/or the need to increase production lead to a shortening of fallow periods, thus provoking a downward spiral of lower productivity and still shorter fallows.

The improvement of shifting cultivation systems should aim at retaining its advantages while reducing or eliminating the disadvantages.

The management and improvement of shifting cultivation systems

Fundamental to an understanding of small farmers practising shifting cultivation in the tropics is that most cultivators are small subsistence farmers who cannot afford to take risks. The system has to ensure that the subsistence farmer always has some food of some sort, and to minimise the risk of total failure. The innovator comes with the sometimes alien idea of increasing yields per hectare of ground, whereas the shifting cultivator might prefer to spread his risk by having several low-yielding patches scattered about rather than one high-yielding one that might fail. Often he cannot find the money needed to pay for inputs to raise yields, and is more interested in returns per day of work than per area of land.

Since the main disadvantages of shifting cultivation are the generally low yields and the high proportion of the total land that is under fallow, the overall productivity of the system can be raised, first, by increasing yields and, secondly, by lengthening the cropping period. Both of these are influenced by numerous factors apart from soil factors.

A third way of raising the productivity of the system is to find ways of obtaining the beneficial effects of the fallow in a shorter time, or with planted fallows if they are more efficient than natural ones.

A fourth approach is to eliminate the fallow by a rotation, if such a rotation can be found, that keeps the land constantly productive.

A fifth approach is to plant perennial tree crops rather than cultivate a succession of annuals. In areas where the rainfall distribution favours tree crops, the last approach has proved successful and been widespread. The establishment of cocoa farms in the West African forest zone traditionally began with clearing of high forest followed by planting of

Fig. 17.5 A forest clearing amid high forest, western Liberia.

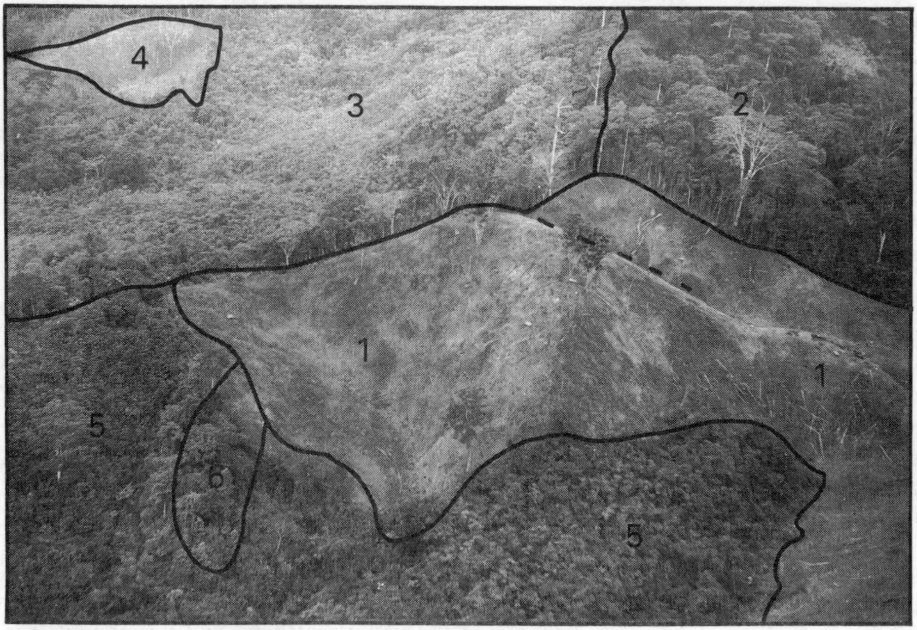

Fig. 17.6 Stages in shifting cultivation in a hilly forest area.
1 and 4 Newly cleared areas. 2 and 6 Original forest. 3 and 5 Forest regrowth vegetation (fallow).

food crops, particularly plantains and cocoyams (*Xanthosoma sagittifolium*), both of which have large leaves and provide shade for the cocoa seedlings planted between the food crops. After three to six years, depending on soil conditions and the variety of cocoa planted, the cocoa began to yield and the food farm gave way to a permanent tree crop which was established with very little extra effort above that which had already been expended on establishing the food farm. In a comparable way other tree crops can be established amid annuals, to take over from them as they mature.

Increasing yields within the cropping period without altering the basic cycle of cropping and fallow has been successful where improved practices are within the resources of the farmers concerned. Improved practices include the use of improved varieties, better planting densities, and more timely planting, as well as the use of fertilisers and insecticides.

The use of fertilisers can increase yields only where soil fertility is the limiting factor and where the plant variety used is genetically able to respond to an improved nutrient supply. Many other factors may limit yield responses. These include poor physical soil properties, water shortages, inadequate management (including weeding), and poor husbandry methods. Farmers will use fertilisers only if the increase in yield is profitable, and this depends on the selling price of the extra yield produced in relation to the local cost of the fertiliser used to obtain it.

According to some investigators, possibly the most important single factor responsible for low yields in many parts of the tropics, and particularly in Africa and South America, is wrong time of planting, which is usually too late. Late planting shortens the growing season, makes less use of the available water, and reduces yields for other reasons, including lessened benefit from the nitrogen flush that is sometimes associated with the first rains or with rains following a dry spell (Chapter 9). Late planting may be due to caution, especially where the date of the onset of the wet season is uncertain, and partly to the fact that at planting time the farmer is too busy to do all that has to be done on time.

Other factors responsible for low yields are wrong planting densities, which are often too low, and delayed or inadequate weeding. In an investigation of maize yields on small farms in western Kenya it was found that good husbandry was more important in raising yields than were fertiliser applications. Good husbandry (GH) was defined as early planting at the start of the rains, a density of 36 000 plants/ha, and clean weeding up to tasselling time. Bad husbandry (BH) was described as planting four weeks late with a low population density of only 18 000 plants/ha and only one late weeding. These two types of husbandry were investigated with a local maize variety and using no fertiliser (LM, no F), and with a hybrid maize with fertiliser (H + F). Average yields for the four combinations were as follows:

BH + LM, no F	1920 kg/ha
BH + H + F	3270
GH + LM, no F	4880
GH, H + F	8020

These results indicate that with good husbandry but no fertiliser, considerably higher yields were obtained than where fertiliser was used but husbandry was bad. In the circumstances investigated, therefore, correct date of planting and planting densities and better weeding gave a much larger yield increase than did the use of hybrid maize and fertilisers without these good husbandry practices.

Shortening the fallow period must be considered in relation to the soil changes during fallow (see Chapter 7). A desirable length of fallow was described as one that continues as long as the curve of net organic matter gain is still steep. The steepness of the curve and hence the rate of gain is related to the type and rate of growth of the fallow vegetation. Many attempts have been made to plant leguminous species that are thought capable of fulfilling the functions of the fallow more quickly and efficiently than spontaneous natural fallow growths, but artificial fallows are very little used as yet. Natural fallows vary so much in floristic composition that generalisations are difficult to make, but in general the species that invade abandoned land are the most vigorous and aggressive colonisers locally available. Even if a pure stand of a plant such as a fast-growing legume were shown to be more efficient at building up organic matter than the local natural fallow species, the farmer would have to weigh the cost of planting and establishment against the extra benefits obtained. Small tropical farmers have so far shown little interest in deliberately planting a fallow merely to improve soil productivity, though possible additional benefits, such as increased fodder supplies or timber production, might make planted fallows more attractive.

It has in practice proved difficult in many shifting cultivation areas of the tropics to produce arable crops year after year without a fallow period. In these areas there are two proved alternatives to shifting cultivation, namely swamp rice, generally in valley bottoms or on terraces, and permanent or semi-permanent tree crops. An integrated landscape using different parts of a soil catena in different ways might have permanent irrigated rice or bananas in the valley bottoms, and permanent tree crops on upland areas which, because of their steep slopes or for other reasons, require protection or are not suited to arable cropping. Annual crops would thus be confined to those parts of the catena most suited to them. This concept can be applied on a broader scale to a country or region with different ecological zones. In West Africa, for example, many of the coastal states have a southern zone of forest with areas of savanna in the interior. An ecologically desirable land use policy in these cases would be to emphasise tree crops in the wet forest areas of generally hilly relief, and annual crops in the more seasonal rainfall areas of the interior where gradients are generally gentle and the risks of accelerated erosion lower.

In a broad context, the forces that progressively modify shifting cultivation are related to population densities and the pressure on the land to produce crops. In areas of low population density where land is relatively plentiful, small farmers may feel little incentive to increase yields per unit area or to look for ways of shortening or eliminating fallow. In areas of very dense population, pressure on the land will progressively shorten fallows, and higher food prices will pay for the inputs needed to intensify agriculture and raise yields.

18 Soil management in multiple cropping systems

Multiple cropping systems in the tropics

In the majority of tropical countries most of the food produced comes from small farms. In tropical Asia, 75% of farms are less than 2 ha in area, and in Central America about 70% are less than 5 ha. The average farm size for 20 tropical African countries in 1973 was about 5 ha.

An important and widespread feature of small farms in the tropics is the practice of multiple cropping – the growing of more than one crop on a piece of land in a year. Multiple cropping has received increasing attention from agricultural scientists in recent years.

Multiple cropping has many forms, reflecting the different ways that two or more crops may be combined (intercropping systems), or succeed each other on the same piece of land in the same year (sequential cropping systems).

In **intercropping systems**, two or more crops are grown at the same time. In **mixed intercropping** the crops are mixed together in a random, irregular way. In **row intercropping**, one or more of the crops is in regular rows. In **strip intercropping**, crops are grown in strips that are narrow enough for the crops to influence each other. In **relay intercropping**, two crops overlap in time, as when a second crop is interplanted among a first crop when it has reached an appropriate stage of its development but before it is harvested.

In **sequential cropping systems** crops are not grown at the same time, as in intercropping systems, but follow each other on the same land during one year. Depending on the climate and on the length of growing season of the crops concerned, two, three or four crops may be grown in sequence in one year, to give double, triple or quadruple cropping, but each crop is harvested before the next one is planted, so there is no overlapping of crops as in relay planting. **Ratoon cropping** is a form of sequential cropping in which the harvested crop is

Fig. 18.1 Mixed cropping of sorghum and groundnuts.

Fig. 18.2 Intercropped sorghum and pigeon pea at the ICRISAT centre, Hyderabad, India. This is one of many crop combinations tested by ICRISAT for use in the semi-arid tropics. Pigeon pea has also proved its value as an alley crop.

allowed to grow again after harvest, as with rice, sorghum or sugarcane, to produce further harvests from the same plants.

In most small farms of the tropics, other than flooded rice farms, crop mixtures rather than single stands are dominant. Intercropping systems sometimes produce greater yields than an equal area of land devoted to single stands of the same crops because the mixing of crops gives a better utilisation of space, time, and the resources of sunshine, water and soil nutrients. The ways the crops are arranged when they are grown together can be such as to minimise competition for these resources, as when tall crops or trees are mixed with shorter, shade-tolerant ones, or when the root systems of the crops are such that to some extent they exploit different soil layers. Nevertheless, the often greater total yields associated with mixed cropping are often probably of less concern to many tropical subsistence farmers than is risk avoidance. They diversify in order to minimise the risk of total crop failure due to climatic conditions, pests or diseases.

The most intense and complex intercropping systems in the tropics are often found in areas where rain and temperatures permit crop growth all the year, but multiple cropping is also practised in those extensive areas of the tropics with seasonal rainfalls, as well as in dry areas with irrigation.

In shifting cultivation, a random mixture of crops is generally planted, with different crops maturing and being harvested at different times or over an extended period, so that the ground is never bare or completely exposed to accelerated erosion.

In the Amazon area of South America, farmers may plant rice, cassava and plantains at the same time, together with vegetables. When first the rice and later the cassava are harvested, other crops are planted amid the remaining plantains.

In the forest areas of West Africa, farmers may plant a complex mixture of tall plantains, cocoyams (*Xanthosoma sagittifolium*), and various smaller crops such as groundnuts, peppers, okra and other vegetables, generally all on the flat. In other areas, as in Nigeria, different crops may be planted on different parts and positions on mounds and ridges, so that cassava and yams may be on the upper parts and maize and other crops on the sides, with crops more tolerant of poor drainage, such as rice, between the mounds. Many hundreds of different intercropping systems are found, some of them complex.

In areas of more seasonal rainfall and cropping, intercropping patterns include maize and beans – a particularly widespread combination in South America but also in East Africa and elsewhere – maize and upland rice, and maize with pigeon pea, millet and sorghum. In such cases crops are often in rows rather than randomly arranged. In the drier areas of Kenya with two wet seasons, pigeon pea is interplanted with the first-season maize but matures during the second wet season when maize is often planted again. If the second rains are inadequate and the second-season maize fails, the deep-rooted pigeon pea provides some yield from the plot.

Many successful crop combinations feature crops that have different lengths of growing season so that the maximum demand for light, water and nutrients may occur at different times. In the Philippines, maize and mung beans (*Phaseolus aureus*) are planted at the same time but the beans flower 35 days after planting, before the maize is tall enough to shade them, and are harvested after 60 days, when the maize requirements are at their highest. Other examples include 128-day maize with 168-day pigeon peas in Trinidad, and 75-day millet with 195-day sorghum in Nigeria.

When crops are not planted at the same time, but the second crop is planted between the first at a later stage of its development but before it is harvested, the two crops overlap to give relay cropping. The very widespread maize and beans intercropping of much of South and Central America often forms an example of this. The climbing beans are interplanted among the maize when the cobs have formed but are not yet mature. The maize cobs are bent so that they hang down but mature normally. The beans climb the maize stalk, making use of stored soil moisture, and mature in a later dry period when diseases are less prevalent than in the main wet season. Both crops are harvested together in the early part of the dry season. In Taiwan, rice is relayed with melons, cabbages and maize, and in East Africa pasture is planted between growing maize to give a maize–pasture relay.

Another form of intercropping features the growing of annual crops with perennials. When the perennials are first planted, tall crops such as plantains, cocoyam, maize or cassava may be used to shade the seedlings of such trees as cocoa, coffee or rubber, and to provide an income before these crops come into bearing. A range of crops is sometimes grown interplanted among young sugarcane. When the tree crops are fully grown the situation is reversed, and it is the trees that now shade the plants grown between them. These plants may be shade-tolerant perennials, or pasture, or annual food crops. In some complex cases the crops may be a combination in three layers of tall trees, medium-sized intermediate plants such as cocoa or pawpaw, and low crops such as groundnuts or pineapples. Such combinations may resemble the strata of some natural forests, and make good use of space and light.

The combination of tree crops with annuals is one aspect of agroforestry, a term coined relatively recently which deals with the study of a very old practice (Chapter 19).

Fertiliser use in multiple cropping systems

Most of the information available on fertiliser use in the tropics relates to the growing of single stands of crops, or monocultures, particularly of commercial crops. Much less is known about soil management and fertiliser use in relation to the many types of multiple cropping practised. This is particularly true in relation to crops grown on small farms for domestic consumption.

By combining different crops at different times, and with different spacings and arrangements, the agronomist can minimise competition for light. The competition for water and for nutrients is usually more severe, and more of a limiting factor for crop production. The development of the root systems of two or more crops grown together can be considered both in space and time. In space, the root systems of different crops may tend to avoid each other, particularly below the topsoil, even if they intermingle at the surface. In an investigation of row intercropping of maize and groundnuts it was found that the groundnut roots were more in the upper soil layers and that the maize roots were at greater depths. In time, competition may be lessened if a second crop is planted later than the first. Most root systems cease to develop after flowering, after which the uptake of water and nutrients decreases until harvest. A second crop which develops its root system after the first has flowered would therefore face less competition than if they both developed and reached the stage of maximum intake at the same time.

The total nutrient uptake of two or more crops grown together usually exceeds the nutrient uptake of single stands of equivalent area of the same crops. However, the nutrient demands of the crops forming the mixture may differ from each other. Both these considerations affect the amount and type of fertiliser it may be profitable to apply.

Trials with maize and pigeon pea in Trinidad showed that one hectare of the two crops grown together accumulated 30–63% more nutrients than two half hectares of single stands. The increase was greater for row intercropping than for mixed intercropping. The ratio of intercropped yields to yields of the same total area of single stands is known as the relative yield total, or RYT. The RYT for dry matter production, and also for the major nutrients nitrogen, phosphorus and potassium, was about 1.5 for mixed intercropping and 1.8 for row intercropping. In a row intercropping trial of short-season millet and long-season sorghum in northern Nigeria, the RYT for nitrogen uptake was also about 1.8.

When nutrient uptake is increased by intercropping, this may be because the roots of mixed cropping systems draw nutrients from a larger total soil volume, or because a similar soil volume is exploited more intensively and effectively. If some of the roots in mixed stands go deeper than in single stands, these deeper roots may take up nutrients moving downwards in soil water that would otherwise be leached beyond root depth and therefore lost. The increased nutrient uptakes commonly found in mixed crop stands are normally no greater for the more mobile nutrients nitrogen, sulphur and potassium, than they are for those nutrients that move less easily in the soil – phosphorus, calcium and magnesium. This suggests that intercropping results in the exploitation of an increased total soil volume.

Where non-legumes are intercropped with nodulating legumes, additional benefits may be obtained from the added nitrogen fixed by the latter. In trials at Ife, Nigeria, maize was intercropped with shade-tolerant legumes such as *Calopogonium muconoides*, cowpeas and mung beans, which were incorporated in the soil after both crops were harvested. For four successive crops the *Calopogonium*-intercropped maize produced yields comparable with those of maize grown with fertiliser, thus saving the cost of the latter. In India, intercropping sugarcane with rows of *Crotolaria juncea*, later incorporated as green manure, was found to have the same effect on sugar yields and juice quality as an application of of 168 kg/ha of nitrogen. Intercropping with shade-tolerant legumes can increase crop yields, but the legume must not compete seriously with the main crop for water and nutrients and it must have a growing period long enough for it to fix appreciable amounts of nitrogen.

If intercropping results in the more efficient use of soil nutrients, it may also result in the more efficient use of applied fertilisers. This was found to be the case with nitrogen fertilisers applied at rates up to 180 kg/ha to a maize and upland rice row intercrop in the Philippines. However, much depends on the relative responses to added nutrients of each crop in the mixture. At Turrialba, in Costa Rica, large numbers of different crop combinations have been carried out. In the trial involving the intercropping of maize, cassava and beans followed by sweet potatoes, increasing applications of a mixed NPK fertiliser increased the cassava and climbing bean yields, but lowered those of sweet potatoes. The yields of soya beans planted instead of sweet potatoes were not increased. Crops planted together may differ, therefore, in their responses to a fertiliser application. Responses will be influenced by the crops and crop varieties grown, by the length of growing season, and by the local soil and climate.

In sequential cropping systems the second crop is planted after the first is harvested. The crops are single stands, and follow each other much as they do in an annual rotation except that the time interval between crops is shorter. The influence of a crop on the succeeding crop may be greater for this reason.

In irrigated lowland rice farms in the tropics, double or triple cropping of rice transplanted from a nursery is widely found. Double or triple cropping is also found in some shifting cultivation areas, as in the rice–maize–soybean sequence in the Amazon area of Peru. Without irrigation, such systems are found in areas that have rain all or most of the year. With irrigation, crops grown in the dry season may benefit from higher sunshine levels at that time of year.

The influence of a crop on a succeeding crop may be beneficial or harmful. The influence can take many forms, since it may involve soil nutrient levels, soil water storage and the carry-over of moisture from one crop to the next, the modification of soil physical factors such as structure, porosity and aeration, or pest and disease factors.

In areas where water is limiting, the extent and depth to which a crop depletes the soil profile of water may be particularly important to the success or failure of a succeeding crop. Some deep-rooted natural savanna fallows deplete soil moisture more thoroughly than do annual crops. Stored water may be higher therefore after cropping, particularly in the case of a shallow-rooted crop such as beans or potatoes, than after a natural fallow or a deep-rooted crop.

The residual effects of a fertiliser on a succeeding crop are of particular interest in sequential cropping systems. Residual effects will vary with the rate and type of fertiliser applied to the first crop and the amount recovered by it, the rainfall pattern and possible leaching, and the extent of nutrient immobilisation or denitrification. It is generally thought that the relatively immobile nutrients – phosphorus, calcium and magnesium – are more likely to have a residual effect than is a relatively mobile nutrient such as nitrogen, though some cases of residual nitrogen effects have been observed.

19 Soil management for tropical tree crops and for agroforestry

Advantages of perennial tree crops

In a natural forest, nutrients circulate in a closed cycle from trees to soil and back again (Chapter 17). Losses from such a system may be very small since little is exported and the trees protect the soils from accelerated erosion. The stock of nutrients in circulation between soil and vegetation may even be progressively built up by additions from the atmosphere and from the deeper layers of the soil.

Under a perennial tree crop cover, conditions may sometimes approach those under an original natural forest except that there is a more systematic export of nutrients in the harvested crop. Nutrients may be tapped by deep roots from the lower soil layers where they are released by the decomposition of minerals in the parent material or where they have been washed down by leaching. The tree crop, though it may not have the multilayered structure of some natural forests, may nevertheless provide adequate protection against erosion by breaking the fall of the rain. As in a natural forest, the tree cover also provides shade, lowers soil surface temperatures and modifies the microclimate beneath the canopy.

These effects vary between tree crops according to the configuration and spacing of the trees, and their age. Oilpalm or coconut palms spaced at 60 trees/ha require several years before the crowns are large enough to give a fairly complete cover, and during the early years a cover crop or food crop can be grown between the palms. The tea plant, in contrast, is planted much more closely and from about the third year provides a complete cover, effectively protecting the soil from erosion, even on steep slopes. In many respects, tea is the ideal land use in steep areas of heavy rainfall and illustrates well the advantages of perennial tree crops. The dense tea bushes allow no rain to reach the ground directly. The ground surface is shaded not only by the bush itself but by the usually dense layer of dead leaves, twigs and prunings which forms a thick mulch at the soil surface, and which returns nutrients and supplies organic matter to the soil as it decomposes. Under a well-fertilised tea cover, soil fertility can be progressively built up, and the system forms an excellent example of sustainable agriculture.

Exports from the soil–plant system under perennial tree crops are mainly in the harvested produce, and can thus be estimated when the weight and approximate composition of the harvest is known. Nutrients exported in this way vary very greatly between crops, being relatively high, for example, for cocoa and coffee, and remarkably low for rubber (see Chapter 16).

Soil fertility maintenance under perennials

The maintenance of soil fertility under semi-permanent perennial crops is in some ways easier than under arable annual crops, and farmers may be more inclined to take a long-term view and more ready to invest in maintaining fertility. As discussed in Chapter 16, soil fertility has to be considered in relation to the phases of (1) the nursery, (2) crop establishment in the field, (3) the growing period between establishment and maturity, and (4) the production phase when nutrients are regularly exported from the soil–crop system.

For most tree crops, fertile topsoil reinforced with fertilisers and manures as necessary is normally used in the nursery, and considerable expense and effort may be devoted to digging adequately large planting holes and putting in them manure and fertilisers designed not only to stimulate initial rooting but also to serve as a long-term reserve. In both these respects, the importance of phosphatic fertilisers is often emphasised.

Thereafter, fertility maintenance requirements during first the growing period and then the production period vary widely between tree crops, and even between varieties of the same crop, and in relation to different soils and climates, as outlined for a range of the more important tree crops in Chapter 16. For coconut palms, for example, nitrogen is emphasised prior to bearing, but once the palms are in production and vegetative growth is slower, potassium is often considered the main or only need. Similar considerations apply to oilpalm. For tea, in contrast, nitrogen continues to be the main nutrient supplied throughout the productive period, since it is the leaf that is harvested, and not the seed. Coffee is commonly supplied with phosphorus during establishment and the growing period, but this nutrient is less often given to the crop when it is in production. The nutrient needs of rubber were traditionally considered to be very low because the latex exported is a hydrocarbon containing very little in the way of nutrients, but current thinking now emphasises that it may be profitable to fertilise immature trees, particularly with nitrogen and phosphorus, in order to bring them more quickly to tapping girth.

Leaf analysis is particularly valuable for perennial crops for monitoring nutrient deficiencies and to enable them to be corrected before they have an appreciable effect on yields (Chapter 13).

Agroforestry principles and practices

Agroforestry is a relatively new term denoting initially a combination of agriculture and forestry, but now somewhat widened to include any combination of trees with annual crops and/or pasture in which there is both an economic and an ecological interaction between the tree and the non-tree components. To its exponents agroforestry is 'a sustainable land management system that increases the overall yield of land', and it includes all agricultural systems where trees are added to annual or perennial crops and grasses or combined with animals. The term as now employed by its advocates thus includes a very wide spectrum of land use combinations (Table 19.1), including combinations that have been in use for a long time, although there is also an emphasis on finding and improving useful trees little used or known at present, and new combinations. Others emphasise agroforestry as a desirable replacement or improvement of systems that are degrading under the pressure of increased population densities in areas with low inherent potential for intensive agriculture. Many of the combinations studied by agroforesters, including those involving the use of perennial tree crops with annuals, are examples of mixed cropping, a subject that has attracted increased research attention in recent years (Chapter 18).

Table 19.1 Some agroforestry practices

Improved shifting cultivation	Trees as shelter for crops, animals and homesteads
Planted tree fallows	Trees for soil conservation
Taungya	(bunds, terraces, strips, hedges)
Alley cropping	Trees in water management
Boundary planting	Plantation crop
Live fences	combinations:
Multipurpose trees on cropland, range and	– with lower-storey trees/ shrubs
improved pastures	– with herbaceous crops
Woodlots (multipurpose	– with upper-storey trees
management)	– with pastures, livestock
Dune fixation	Livestock under trees
Aquaforestry	Apiculture with forestry

Adapted from the International Council for Research in Agroforestry agroforestry systems inventory.

The mechanisms by which trees may improve and protect the soil have been looked at afresh by agroforesters. They are fundamentally those dealt with in Chapter 17 on shifting cultivation and the role of fallow vegetation in restoring soil fertility and building up favourable physical soil properties by recycling nutrients and increasing the soil organic matter content. A recent additional hypothesis concerns the behaviour of litter when it is derived from a mixture of sources, both woody and herbaceous. It is conjectured that litter of mixed origin might display a wider range of desirable properties and qualities than litter from a single source, though this remains to be proved and quantified.

Agroforestry in the narrow sense was at first equated with agrisilviculture as exemplified by *taungya* farming, a system of forest establishment that has been successfully practised in areas of the humid tropics, originally for teak establishment in South-east Asia and more recently with *Gmelina arborea* plantings in West Africa.

The *taungya* method consists essentially of clearing or partially clearing forest and then growing tree seedlings in combination at first with food crops which are later eliminated when the tree seedlings have reached a certain size. The advantage of the method is partly one of cost since the areas to be planted are divided into plots. These are given to farmers who clear, plant and tend the seedlings in return for the food crops they grow between them. Alternatively, hired labour is employed to carry out the operation and the food crops are grown to offset part of the establishment costs.

Recent studies in Nigeria, Suriname, Malaysia and Costa Rica have found that *taungya* systems substantially cut the cost of reafforestation, both where the farmers took the food they produced and where they were paid for their work.

The *taungya* system forms an example of perennial/annual intercropping that is comparable to purely agricultural combinations such as oilpalm or coconuts with annuals. Agroforestry as now understood and defined by its advocates and practitioners forms an updated, expanded and sometimes more intensive version of the original *taungya* concept widened to embrace the study of all combinations of trees with annuals or short-lived perennials or pasture in a wide variety of ecological situations from the wettest tropics to the Sahel. As such, the subject has attracted to its ranks both foresters and agronomists.

In West Africa, cocoa and coffee farms have been traditionally established by planting tree seedlings in the shade of mixed plantains and cocoyams (*Xanthosoma sagittifolium*) which are progressively eliminated as the cocoa or coffee grows. The food produced in the first two to three years may be regarded as a by-product of the cocoa or coffee establishment.

There are many other combinations of annuals and perennials, and some of these are relatively long lasting. A well-known example of an intensive tree-cropping system is found in the Kandy gardens of Sri Lanka. Here, tall trees, including coconuts, kitul and betel palms, are associated with other trees (cloves, cinnamon, nutmeg, citrus, mango, durian, jackfruit, rambutan and breadfruit) which are less tall. A lower storey is formed by bananas and pepper vines. Particularly at the edges, crops

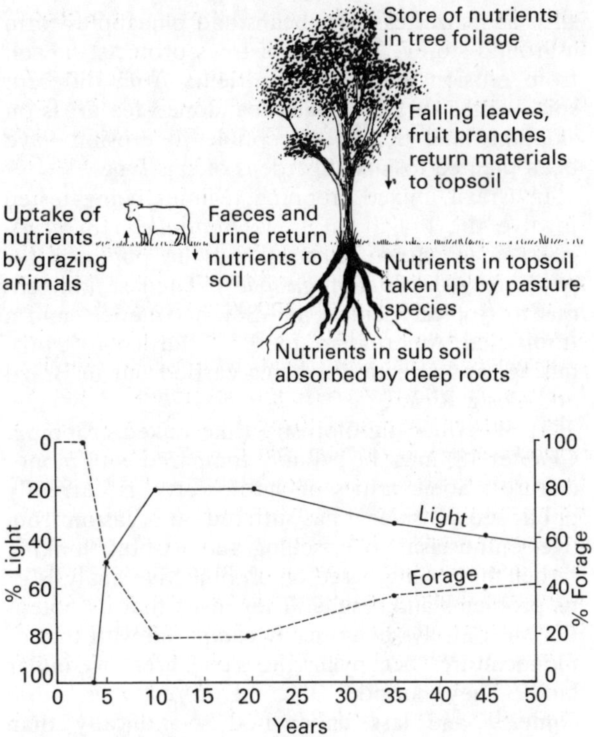

Fig. 19.1 Agroforestry practices listed in Table 19.1 include plantation crop combinations with pastures and livestock. This diagram represents components of a livestock–pasture–tree crop system (above) and the availability of forage in a rubber plantation (below).

In the lower diagram the solid line represents the estimated percentage of light reaching the ground, and the broken line the estimated quantity of forage available as a percentage of forage available on open land under similar conditions. The reciprocal nature of these two curves shows how forage production is related to the amount of light intercepted by the tree crop. Source: **Payne, W. J. A.** (1985) 'A review of the possibilities for integrating cattle and tree crop production systems in the tropics', *Forest Ecology and Management* **12**, 1–36.

such as maize, cassava, beans and pineapples form a ground storey. The mix of trees protects the soil from erosion and cycles nutrients from different soils depths. In Sri Lanka, abandoned tea areas on slopping land that is susceptible to erosion have been planted to mixed gardens of this type.

In Brazil, mixed cropping regimes being tested involve the brazil nut tree (*Bertholletia excelsa*), guarana (*Paullinia cupana*) and the pupunha (or pejibaye) palm (*Bactris gasipaes*). The last, indigenous to tropical Central and South America, and a traditional Amerindian food, is particularly promising, and selection is being carried out in Brazil and Costa Rica.

To its critics, agroforestry (like mixed cropping, Chapter 18) must be suitably compared with monocultures. Some critics of agroforestry as currently publicised consider it has suffered on occasion from over-enthusiasm, over-selling and wishful thinking rather than being based on an objective analysis of its problems and results. They insist that its potential must always be critically compared with that of monocultures occupying the same area of similar land. They consider that agroforestry is more complex and less understood scientifically than monocultures, and that its scientific testing is correspondingly lengthy, difficult and costly. However, many of the soil management principles discussed in relation to tree crops in the earlier part of this chapter and, in relation to mixed cropping, in Chapter 18, apply to most agroforestry situations.

Agroforestry management problems

The variety and complexity of some tree/annual crop combinations found in the tropics has itself posed research and management problems. Some investigators consider that management can deal competently only with a small number of main crops, thus simplifying the provision of high-yielding material, the development of nursery techniques and of planting, maintenance and harvesting schedules.

Where perennial tree crops are combined with annuals, care is particularly needed in the establishment phase, and if the main crop is not to suffer from the intercropping, special care must be given to

1 the degree of shade permissible,
2 the use of fertilisers, and
3 weed control.

The soil must be capable of responding to the increased demands put upon it. On poor soils, during the years of tree crop establishment sufficient fertilisers must be supplied both for current production and for immobilisation in the growing tree crops before they reach the production stage.

To get satisfactory food crop production in inter-row areas without damage or slowing down of the tree crop the secondary crops must be kept well away from the trees and fertiliser applied as necessary to both these and to the tree crop, which must be vigorous enough to suppress weed growth once the developing canopy prevents further food cropping.

Intercropping perennial tree crops with annuals

Results of intercropping perennial tree crops with food crops vary considerably with the nature of the main tree crop and the soil.

Rubber has low nutrient demands and is often grown on poor leached soils in areas of high rainfall (Chapter 16) but intercropping with food crops has given doubtful economic results where it has delayed the time taken for the tree to reach tapping size. Food crop intercropping has rarely been favoured in large-scale rubber establishment, but has been a traditional practice in rubber smallholdings. Intercrops include pineapples, bananas, cucurbits, maize and upland rice, but in the absence of fertilisers most of these yield well only on soils such as those newly cleared from forest which are fairly fertile. In seasonal rainfall areas, intercropping may be confined mainly to the wet season. In a study of rubber intercropping in Indonesia it was found that yields of maize and green gram (*Phaseolus aureus*) were very disappointing and that the crops also depressed rubber yields, so that, without fertiliser, there was no economic advantage. In Sumatra, rice intercropping of the rubber inter-rows has been more successful. One danger is always that the perennial main crop is neglected in the early years for the food intercrop, as occurred with too heavy intercropping of rubber in the Ivory Coast. On poor sandy soils, farmers insisted on intercropping with vegetable crops, maize and even sugarcane, with the result that rubber growth was poor and the time to tapping was extended, again with no net economic gain. Similarly disappointing results were obtained when rubber in north-east Brazil was intercropped with coffee, maize, beans, rice and pineapples. Extreme competition in the absence of fertilisers resulted in the short-term advantages being offset by long-term losses in rubber productivity, and the fact that the onset of tapping was postponed to eight or nine years.

These experiences suggest that despite the fact that rubber is regarded as having low nutrient demands, it can suffer in competition with food crops if soil deficiencies are not made good by adequate fertiliser applications. This supports the conclusion in Chapter 16 that one important aspect of fertiliser application to rubber is to ensure that the tree reaches tapping girth as early as possible in order to get an earlier return.

Results of intercropping with oilpalm have been more encouraging than those with rubber. Intercropping with coconuts has received considerable attention and, depending on soil conditions, has sometimes proved very satisfactory.

Although most **oilpalm** is grown in monoculture, often with heavy fertiliser programmes to ensure high yields, intercropping of oilpalm in West Africa has been shown to be practicable. In one experiment in which oilpalm was intercropped with yams, maize and cassava followed by cocoyams (*Xanthosoma sagittifolium*), the intercropping stimulated palm growth and yields were increased. Attention needs to be paid to soil fertility levels, because in another experiment on degraded soils, in eastern Nigeria, intercropping with maize and cocoyams caused a decline in long-term oilpalm yields, apparently due to an earlier exhaustion of the soil in the intercropped areas as compared with that which was not intercropped. In West Africa, oilpalm intercropping appears feasible on the more fertile soils, and is compatible with local traditions, but on poor soils and in the absence of fertilisers it increases the risk of soil exhaustion and poor yields.

Coconut production is now increasing due to the use of higher-yielding hybrids and adequate fertilisation (Chapter 16). Coconuts are well suited to intercropping by the small farmer. Both in the early years after planting and again later when the palms grow tall to let in more light, the palms do not fully utilise all the available light. Coconut intercropping is traditional in the Philippines and in parts of Indonesia, Malaysia, Sri Lanka and India. Recent studies on the intercropping of coconuts in India have shown that root crops (tubers and rhizomes) did better than pulses, while among perennials, cocoa, cinnamon, nutmeg and cloves all did well, forming an understorey to the taller coconut palms, with cocoa showing the greatest promise.

In Malaysia cocoa–coconut combination is considered to be particularly profitable. This combination can succeed only with fertile soils. Of the two crops, coconut has the wider distribution and is tolerant of a very wide range of soils, including beach sands on which little else can be grown profitably. Thus there are soils that are well suited to coconuts on which cocoa, a relatively demanding crop requiring fertile, base-rich soils (Chapter 16), has no hope of succeeding. A coconut–cocoa combination can only be expected to succeed where cocoa alone would grow. In Malaysia, the soils used for the cocoa–coconut combination are generally fertile marine clays.

Where coconuts are grown on poor sandy soils, then the middle-storey crops with which it can be associated must also be adapted to poor soils. Suitable crops may then include cashew nuts, jackfruit and guavas, and some pasture between the trees may allow cattle and sheep to be grazed as well. With better soils, or with increased fertiliser use, other crops can be grown and cropping intensity increased.

Alley cropping

Alley cropping is an agroforestry system involving the growing of crops between lines of closely spaced trees or shrubs usually cut to form woody hedges. It has long been practised in Indonesia, Hawaii, Fiji and the Philippines where maize or grass was planted between strips of leucaena (*Leucaena leucocephala*). For maize cultivation, the leucaena hedges are cut and the green manure obtained incorporated in the soil, and subsequently cut again after the maize has

A

B

Fig. 19.2 Agroforestry: examples of (a) hedgerows farming terraces (India), and (b) alley cropping (Kenya).

been planted to avoid competition for light and to provide a mulch.

Trials on alley cropping at the International Institute for Tropical Agriculture (IITA), Ibadan, Nigeria, have included several shrubs. On the fairly fertile Entisols at the IITA station, with a pH of 6.0, leucaena was generally the best of those tried. An alley width of 4 m was found preferable, and pruning every five or six weeks is advised.

Continuous alley cropping for six years showed that maize yields could be kept at 2 t/ha with leucaena prunings alone. The plots that received the prunings had a higher nutrient status than plots not receiving prunings, and twice the organic matter content.

Attempts to establish similar alley cropping experiments elsewhere have sometimes been less successful, particularly on soils that are poorer and more acid than those used by IITA at Ibadan.

An alley cropping experiment on sandy soils with a pH of 4.2–4.3 at Yurimaguas, in the Amazon basin area of Peru, showed that leucaena grew poorly on these acid soils because of its susceptibility to aluminium toxicity and the low soil base status.

It must be concluded that alley cropping with leucaena can work well in some situations on moderately fertile soils, but that much further study is needed before it can be considered widely applicable in the tropics. On acid and infertile Ultisols and Oxisols it may be necessary to use inputs such as lime and phosphorus to allow leucaena establishment. Other species of quick-growing leguminous shrubs and trees such as *Glyricidia sepium* and *Inga edulis* appear better suited than leucaena in very wet areas of leached acid soils.

20 Rice soils and their management

Rice is the most important food crop in the tropics. It differs from other important crops in that it can tolerate flooding. Rice in the tropics is grown in a wide variety of climates, on many different types of soil, and using several different cropping systems. In some areas it is grown continuously flooded, and in others it is intermittently flooded, while the so-called upland rice is grown as a normal rainfed crop without flooding.

When rice soils are flooded, certain important soils changes occur. These changes make flooded rice-growing different from the growing of non-flooded crops.

Traditional rice varieties have not been very responsive to fertiliser use because the application of nitrogen has usually resulted in lodging – the bending over of the rice stalk – which reduces yields and makes harvesting difficult. The introduction in the 1970s of short, stiff-strawed rice varieties which respond well to nitrogen fertilisers has resulted in higher yields. The emphasis now, for flooded rice, is on better water management in relation to the use of nitrogen and other fertilisers. Rice yields are also being increased by the selection and use of varieties adapted to particular situations, including adverse soil conditions such as salinity or low phosphorus levels.

Rice cropping systems

Rice in the tropics is grown in five main cropping systems, as follows:

1 Irrigated paddy rice (about 20% of the total area).
2 Rainfed paddy rice (about 45% of the total area).
3 Direct seeded irrigated rice (about 1% of the total area).
4 Deep-water floating rice (about 9% of the total area).
5 Rainfed upland or hill rice (about 25% of the total area).

Fig. 20.1 The watery world of the rice field: paddy rice on alluvial soils in central Thailand. In the right middle distance is the nursery from which seedlings are transplanted by hand. The fields show newly transplanted rice in the foreground, and older rice elsewhere. Behind the rice paddies are rolling well-drained upland soils planted to trees, arable crops and pasture.

About 90% of the rice grown in the tropics is grown in Asia, with about 7% in tropical America and 3% in Africa. In Asia, half of the rice is grown in rainfed paddy without irrigation, and only 20% is grown with irrigation and full water control. A further 20% is upland, rainfed rice. The remaining 10% of the rice grown in Asia is deep-water floating rice, mostly along the Mekong and other large rivers. In tropical America, most of the rice is upland rice, but 20% is direct seeded irrigated rice. In Africa, most of the rice grown is upland rainfed rice.

1 Irrigated paddy rice

Paddy rice is often called lowland rice because it is usually grown in flat, small fields, though not necessarily at low elevations. Typically the essential features of this system are that a field is surrounded by low mud dikes or bunds to hold water and, when well saturated with water, is first ploughed to turn in stubble and weeds and then harrowed or worked in various ways at progressively lower soil moisture contents to puddle the soil.

The term 'puddling' refers to the working of the soil when wet to break down the soil structure and to produce a relatively uniform and impermeable mud. Puddling is often performed in Asia with animal-drawn implements and by trampling. During puddling the field is also levelled, and a basal fertiliser dressing may be incorporated.

At the same time as the field is being prepared, rice seedlings are grown in a separate nursery, usually from soaked, pre-germinated seeds. The nursery is flooded and the water level carefully controlled. The rice seedlings are transplanted from the nursery into the field in groups of 3–6 seedlings, usually 15–30 days after seeding. The transplanted seedlings may be left a few days without flooding, but are then flooded to a depth of 5–10 cm until two to three weeks before harvesting. Harvesting is usually 100–150 days after seeding. Nitrogen may be top-dressed on the flooded soil at the panicle initiation stage.

The use of a separate nursery saves time by shortening the period that the rice occupies the paddy field. This may allow a second or even a third crop to be fitted in during the year. About half of the South-east Asian irrigated paddy rice fields are double cropped.

2 Rainfed paddy rice

Rainfed paddy rice is more extensive than irrigated paddy, and is the main rice growing system in the tropics. It differs from irrigated paddy rice mainly in the fact that the farmer does not have control over the water supply but relies on rain to flood the fields. As a result, both the timing and the extent of the flooding can be very variable. Land preparation and puddling may have to be delayed until well after the start of the rains. Flooding during the growing season may be intermittent, depending on the local rainfall in that year. Intermittent flooding may result in periodic water stress, to which the rice plant is relatively susceptible. In areas of seasonal rainfall, only one crop a year can usually be grown. This system can be described as one of intermittent flooding, in contrast with the continuous flooding practised in irrigated systems.

3 Direct seeded irrigated rice

Direct seeded irrigated rice is a relatively advanced mechanised form of rice growing. The fields are usually large. Dikes are constructed, the fields are carefully levelled and the land is prepared for sowing by machinery. Pre-germinated seeds are either broadcast onto standing water from aeroplanes or drilled into the dry soil. The soil is flooded from 20 days after seeding to 20 days before harvest. The system was developed in the United States and in Australia. In the tropics it is found mainly in Latin America.

4 Deep-water floating rice

Deep-water rice varieties have the ability to elongate their stems very rapidly, as flood water rises. They are usually grown in low-lying areas where the depth of water reaches 1.5–5 m or more. Even uprooted plants can float, grow and produce panicles. Deep-water rice is found in association with some of the major rivers of South-east Asia, particularly in the Mekong delta area of Vietnam and in Bangladesh, India and Burma. It is also found along the Niger in West Africa.

5 Upland rice

Upland rice is also known as 'hill rice' in English, *riz pluvial* in French and *arroz secano* in Spanish. It is essentially rice that is rainfed and grown without flooding or transplanting, so that it is comparable with other rainfed crops. Most of the rice in tropical America and in Africa is grown in this way, sometimes in a shifting cultivation system (Chapter 17). In Liberia, rainfalls are heavy and rice is grown as the first crop after the forest is cleared, often followed by cassava. Upland rice yields are generally lower than those associated with more intensive systems using flooding because flooding eliminates water stress, controls weeds and may produce beneficial chemical changes in the soil.

Soil aspects of paddy rice production

Puddling and its effects

Puddling, which is widely used in paddy rice production, is the process of working the soil in a wet condition with the aim of breaking down soil structure almost completely to form an impermeable mud. Puddling is usually accompanied by a series of tillage operations which begin when the soil is very wet, above saturation, and continue at progressively lower moisture contents down to about field capacity (Chapter 4).

Optimum puddling occurs with high soil moisture contents of above field capacity but below saturation. At these moisture contents, cohesion of soil particles within the soil aggregates is relatively low because cohesion becomes less as soil moisture increases. The aggregates become soft and disintegrate relatively easily. Cohesion between aggregates, on the other hand, is higher at field capacity than in a drier soil because as the soil gets wet the aggregates swell, causing increased contact and cohesion between them. At water contents above field capacity, cohesion between aggregates decreases due to the increased lubricating effect of the additional water. Puddling is therefore easiest when the soil water content is between field capacity and saturation, since cohesion within aggregates is then low compared with less wet soils, but cohesion between aggregates is relatively high. Thus aggregates are broken down because of their low internal strength and the relatively high friction between them.

The ease with which soils can be puddled is influenced by the amount and type of clay they contain, and by their structural stability. Clay soils are easier to puddle than sandy soils. Soils containing montmorillonite are puddled more easily than those in which the clay fraction is high in kaolin, or in iron and aluminium oxides. High organic matter contents make soils more difficult to puddle, but sodium-saturated soils are easily puddled because the sodium disperses the clay. Soils with stable structures or microstructures, particularly the Oxisols of the US Soil Taxonomy (Chapters 8 and 15) are difficult to puddle because of the difficulty of breaking down the aggregates.

The effects of puddling include the following:

1 Destruction of soil aggregates.
2 The almost complete elimination of non-capillary pores and an increase in capillary porosity.
3 An initial decrease in bulk density followed by an increase during flooding, and a more marked increase when soils dry out and shrink.
4 Reduced moisture losses due to lower percolation losses.
5 A slower rate of drying than in well-structured soils.
6 The prevalence of reducing conditions as soon as the soil is puddled (whether flooded or not) until cracks form and let in air.

The effects of puddling can be reversed and the original structure formed again if soils are first dried and then alternately wetted and dried. Structural regeneration can be helped by tilling at water contents of below field capacity. In some cropping systems where other crops follow paddy, the ease or otherwise with which structure can be regenerated is important. Regeneration is relatively easy with lighter-textured soils, soils high in iron and aluminium oxides, and soils high in organic matter.

The main purpose of puddling is to reduce water losses by percolation when the paddy is flooded. In doing this, puddling also reduces the loss of nutrient ions through leaching.

The effects of flooding

Rice generally yields better under flooded conditions than unflooded conditions, even though very high rice yields are sometimes obtained without flooding. The main reasons for this are as follows:

1 Water stress is eliminated.
2 Weed control is easier because rice tolerates flooding better than do the weeds.
3 Flooding brings the soil pH to near neutral, thus making most nutrients more available, particularly phosphorus.

Chemically the important result of flooding is the change from a well-aerated soil environment to one that is lacking air oxygen in which reducing conditions prevail. Not all the soil profile is reduced on flooding, since a thin upper layer of the soil 1–10 mm thick remains in an oxidised state because it is in contact with the floodwater and in equilibrium with the oxygen dissolved in it. This thin upper soil layer is easily identified because it retains the original soil colour whereas the reduced soil below it is grey to blue or black in colour, though it may contain 'rusty root channels', i.e. reddish-brown stains along root channels caused by the oxidising effect of active roots on their immediate rhizosphere.

The change of the soil from the oxidised to the reduced state is due to the activities of anaerobic bacteria which obtain their energy by reducing oxidised soil components. When a soil is flooded the oxygen supply quickly falls to zero. The aerobic bacteria die or become dormant, but the anaerobes multiply rapidly. The oxidation–reduction potential (E_h values corrected to pH 7.0) is a useful measurement of the intensity of soil-reducing conditions, and a sequence of chemical reactions takes place as the intensity increases. First, nitrates become unstable and are denitrified by being changed to nitrogen gas (N_2). After all the nitrate has been reduced, any manganese IV (Mn^{4+}) compounds present are reduced to Mn^{2+} compounds, but these are usually minor in amount. The most important reduction reaction that usually takes place in flooded soils is the reduction of iron III (Fe^{3+}) to iron II (Fe^{3+}) ions because iron III compounds are usually more abundant in rice soils than the other compounds that are reduced. In some acid soils high in iron, increases in Fe^{2+} may temporarily reach levels that are toxic to rice. When all the iron III iron has been reduced, the sulphates present are next affected by the reducing conditions.

In general the intensity of the reduction process is related to soil organic matter content and temperature; the more organic matter, the greater the intensity. In practice iron III iron is the most important inorganic reduced product and E_h values are equilibrated by the $Fe^{3+}-Fe^{2+}$ system so that values low enough to reduce sulphate are not usually attained except in soils both high in organic matter and low in iron.

The flooding and consequent reduction of the soil has other important consequences. Whatever the original pH of the soil, after one month of flooding the soil pH is usually 6.5 to 7.2 and it remains in this range until it dries. The effect of submergence is thus to raise the pH of acid soils and to decrease that of alkaline soils. In acid soils the pH rises because of the release of OH^- ions, as when $Fe(OH)_3$ is reduced to $Fe(OH)_2$. In alkaline soils partial pressure of carbon dioxide causes an increase in H^+ ions. Together these mechanisms mean that a remarkable and very important consequence of flooding is a self-liming operation which results in the soil reaching a pH range that is optimum for most nutrients. In acid soils, any aluminium toxicity associated with low pH values is quickly eliminated within two weeks of flooding by precipitation of exchangeable aluminium at pH 5.5.

Nitrates in the reduced soil are reduced to N_2 and lost as nitrogen gas, and considerable amounts of soil nitrate may thus be lost as a consequence of flooding, particularly when nitrate has accumulated in the soil in a previous dry season. In contrast to the nitrate anion, the ammonium cation is stable in the reducing environment of the flooded soil. The mineralisation of organic matter stops at the NH_4^+ stage. As a result ammonium (NH_4^+) ions may accumulate in flooded soils either as exchangeable NH_4^+ or in the soil solution, though ammonium ions may also be lost by leaching where water seeps away from the paddy field. If ammonium-containing fertilisers are added to a flooded soil they should not be left in contact with the thin soil surface layer in which oxidising conditions still prevail, since there they are likely to be first oxidised to nitrate and then reduced to nitrogen gas and lost, but should be mixed into the reduced layer to prevent their oxidation.

Flooding increases the concentration of phosphorus in the soil solution. This is due to the reduction of relatively insoluble iron III phosphates to more soluble iron II phosphates, and to various other chemical reactions in the flooded soil. The importance of this varies very much between soils, but in most soils the increase is enough to raise soil solution phosphorus to a level that is adequate for rice growth.

Organic matter decomposes more slowly in flooded soils than in well-aerated ones. On the other hand nitrogen mineralisation can proceed at higher C/N ratios and it is thus possible to mineralise materials with a high C/N ratio such as rice straw which do not easily decompose in well-drained soils.

All these reactions are reversed when a flooded soil dries and is re-oxidised, though this happens more slowly in clayey, puddled soils than in lighter-textured or well-aggregated soils in which re-oxidation begins again as soon as the soil is drained.

Fertiliser use on paddy rice

Nitrogen

In the great majority of cases rice grows better when it is supplied with extra nitrogen. Exceptions occur on newly cleared land which supplies sufficient nitrogen, and when growth is held back by other limiting factors. The value of added nitrogen in producing extra yields, however, depends on the rice variety and on whether the rice is a tall or short type. Tall rice varieties unfortunately tend to lodge (bend over) because added nitrogen causes a tall, weak stem. Lodging makes harvest difficult and may reduce yields. Responses to nitrogen have, however, become much more widespread and important with the introduction of short, stiff-stemmed rice varieties that are resistant to lodging. This is one of the most important aspects of the so-called 'green revolution': the increase in rice and wheat yields widely obtained in the 1970s from improved fertiliser and other management practices associated with new varieties. Apart from the choice of variety, rice responses to nitrogen are influenced by sunshine hours, temperature, water management and soil fertility.

The new rice varieties that respond to nitrogen applications are characterised by a short stature, erect stems and leaves, high tillering capacity, high grain/straw ratio and resistance to lodging. They respond to added nitrogen differently from the traditional tall leafy indica varieties. The tall varieties increase their height with added nitrogen, which causes lodging and yield losses, but the short varieties show little height change and respond mainly by increasing the number of tillers. In neither case does added nitrogen much change the length of the growing season. Nitrogen uptake is also similar, but in the new varieties with a better grain/straw ratio, more of the added nitrogen is converted into grain.

Nitrogen responses from flooded rice are usually better in the dry season than in the wet, which has more cloud and less radiation, because the greater solar radiation in the dry season gives more photosynthetic energy. Similarly, areas with low solar radiation give poorer nitrogen responses than those areas where it is high.

Nitrogen responses of short rice varieties grown with only intermittent flooding are not as good as with constant flooding, and the traditional tall varieties are relatively better adapted to poor water management.

Ammonium-containing fertilisers are better for flooded rice than those supplying nitrate because of the possibility of nitrate reduction and loss as nitrogen gas (see the section on the effects of flooding, above). The major forms used – ammonium sulphate and urea – have been shown to be equally efficient in constantly flooded soils. If sulphur is deficient, then ammonium sulphate is to be preferred, but in extremely acid soils with particularly low iron contents it may be better to avoid adding sulphur because of the possibility of H_2S formation and toxicity. In soils normally supplied with iron, however, H_2S is not formed because the sulphur is precipitated as FeS before this can occur. Urea applied to flooded soils is hydrolysed to ammonium carbonate in the same way and at the same speed as it is in normal well-drained soils, but before hydrolysis it is not retained by the cation exchange complex of the soil and is mobile and free to migrate. This may occasionally result in some loss.

Nitrogen fertilisers are either incorporated in the soil before seeding or transplanting, or broadcast at different stages of growth, but in this case the fertiliser should be incorporated into the soil at depths below 5 cm to ensure that it is in the reduced layer of the soil and is not re-oxidised to nitrate with the possibility of subsequent reduction to nitrogen gas. For intermittently flooded rice a deeper incorporation is needed to prevent this.

A relatively new innovation is the use of urea 'supergranules' – large granules of 5 g or more each, which can be placed more easily than conventional urea.

The timing of nitrogen applications to rice should be such as to ensure a good supply at the two periods when the nitrogen supply is most critical to the plant, namely at tillering and at panicle initiation. In practice, however, where flooding is constant, a basal dressing well incorporated before seeding or transplanting is normally sufficient. It is only when the flooded soil has high percolation rates that it is better to split the nitrogen applications into two or more in order to reduce seepage losses. With intermittently flooded rice and with rainfed rice the situation may be quite different, since more of the basal dressing is likely to be lost and the advantages of splitting applications and of applying nitrogen at the panicle initiation stage are likely to be much greater.

The optimum amounts of nitrogen fertiliser that can be applied are related primarily to the type of rice grown and the possibility of inducing lodging in the traditional varieties, but also to radiation and temperature, and to water management. The ability of the soil to supply nutrients other than nitrogen has also to be considered. With traditional tall varieties it is rarely possible to apply more than 30–40 kg/ha of nitrogen without inducing lodging and drastically reduced yields. Much greater amounts can be profitably applied to the new shorter varieties with greater straw strengths, and in optimum yields may be associated with dressings of up to 120–160 kg/ha, with 90 kg/ha often considered a practical minimum. For many varieties intermediate in lodging characteristics, dressings of the order of 60–80 kg/ha are recommended. For high-yielding nitrogen-responsive varieties, yield responses of 20 kg/ha of rice per kg of nitrogen supplied are common.

Under favourable conditions some strains of blue-green algae can fix some nitrogen. This can contribute to the maintenance of yields when no fertiliser is given, particularly with traditional varieties, but since they are thought at best to be able to fix and supply only some 15–20 kg N/ha, fertiliser nitrogen is required for optimum yields of the new short nitrogen-responsive varieties.

Phosphorus

Soils used for rice differ widely in their phosphorus status under rainfed conditions, but despite this, responses of paddy rice to phosphorus applications are not common. This is partly because rice needs less phosphorus than other cereals such as maize, and some rice varieties are much more tolerant of low phosphorus levels than others, but is mainly due to the fact that flooding, as discussed in an earlier section, increases phosphorus availability and uptake in most soils. Exceptions include some highly weathered soils that are deficient in phosphorus and on which rice does respond to phosphatic fertilisers. The fact that phosphorus availability is increased on flooding means that normal routine soil tests designed to measure plant-available phosphorus are not applicable to flooded soils since they do not measure what is likely to be released when the soil is flooded.

Single superphosphate, triple superphosphate and the ammonium phosphates are all suitable for use on paddy rice. In soils with a low soil pH before flooding, rock phosphates have sometimes proved as efficient as the much more expensive soluble phosphatic fertilisers. They may also be used on soils containing pyrites which may oxidise to sulphuric acid.

The best method of application for paddy rice is to incorporate the phosphatic fertiliser into the puddled layer of the soil before transplanting. Early applications contribute to a good root system. Later applications have much less effect.

Flooding leads to better utilisation of applied as well as of native phosphorus, so that even on phosphorus-deficient soils, optimum dressings of phosphatic fertilisers are generally low, and often of the order of only 20 kg P_2O_5/ha and seldom higher then 40 kg P_2O_5/ha for traditional varieties or 60 kg P_2O_5/ha for high-yielding ones.

Potassium

The rice plant takes up more potassium than either nitrogen or phosphorus (Table 20.1) but much of this goes into the straw, and if the straw is returned

Table 20.1 Nutrient removal of a rice crop (kg/t of straw or grain) grown in the Philippines in the 1979 dry season. The variety was IR8. The yield was 7.9 t/ha of grain and 7.0 t/ha of straw.

	Element										
	N	P	K	Ca	Mg	S	Fe	Mn	Zn	Si	Cl
Straw	5.3	0.8	13.6	3.9	2.6	0.7	0.2	0.6	0.03	74.0	1.8
Grain	10.9	2.0	3.1	0.5	1.1	1.0	0.04	0.05	0.01	16.8	1.6

Source: **Datta, S. K. de** (1981) *Principles and Practices of Rice Production*. IRRI: Los Banos.

to the field, only about 5% of the potassium taken up by the crop is exported. Potassium is obtained from the soil, from decomposing rice straw, and from the irrigation water. For the low-yielding traditional indica varieties where the straw is returned, these three sources are generally sufficient without added fertiliser potassium. High-yielding varieties in the presence of sufficient nitrogen and phosphorus have, however, sometimes responded to fertiliser potassium, particularly on sandy soils. Trials in Burma, India, Sri Lanka and Madagascar have shown responses under these conditions, and yield increases were mostly of the order of 7–13 kg of paddy per kg K_2O applied. In Taiwan the average fertiliser recommendation is 98 kg N, 49 kg P_2O_5 and 65 kg K_2O per hectare, and potassium is also included in standard fertiliser recommendations in South Korea and in Japan. Numerous experiments in Bangladesh showed the need for potassium and for a balanced NPK fertiliser dressing for high-yielding varieties. In Malaysia potassium is recommended for the lighter soils and for high-yielding varieties but not for traditional varieties on other soils.

Silicon and minor elements

Silicon is taken up in considerable quantities by the rice plant (Table 20.1) and although not regarded as an essential element it seems to play a physiological role by silicifying the cell walls. This may increase resistance to some pathogenic fungi. Available silicon may be deficient in highly weathered soils, sandy soils and organic soils. Experiments have sometimes shown yield responses to silicon applications in the form of calcium magnesium silicate, and according to one estimate about a quarter of Japanese rice soils show some response to this element. Silica is supplied in rice hulls and rice straw or their ash.

Iron-deficient rice soils (Akiochi soils) are well known in Japan, and are mainly sandy soils derived from granites and similar rocks. Dressings of red, iron-rich clay soil are often given, sometimes with additional manganese as manganese sulphate.

Zinc deficiencies have recently been recognised as being very widely deficient in rice soils and as an important factor limiting yields. They are found in organic soils, sodic and calcareous soils, in very sandy soils and in soils that are wet for long periods, and can be corrected with small applications of zinc sulphate.

Fertiliser use on upland rice

Rice that is grown on upland soils without flooding or irrigation is more comparable to other cereals under similar conditions than to flooded paddy rice. Yields are generally less than for lowland flooded rice. Upland rice is produced in large quantities in Brazil, as well as in parts of Indonesia, Malaysia and the Philippines, and in Africa. It is often the first crop in shifting cultivation agriculture of the wetter areas.

Upland rice varieties are generally tall, leafy and prone to lodging if given more than modest nitrogen dressings, and new short, stiff-strawed upland varieties are needed if yields are to be raised substantially. Experiments have sometimes shown that some of the new varieties developed for flooded conditions have also outperformed traditional upland varieties when grown under rainfed conditions. Other causes for generally moderate yields are unevenly distributed or inadequate rainfall, and poor control of weeds and of pests and diseases.

In Senegal, the Ivory Coast, Ghana, Dahomey, and the Central African Republic, numerous experiments have shown that yields can be profitably increased by relatively small nitrogen applications, usually of 20–40 kg/ha with or without phosphate. In Brazil, where about 80% of the total rice area is upland rice, soils are widely deficient in phosphorus, and modest dressings of nitrogen and phosphorus are beneficial.

21 Soil management for pasture and forage production

Ruminant livestock in the tropics are raised mainly on relatively poor soils or where the low or uncertain rainfall is a serious limiting factor for crop production. For these reasons, although about half of the pasture and rangeland and half of the cattle population of the world are in the tropics, the productivity of tropical livestock is low relative to that in temperate areas. This low productivity is partly due to environmental and health aspects, but is mainly due to the quality of the feed supply. Feed supply and quality are in turn related to soil properties that limit pasture production.

There are three main systems of pasture production in the tropics. These are:

1 Extensive grazing on natural grasslands.
2 Extensive grazing on improved grass–legume mixtures.
3 Intensive pasture or intensive forage production using fertilisers.

Other systems of animal production include nomadic grazing in dry areas, and animal production based on crop residues, practised in parts of Asia in particular.

Of these systems, the first – extensive grazing on natural grasslands – is the most important. This system is particularly extensive and dominant in Africa, Australia and South America.

The productivity of the three main systems listed varies between wide limits. Extensive grazing of natural grasslands usually results in annual liveweight gains of only 20–50 kg/ha, and where rainfall is very seasonal much of the weight gained in the wet season may be lost again in the following dry season. Improved grass–legume mixtures with only low fertiliser inputs can give considerably higher liveweight gains in the range 100–300 kg/ha a year, whereas intensively fertilised grass pastures may give annual liveweight gains of 500–1000 kg/ha or more.

Natural grasslands and their management

Factors affecting productivity

Most of the natural grasslands of the tropics are found in areas of low or seasonal rainfall. Forest vegetation is generally associated with moderate to high rainfalls without marked dry seasons, but the savannas of Africa, South America, South-east Asia and Australia generally have rainfalls that are low and/or which have a marked dry season.

Grasses are well adapted to dry seasons because annual grasses set seed, die and then grow afresh from seed the following wet season, while perennial grasses die back but then resume growth when the next rains begin. In some areas, particularly in South America, savanna may be found in areas that are climatically suited to forest, but because of soil conditions, such as a shallow soil depth to an impermeable layer, the soils may be waterlogged in the wet season and dry out at other times of the year, and thus their vegetation may be limited to grasses and sedges.

The productivity of the natural grasses of the tropics is usually limited not only by the low or seasonal rainfall but also by low soil fertility. The soils of tropical savanna areas generally have low organic matter contents, due in part to the limited vegetation they support and in part to the effects of annual dry season grass burning. Annual burning contributes to the fact that the savanna soils of much of Africa are low in nitrogen and sulphur. The savanna soils of South America are frequently limited by high soil acidity as well as by low nutrient status.

The frequent combination of seasonal rainfall shortages and low soil fertility results in the fact that even if pasture production is relatively plentiful in the wet season, it is followed by a dry season when forage is low in quantity and poor in quality.

In the rains, pasture is relatively plentiful and grows rapidly. The young leafy material is relatively digestible, and high in crude protein. Cattle normally gain weight during this period. However, since the size of the herd in most cases is controlled to some extent by the dry season carrying capacity, there are sometimes not enough animals to keep the grass grazed sufficiently. As the grasses approach maturity the leaf/stem ratio goes down, and the plants become progressively lower in proteins, minerals and soluble carbohydrates, and higher in fibre and lignin. The grass thus becomes less palatable, less digestible, and of lower nutritional value. The animals grazing it obtain less energy and less protein. Whereas immature grasses contain 8–10% crude protein by weight on a dry matter basis, protein may fall at maturity to only 3–5%, which is critically low, and the grasses become increasingly siliceous. Phosphorus and calcium may also be deficient.

When the nutritional value of the forage falls to below what is needed to maintain the weight of the animal, liveweights fall, fertility declines, and the time taken to maturity increases. These factors result in alternate liveweight gains and losses with the season, and in a zigzag or sawtooth growth curve showing weight increases followed by weight losses. Animals take much longer to reach maturity than they would if they had a better, less fluctuating food supply.

Dry season grass burning

In Africa and in South America, large areas of savanna are burnt in most dry seasons. The burning is usually deliberate, partly to help in the hunting of game, but mainly to clear the ground of the dead grass and ticks, and to encourage a new growth of grass the following wet season. If the grass is set on fire early in the dry season before the dead plant material is thoroughly dry, the fire is less complete than later in the dry season. A burn at the end of the dry season is more complete, hotter, and more harmful to any shrubs or trees present. Typically a narrow curtain of fire sweeps across the grassland, moving in the direction of the wind, but advancing too rapidly to cause more than a brief and superficial heating of the soil.

During the burn most of the carbon, nitrogen and sulphur in the dead plant material is lost to the atmosphere (though some may be returned later in the rain) but the other plant elements, including phosphorus, calcium, magnesium and potassium mostly remain in the ash and have a fertilising effect on the new growth.

The effects of repeated annual burning on the savanna vegetation have been widely studied, particularly in Africa where savanna that is the product of human activities extends over large areas on both sides of the equator. In a number of locations in the savanna areas of West Africa, plots have been completely protected from burning, and others burnt only at the beginning of the rains when the burn is less fierce. The vegetation of these plots has then been compared over a period with that of plots subjected to normal late burns each year. Complete protection from burning resulted in an increase in shrubs and trees, and in a change in the species of grass found.

One of the best known of these experiments is located near Bouaké in central Ivory Coast, where a plot protected from burning since 1940 now has a dense savanna woodland vegetation, in contrast to the savanna of the normally burnt plot. In a similar experiment in northern Ghana, the less severe early dry season burn resulted in an increased growth of the palatable *Andropogon gayanus*, whereas the hotter burns of the late dry season promoted *Loudetia acuminata*, which is not palatable. Comparable results were obtained in Tanzania.

In the Ngong Hills area of Kenya, near Nairobi, annual burning was found to be beneficial since it maintained the dominant indigenous grass *Themeda trianda* in good grazing condition. Eight years of protection from burning had the undesirable effect that the *Themeda* was replaced by the less valuable species, *Digitaria abyssinica*.

Fertiliser use on natural grasslands

In addition to the adoption of improved burning and grazing practices, native tropical grasslands can be improved by

1 the use of fertilisers, and
2 oversowing legumes and using fertilisers.

The direct use of fertilisers on natural savanna grasslands of the tropics has given generally poor results, or results that were not economic. This appears to be due mainly to the fact that the indigenous grasses are adapted to the local conditions of low soil fertility, which often means that they grow slowly and have relatively modest nutrient demands, and are not able to respond well enough to an added supply of nutrients in the form of fertiliser. Only a small proportion of the fertiliser nitrogen or phosphorus may be taken up by the vegetation, resulting in a low efficiency of fertiliser use. In trials in northern Australia, for example, the local *Themeda* grasses recovered less than 10% of the nitrogen added in the fertiliser trials, compared with a recovery of 40–70% by improved grass species.

Most tropical savannas are low in available nitrogen and contain few ground legumes. The most successful method of improving natural tropical pastures has been to oversow them with legumes to produce a mixture of legumes and the original grasses. Some phosphate is often given to help the establishment of the legumes.

In the Kitale area of Kenya, improvement of the natural *Hyparrhenia* grasslands has been achieved by broadcasting *Desmodium uncinatum* and applying phosphate at the rate of 25–50 kg/ha of phosphorus. The use of the phosphatic fertiliser was essential. It increased the percentage of legumes in the cover, doubled total dry matter production, and tripled the production of protein.

In considerable areas of the grasslands of northern Australia the soils are deficient in nitrogen, sulphur, phosphorus and molybdenum. The indigenous savanna is characterised by spear grass (*Heteropogon contortus*) and is usually burned during the dry season. These areas have been improved by drilling in or broadcasting seeds of Townsville stylo (*Stylosanthes humilis*) mixed with molybdenum-enriched single superphosphate applied at 220 kg/ha to give 22 kg phosphorus, 22 kg sulphur and 1 kg molybdenum per ha. The use of fertilisers is essential for the establishment and persistence of the legume, and maintenance applications are given at one-quarter to one-half of the establishment rates annually. After the initial germination of the stylo, the pasture is grazed heavily to reduce the grass and to give more light to the legume.

Improved grass–legume mixtures

Since most native grass species have a low productivity, and do not respond well to added fertilisers, to increase carrying capacities of natural grasslands to one or more animals per hectare it is usually necessary to replace the indigenous grasses with better, introduced species. At the same time soil fertility has often to be raised by using fertilisers, though suitable legumes mixed with the grasses can become the main source of nitrogen.

Most improved tropical pastures consist of grass–legume mixtures rather than stands of a grass or a legume alone. Both the grass and the associated legume must be adapted to the local soil and climate. Successful grass–legume combinations may therefore suit only a specific range of soil and climatic conditions. Tall bunched grasses can combine well with low-lying spreading legumes such as *Stylosanthes*, but grasses that cover the soil with a thick mat may smother legumes planted with them.

Both grasses and legumes differ widely in their degree of adaptation to, and tolerance of, specific adverse soil conditions such as high acidity and exchangeable aluminium, and low available nutrients such as phosphorus and calcium.

Grass species adapted to seasonally dry, ustic soil moisture regimes that are thought to be tolerant of both acid soil conditions and low available phosphorus include *Hyparrrhenia rufa, Melinis minutiflora, Paspalum plicatulum* and *Brachiaria decumbens*.

Grasses adapted to wet, udic moisture regimes that are thought tolerant of the same conditions of both acid soil conditions and low available phosphorus include Guinea grass (*Panicum maximum*) and Pangola grass (*Digitaria decumbens*). *Pennisetum purpureum* (Elephant grass or Napier grass) is also adapted to wet climates and to soils with a high aluminium content, but is more demanding of nutrients than is Guinea grass or Pangola grass.

A number of tropical legumes are now widely used in grass–legume mixtures. The nutritional needs and soil tolerances of these legumes are generally better known than those of many tropical grasses. Table 21.1 lists 14 commonly used tropical legumes and gives information as to whether they are tolerant or susceptible to high soil aluminium, high manganese, low calcium and magnesium, and

Table 21.1 Commonly used tropical legumes showing tolerance to high soil aluminium and manganese, and to low soil calcium and magnesium phosphorus and potassium, and the type of rhizobium used.

Scientific name	Common name	Tolerance (1) to High Al	High Mn	Low Ca + Mg	Low P	Low K	Rhizobium type
Stylosanthes guyanensis	Stylo	T	T	T	T	?	Cowpea
Stylosanthes humilis	Townsville lucerne, Townsville stylo	T	T	T	T	T	Cowpea
Pueraria phaseoloides	Kudzu, Puero	T	T	T	T	?	Cowpea
Centrosema pubescens	Centro	T	T	T	T	T	specific
Desmodium intortum	Greenleaf desmodium, pega-pega	T	M	T	S	M	Desmodium
Desmodium uncinatum	Silverleaf	T	S	T	S	T	Desmodium
Calopogonium muconoides	Calopo	T	T	T	T	?	Cowpea
Macroptilium lathyroides	Murray lathyroides	T	M	T	S	T	Cowpea
Lotononis bainesii	Miles lotononis	T	T	T	T	S	specific
Phaseolus atropurpureus	Siratro	T	S	?	T	T	Cowpea
Leucaena leucocephala	Ipil-ipil, Leucaena	S	S	?	?	?	specific
Glycine wightii	Perennial soybean	S	S	?	S	M	Cowpea
Trifolium repens	White clover	S	S	?	S	S	Clover
Medicago sativa	Alfalfa, lucerne	S	S	?	S	S	Alfalfa

(1) T = tolerant; M = moderately tolerant; S = susceptible
Source: various authors as summarised in Sanchez, 1976.

low phosphorus and potassium. In the case of most of these 14 legumes, the associated *Rhizobium* is the slow-growing cow pea type which is widely distributed in acid tropical soils, so that these legumes do not normally require *Rhizobium* inoculation. It is thought that slow-growing strains of *Rhizobium* such as the cow pea type release alkaline substances which appear to favour tolerance of acid soil conditions. Fast-growing strains typical of the temperate areas release acid substances and are less suited to acid soils.

The first 10 of the legumes listed in Table 21.1 are relatively tolerant of acid soils, but the last 4 are better adapted to non-acid, high base status soils. These include *Leucaena leucocephala*, a fast-growing tree that has attracted considerable attention in recent years for agroforestry systems and for browse production. Pelleting the seeds with lime has proved beneficial for the establishment in acid soils of Leucaena, alfalfa and clovers. In contrast, lime pelleting has had no effect or even an adverse effect on those legumes that form an association with the cow pea type of *Rhizobium*, which is thought to exude acid substances.

The establishment of improved pastures

The establishment of improved pastures is influenced by the nature of the previous vegetation, and whether the pasture is replacing forest, savanna, arable land or a less productive pasture.

The ash provided by the burning of tropical forest contains a wide range of nutrients (though little or no nitrogen or sulphur) which help to establish pasture. The clearing of tropical forests for pasture establishment has been rapid in recent years, particularly in the Amazon area of South America. In this area, most of the cleared forest has been planted either to Guinea grass (*Panicum maximum*) in the wetter areas, or to Jaragua grass (*Hyparrhenia rufa*) in the areas of more seasonal rainfall.

The grasses are planted immediately after burning, using seed or, more usually, planting material. Legumes are not normally planted with the grasses but some, such as *Centrosema pubescens*, often appear spontaneously and supply nitrogen.

In the wetter areas of the Amazon the Guinea grass may persist for 20 years or more if planted after burning and if management is good. In areas of more seasonal rainfall the Jaragua grass may degenerate after only four or five years if established after mechanical clearing of the vegetation. This is thought to be due to soil compaction on clearing, and to low soil fertility. In some cases the decline has been prevented by mixing legumes with the grass and applying single superphosphate at rates corresponding to 10–50 kg P/ha. Both Guinea and Jaragua pastures may decline by being replaced by coarser species such as *Paspalum conjugatum*.

Forest areas in Hawaii were converted to pasture by killing the trees with herbicides, burning the dead wood and then broadcasting from the air a mixture of Guinea grass and *Desmodium intortum* seeds and granulated superphosphate.

Tropical savanna areas have been converted to improved pastures by burning late in the dry season and then incorporating into the soil the ashes, fertiliser and seed by using a disc harrow.

In some areas the cost of pasture establishment can be lowered by planting an initial crop and then interplanting the grass. In East Africa, grasses have been successfully planted between maize after the last maize weeding.

When establishing improved pastures on old pastures that have run down, soil compaction of the old pasture by grazing animals may make establishment more difficult by reducing soil porosity and aeration. Soil cultivation may be needed to reduce this condition. In some overgrazed, compacted soils in Kenya, Star grass (*Cynodon plechtostachyum*) was successfully established by contour cultivation. This doubled water infiltration rates into the soil.

The establishment of acid-tolerant tropical legumes from seed is often helped by pelleting the seed with superphosphate or with rock phosphate. Single superphosphate supplies sulphur in addition to phosphorus and calcium, and this may be beneficial. Many legumes have higher phosphorus needs at establishment than for later growth, and most legumes are thought to have relatively high sulphur needs.

The ability of some pasture grasses and legumes to grow on soils with low available phosphorus may be due to mycorrhiza on the roots. Mycorrhiza inoculations have increased the growth of some legumes in Colombia and Australia. Mycorrhiza inoculations require further research and may become more important in the future.

Fertiliser use on improved grass–legume mixtures

Once grass–legume mixtures are established and grazed, nutrients are cycled through the grazing animals. Animals excrete in their dung and urine over 80% of the nitrogen, phosphorus and potassium they take in by grazing, though if these nutrients are to be adequately distributed over the pasture the animals must be allowed to move freely.

Losses from the system occur when the animals are removed, when ammonia is lost by volatilisation from dung, and when nutrients are leached away by rain. Fertilisers are needed to replace these losses and to maintain a stable nutrient supply cycling between soils, plants and animals.

Nitrogen losses may be made good by the nitrogen fixed by legumes in the mixture. Nitrogen applications as fertiliser may have undesirable effects, since high available nitrogen in the soil results in less nitrogen fixation by the *Rhizobium*. Nitrogen applications may stimulate the growth of the grasses to such an extent that they outshade the legumes and cause their proportion in the mixture to decline. Nitrogen applications therefore are often confined to small quantities at establishment.

Phosphorus needs for pasture maintenance depend on the phosphorus requirements of the species concerned, on the nature of the soil and its phosphorus fixation capacity, and on the amounts of phosphorus initially supplied at establishment.

For soils with high phosphorus fixation capacities

(Chapter 10) both the phosphorus at establishment and subsequent maintenance applications have to be higher than for soils with low fixation capacities. In acid soils, the cost of the phosphatic fertiliser used at establishment can sometimes be reduced by using mainly cheap rock phosphate mixed with small amounts of the more expensive soluble phosphatic fertilisers such as the superphosphates.

For soils with low phosphorus fixation capacities, annual phosphate maintenance applications can be low, particularly for pasture species that are tolerant of low phosphate soil levels. In much of tropical Australia, where these conditions apply, annual phosphate maintenance levels of only 5–10 kg P/ha have proved adequate.

Potassium requirements of tropical grass–legume mixtures seem in many cases to be low. This may be partly because some legumes have relatively low potassium needs, partly because many savanna soils appear able to provide adequate potassium for considerable periods (Chapter 10), and partly because about 80% of the potassium taken in by grazing animals is returned in the excreta. However, some highly weathered tropical soils, and some sandy soils, appear to have low potassium reserves, and more research is needed.

Sulphur responses have been widespread in tropical pastures, particularly in some West African savanna areas, Australia and Brazil. These responses seem to be related to the generally low organic matter levels in savanna soils, and to low sulphur levels. Sulphur, like nitrogen, is lost when vegetation is burned. Legumes in general have relatively high sulphur needs. *Stylosanthes* and alfalfa are pasture legumes which have shown particularly good responses. Sulphur is often conveniently supplied with phosphorus by using single superphosphate.

Liming has proved beneficial on acid soils in South America with high exchangeable aluminium, using small amounts of up to 1 t $CaCO_3$/ha. These amounts were not enough to raise the soil pH appreciably, so it is assumed that the benefit was from the calcium itself (or from both the calcium and the magnesium in the case of dolomitic limestone) and not from any change in soil reaction. The lime needs of acid-tolerant legumes may in many cases be met by small maintenance applications of single superphosphate. Acid-intolerant legumes, on the other hand, have much higher lime needs if they are to be grown on acid soils. On acid soils of New Guinea, for example, liming at 10 t/ha was needed for maximum growth of *Leucaena leucocephala*, making this practice of doubtful economic viability.

Trace element deficiencies in tropical pastures occur patchily, depending on soil properties (Chapter 12). A very well-known example is that of molybdenum deficiency in Australia, corrected by adding less than 1 kg/ha of molybdenum in the superphosphate applied. Boron and molybdenum have been reported deficient in the Amazon area of Peru, and iron in calcareous soils of Puerto Rico. Trace element deficiencies can usually be detected by soil, plant and animal analyses, and by the use of pot experiments and field trials.

Intensive grass production using fertilisers

The highest forage yields are obtained from the use of fertilisers on high-yielding grasses such as Elephant grass or Guinea grass. Yields can be higher than for grass–legume mixtures because legumes cannot usually provide enough nitrogen for grasses to reach their highest yield potential. Heavily fertilised high-yielding grasses have produced up to 45 t/ha of dry matter a year commercially in Puerto Rico. This gives a carrying capacity of 10 animals/ha if the grass is cut for forage and 5/ha when grazed, as compared with the typical carrying capacities of less than 1 animal/ha for most grass–legume mixtures, and one-third to one-fifth of an animal/ha for improved natural grasslands.

The use of large quantities of fertilisers to achieve really high yields of forage per hectare, however, is only profitable where the cost of fertiliser is low in relation to the selling price of beef or milk. High-yielding fertilised pastures, with or without irrigation, are also used for fattening purposes. Irrigated pastures may be profitable for fattening during the dry season.

The soil management of pastures that are cut and fed to animals elsewhere is different from that of pastures which are grazed. Where animals graze the pasture, nutrients are recycled and fertiliser needs are lower than if the grass is cut and taken away. In Puerto Rico it is considered that pastures that are cut for forage need twice the amount of fertiliser required by grazed pastures. One recommendation made in Puerto Rico for Oxisols and Ultisols (Chapter 8) is 370 kg N/ha, 53 kg P/ha and 197 kg K/ha where the forage is cut, but only half of these amounts where animals graze the pasture and thus return nutrients in their excreta.

In Uganda, in trials involving Elephant grass, it was found that there were good responses to applications of potassium and magnesium when the grass was cut, but no response to these two nutrient elements when applied to pasture that was grazed.

Other differences between grazed and ungrazed pastures that affect soil management include the fact that grazed pastures are liable to animal trampling. Since animals that are allowed to graze freely feed selectively on the younger, more protein-rich plant parts, grazed pastures produce more crude protein. Cut pastures produce a higher total amount of forage with more digestible energy but with a lower protein percentage.

Nitrogen Some grasses respond much better to nitrogen application than do others. Grasses with good responses include the following:

Elephant grass (*Pennisetum purpureum*)
Guinea grass (*Panicum maximum*)
Pangola grass (*Digitaria decumbens*)
Star grass (*Cynodon plectostachyum*)
Para grass (*Brachiaria mutica*)
Congo grass (*Brachiaria ruzizensis*)
Kikuyu grass (*Pennisetum clandestinum*)
Setaria grass (*Setaria sphacelata*)
Rhodes grass (*Chloris gayana*)
Dallis grass (*Paspalum dilatatum*)

Experiment in various parts of the tropics suggests that these grasses need nitrogen applications of 400–900 kg/ha to reach maximum yield when cut as forage, though less than this when grazed. Since the maximum nitrogen supplied by a legume in a grass–legume mixture is probably in the range 100–300 kg/ha, the fact that the responses to nitrogen of the grasses listed above are linear in many cases up to 900 kg/ha indicates that maximum yields cannot be obtained with grass–legume mixtures because not enough nitrogen is supplied.

The responsive grasses listed are relatively efficient users of applied nitrogen and can produce 40–50 kg of dry matter for each kg of nitrogen supplied. Less responsive grasses, such as indigenous savanna grasses of the Andropogon family, are less efficient converters of applied nitrogen and may produce only up to 20 kg/ha of dry matter per kg of fertiliser nitrogen.

The response obtained to nitrogen applications is influenced by the local soil and climate, as well as by the adequacy or otherwise of other nutrients. Responses are best in hot wet periods, and fall when temperatures are lower or when moisture is limiting.

High nitrogen applications may require applications of phosphorus or potassium to be fully effective. If nitrogen is applied in the ammonium form, the acidifying effect on the soil might need to be corrected by applications of lime. Urea is often recommended as a source of nitrogen because it has a high nitrogen content and is relatively cheap, and has only a moderate acidifying effect on the soil. If sulphur is needed, ammonium sulphate may be used as a source of sulphur unless it is supplied in single superphosphate.

Nitrogen fertilisers are usually broadcast on the soil. Applications are often split, being applied up to eight times a year after cutting or grazing, but during cool or dry periods applications may be omitted or reduced in amount.

The recovery of added nitrogen by pasture grasses is often better than by crops, with recovery rates of up to 80% being found in some Australian investigations.

Phosphorus Whereas the nitrogen needs of intensively managed pastures depend very much on the responsiveness of the particular species grown, the phosphorus needs depend mainly on the soil properties.

Optimum phosphorus applications for grass grown on Oxisols and Ultisols (Chapter 8) in Puerto Rico are of the order of 75 kg P/ha for most species, though Pangola grass requires less. In Colombia, application rates of 43 kg P/ha have been recommended for Elephant and Pangola grass grown on volcanic and alluvial soils. Where pastures are grazed rather than cut, these rates may be halved because of recycling.

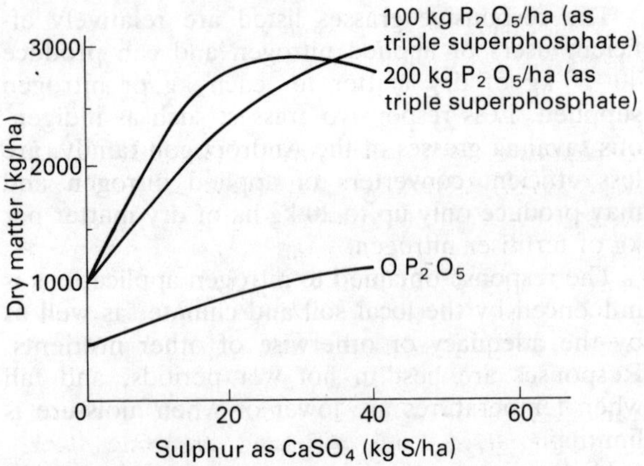

Fig. 21.1 Yields of Bahia (or Batatais) grass (*Paspalum notatum*) on a Brazilian Ultisol as increased by 20–40 kg/ha sulphur (as $CaSO_4$) and 0. 100 and 200 kg P_2O_5/ha. Pasture production was tripled by 40 kg/ha S and 100 kg/ha P_2O_5. Source: **McClung, A. C. and Quinn, L. R.** (1959) *Sulfur and Phosphorus Responses in Batatais grass (Paspalum notatum)*, IBEC Research Institute Bulletin 18, as given in Sanchez, 1976.

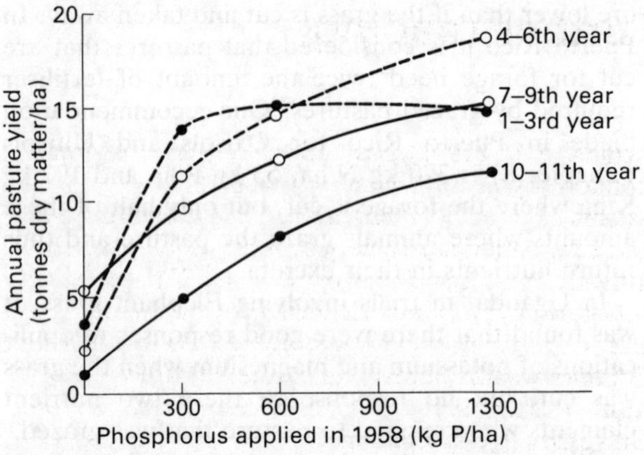

Fig. 21.2 Residual effects of massive phosphorus applications to a Hawaiian Oxisol on grass–legume pasture yields. Source: **Fox, R. I, Hasan, S. M. and Jones, R. C.** (1971) 'Phosphate and sulphate sorption by Latosols', *Proceedings International Symposium on Soil Fertility Evaluation (New Delhi)* **1**, 857–64, as given in Sanchez, 1976.

Potassium Fast-growing, productive grasses remove large amounts of potassium from the soil. Since pasture grasses contain about 1% potassium in their dry matter, the approximate amounts removed can easily be estimated if the dry matter production is known. How much of this potassium, if any, needs to be replaced by fertiliser applications depends on the potassium reserves in the soil and the rate at which the available potassium is replaced from these reserves (Chapter 10). In soils with large potassium reserves, such as many of the soils of East Africa, no response to potassium can be expected for at least several years of production. In many Puerto Rican soils no response to potassium was observed during the first two years, but subsequently fertiliser applications were needed and proved responsive.

If it is assumed that the pasture recovers 75% of the potassium added, then the theoretical annual potassium requirement is about 1.3 times the difference between the potassium removed in the forage and the amount supplied by the soil. In practice recommended applications vary widely with up to 200 kg K/ha being recommended for intensively fertilised pastures on Oxisols and Ultisols in Puerto Rico where the grass is cut, and half of these amounts when grazed.

Liming, sulphur and micronutrients Since most nitrogen-responsive tropical grasses are fairly tolerant of acid soils and of high levels of exchangeable aluminium, they are not often limed except in some cases to counteract the acidifying effects of ammonium-containing fertilisers. The amounts of calcium and magnesium removed by pastures are fairly high, so that in some cases liming may be needed to supply these two elements rather than to raise the pH of the soil. To supply both calcium and magnesium, dolomitic limestone should be used instead of calcium carbonate limestone.

The sulphur needs of intensively managed pastures are relatively high. If sulphur is not supplied in sulphate of ammonia or in single superphosphate, sulphur in the form of gypsum may be needed. Micronutrient deficiencies in intensively managed pastures have not often been reported.

Further reading

Acland, J. D. (1971) *East African Crops*. Longman: London (by arrangement with FAO).
Adams, F. (ed.) (1985) *Soil Acidity and Liming*. 2nd edn. American Society of Agronomy: Madison.
Adland, P. G. and Johnson, J. A. (1983) *Annotated Bibliography: Biomass Estimation – Nutrient Cycling and Organic Matter Relations in Forest Stands*. Commonwealth Forestry Institute : Oxford.
Ahn, P. M. (1970) *West African Soils*. Oxford University Press: London.
Alexander, A. (ed.) (1986) *Foliar Fertilization*. Martinus Nijhoff: Dordrecht.
Alexander, L. T. and Cady, J. C. (1962) *Genesis and Hardening of Laterite in Soils*. Technical Bulletin 1282. US Department of Agricultural, Soil Conservation Service: Washington DC.
Alexander, M. (1977) *Introduction to Soil Microbiology*. 2nd edn. John Wiley: New York and Chichester.
Allan, W. (1965) *The African Husbandman*. Oliver and Boyd: Edinburgh and London.
Alvim, P. de T. and Kozlowski, T. T. (1977) *Ecophysiology of Tropical Crops*. Academic Press: New York.
American Society of Agronomy/Soil Science Society of America (1976) *Multiple Cropping*. American Society of Agronomy: Madison.
____ (1978) *Diversity of Soils in the Tropics*. American Society of Agronomy: Madison.
Arnon, D. I. (1972) *Crop Production in Dry Regions*. Leonard Hill: Aylesbury.
Aubert, G. (1968) *Classification des Sols utilisée par les Pedologues Français*. World Soil Resources Report 32. FAO: Rome.
Baeumer, K. and Bakermans, W. A. P. (1973) 'Zero tillage' in *Advances in Agronomy 25. Academic Press Inc: New York.*
Barber, S. A. and Bouldin, D. R. (1984) *Roots, Nutrient and Water Influx, and Plant Growth*. ASA Special Publication 49. American Society of Agronomy: Madison.
Barnes, A. C. (1974) *The Sugar Cane*. Leonard Hill: Aylesbury.
Bartholomew, W. V. and Clark, F. E. (eds) (1965) *Soil Nitrogen*. Advances in Agronomy, 10. American Society of Agronomy: Madison.
Bates, W. N. (1963) *Mechanization of Tropical Crops*. Temple Press Books: London.
Baver, L. D., Gardner, W. H. and Gardner, W. R. (1972) *Soil Physics*. 4th edn. John Wiley: New York.
Bear, F. E. (1964) *Chemistry of the Soil*. 2nd edn. ACS Monograph, 160. Reinhold Publishing Corporation: New York.
Beckett, P. H. T. and Webster, R. (1971) Soil variability – a review. *Soils and Fertilisers,* **34**, 1–15.
Berger, J. (1969) *The World's Major Fibre Crops: Their Cultivation and Manuring*. Centre d'Etude de l'Azote: Zürich.
Bie, S. W. and Beckett, P. H. T. (1970) The costs of soil survey. *Soils and Fertilisers,* **33**, 203–17.
____ (1971) Quality control in soil survey. *Journal of Soil Science*, **22**, 32–49, 453–65.
Bie, S. W., Ulph, A. and Beckett, P. H. T. (1973) Calculating the economic benefits of soil survey. *Journal of Soil Science*, **24**, 429–35.
Birch, H. F. (1958) The effect of soil drying on humus decomposition and nitrogen availability. *Plant and Soil*, **10**, 9–31.
Black, C. A. (1968) *Soil Plant Relationships*. 2nd edn. John Wiley: New York.
Blackburn, F. H. (1983) *Sugar Cane*. Longman: London.
Blair, G. J. and Till, A. R. (eds) (1983) *Sulfur in South east Asian and South Pacific Agriculture*. The Australian Development Assistance Bureau/ The Sulphur Institute:
Bohn, L., McNeal, B. L. and O'Connor, G. A. (1979) *Soil Chemistry*. John Wiley: New York.
Bornemisza, E. and Alvarado, A. (eds) (1975) *Soil Management in Tropical America*. Soil Science Dept, North Carolina State University (for University Consortium on Soils of the Tropics): Raleigh.
Boyer, J. (1982) *Les Sols Ferrallitiques. Tome X. Facteurs de Fertilité et Utilisation des Sols*. ORSTOM: Paris.

Brady, N. C. (1984) *The Nature and Property of Soils*. 9th edn. Macmillan: New York.

Bridges, E. M. (1970) *World Soils*. Cambridge University Press: Cambridge.

Buol, S. W. and Couto, W. (1981) Soil fertility capability assessment for use in the humid tropics. In **Greenland, D. J.** (ed.) (1981). Oxford University Press: London.

Buringh, P. (1979) *Introduction to the Study of Soils in Tropical and Subtropical Environments*. 3rd edn. Centre for Agricultural Publishing and Documentation: Wageningen.

Chandler, T. and Spurgeon, D. (1979) *International Cooperation in Agroforestry*. ICRAF: Nairobi.

Charter, C. F. (1957) The aims and objects of tropical soil surveys. *Soils and Fertiliser*, **20,** 127–8.

Chijoke, E. O. (1980) *Impact on Soils of Fast Growing Species in Lowland Humid Tropics*. Forestry Paper 21. FAO: Rome.

Child, R. (1964) *Coconuts*. Longman: London.

Childs, E. C. and Youngs, E. G. (1974) Soil physics: twenty five years on. *Journal of Soil Science*, **25**, 399–407.

Cooke, G. W. (1967) *The Control of Soil Fertility*. Crosby Lockwood: London.

____ (1975) *Fertilising for Maximum Yield*. 2nd edn. Crosby Lockwood: London.

Corley, R. H. V., Hardon, J. J. and Wood, B. J. (eds) (1976) *Oil Palm Research*. Elsevier: Amsterdam.

Coursey, D. G. (1967) *Yams*. Longman: London.

De Geus, J. G. (1973) *Fertilizer Guide for the Tropics and Subtropics*. 2nd edn. Centre d'Etude de l'Azote: Zürich.

____ (1977) *Production Potentialities of Pastures in the Tropics and Subtropics*. Centre d'Etude de l'Azote, Zürich.

Dent, D.(1980) Acid sulphate soils: morphology and prediction. *Journal of Soil Science*, **31**, 87–100.

Dent, D. and Young, A. (1981) *Soil Survey and Land Evaluation*. George Allen and Unwin: London.

Doggett, H. (1970) *Sorghum*. Longman: London.

Drosdoff, M. (ed.) (1984) *World Food Issues*. 2nd edn. Cornell University: Ithaca, New York.

Duchaufour, P. (1984) *Abrégé de Pédologie*. Masson: Paris and New York.

Dudal, R. (1963) *Dark Clay Soils of Tropical and Subtropical Regions*. Agricultural Development Paper 83. FAO: Rome.

Faniran, A. and Jeje, L. K. (1983) *Humid Tropical Geomorphology*. Longman: London.

FAO (1974) *Shifting Cultivation and Soil Conservation in Africa*. Soils Bulletin 24. FAO: Rome.

____ (1974) *Approach to Land Classification*. Soils Bulletin 22. FAO: Rome.

____ (1976) *A Framework for Land Evaluation*. Soil Bulletin 32. FAO: Rome.

____ (1977) *Guidelines for Soil Description*. 2nd edn. *Guias para la Descripcion de Perfiles de Suelos. /Directives pour la Description des Sols*. FAO: Rome.

____ (1977) *Soil Survey Investigations for Irrigation*. Soils Bulletin 42. FAO: Rome.

____ (1978–81) *Report on the Agro-Ecological Zones Project*. Vols 1–4. World Soil Resources Reports 48/1–4. FAO: Rome.

____ (1984) *Fertilizer and Plant Nutrition Guide*. Fertilizer and Plant Nutrition Bulletin 9. FAO: Rome.

____ (1984) *Improved Production Systems as an Alternative to Shifting Cultivation*. Soils Bulletin 53. FAO: Rome

____ (1988) *FAO – Unesco Soil Map of the World. Revised legend*. World Soil Resources Report 60. FAO: Rome.

FAO/Unesco (1970–80) *Soil Map of the World 1:5 000 000*. Vols 1–10. Unesco: Paris.

Forbes, T., Rossiter, D. and Van Wambeke, A. (N.D.) *Guidelines for Evaluating the Adequacy of Soil Resource Inventories*. SMSS Technical Monograph 4. Soil Management Support Services, US Dept of Agriculture, Washington DC, and Dept of Agronomy, Cornell University: Ithaca, New York.

Foth, H. D. (1978) *Fundamentals of Soil Science*. 6th edn. John Wiley: New York, London.

Fremond, Y., Ziller, R. and de Nucé de Lamothe, M. (1966) *The Coconut Palm*. International Potash Institute: Berne.

Gibbon, D. and Pain, A. (1985) *Crops of the Drier Regions of the Tropics*. Longman: London.

Godin, V. J. and Spensley, P. C. (1971) *Oils and Oilseeds*. TPI Crop and Product Digest 1. Tropical Products Institute: London.

Gomez, A. A. and Gomez, K. A. (1983) *Multiple Cropping in the Humid Tropics of Asia*. International Development Research Centre: Ottawa.
Gomez, K. A. and Gomez, A. A. (1984) *Statistical Procedures for Agricultural Research*. 2nd edn. John Wiley: New York and Chichester.
Gourou, P. (1966) (translated Laborde, E. D.) *The Tropical World*. 4th edn. Longman: London
Greenland, D. J. (ed.) (1981) *Characterisation of Soils in Relation to their Classification and Management for Crop Production*. OUP: London.
Greenland, D. J. and Lal, R. (eds) (1977) *Soil Conservation and Management in the Humid Tropics*. John Wiley: Chichester and New York.
Grove, T. L. (ed) (1979) *Nitrogen Fertility in Oxisols and Ultisols of the Humid Tropics*. Cornell International Agriculture Bulletin 36. Cornell University: Ithaca, New York.
Hartley, C. W. S. (1977) *The Oil Palm*. 2nd edn. Longman: London.
Hausenbuiller, R. L. (1985) *Soil Science, Principles and Practices*. Wm. C. Brown: Dubuque.
Hillel, D. (1971) *Soil and Water*. Academic Press: New York.
Hodgson, J. M. (1978) *Soil Sampling and Soil Description*. Oxford University Press: London.
Hoekstra, D. A. and Kuguru, F. M. (eds) (1983) *Agroforestry Systems for Small Scale Farmers*. ICRAF: Nairobi.
Humphreys, L. R. (1978) *Tropical Pastures and Fodder Crops*. Longman: London.
ICRISAT (1982) *Proceedings of the International Symposium on Sorghum, 2–7 November 1981*. ICRISAT: Hyderabad.
Ignatieff, V and Page, H. J. (eds) (1958) *Efficient Use of Fertilisers*. Revised edn. FAO Agricultural Studies 43. FAO: Rome.
International Potash Institute (1974) *Le Potassium dans les Cultures et les Sols Tropicaux/ Potassium in Tropical Crops and Soils*. International Potash Institute: Berne–Worblaufen.
International Rice Research Institute (IRRI)
____ (1975) *Major Research in Upland Rice*. IRRI: Los Banos.
____ (1976) *Climate and Rice*. IRRI: Los Banos.
____ (1979) *Nitrogen and Rice*. IRRI: Los Banos.
____ (1979) *Rainfed Lowland Rice*. IRRI: Los Banos.
____ (1980) *Agrometeorology of the Rice Crop*. IRRI: Los Banos.
____ (1985) *Wetland Soils: Characterisation, Classification and Utilisation*. IRRI: Los Banos.
Jacob, A. and Uexkull, H. V. (1963) Fertiliser use. *Nutrition and Manuring of Tropical Crops*. Verlagsgesellschaft fur Ackerbau mbH: Hannover.
Jenny, H. (1941) *Factors of Soil Formation*. McGraw Hill: New York.
Jones, C. A. (1985) *C4 Grasses and Cereals. Growth, Development and Stress Response*. John Wiley: New York and Chichester.
Jones, M. J. (1973) The organic matter content of the savanna soils of West Africa. *Journal of Soil Science*, **24**, 42–53.
Jones, M. J. and Wild, A. (1975) *Soils of the West African Savanna*. Commonwealth Bureau of Soils, Technical Communication 55. Commonwealth Agriculture Bureau: Farnham Royal.
Jones, U. S. (1979) *Fertilizers and Soil Fertility*. Reston Publishing Co: Reston, Virgina.
Kalpagé, F. S. L. P. (1974) *Tropical Soils. Classification, Fertility and Management*. Macmillan Co. of India: New Delhi.
Kassam, A. H. (1976) *Crops of the West African Semi-arid Tropics*. ICRISAT: Hyderabad.
Kassam, A. H. (1980) Agro-climatic suitability for rainfed crops of winter barley, upland rice, groundnut, sugarcane, banana/plantain and oilpalm in Africa. In FAO (1980) *Report on the Second FAO/ UNFPA Expert Consultation on Land Resources for Populations of the Future*. FAO: Rome.
Kategile, J. A. (1985) *Pasture Improvement Research in Eastern and Southern Africa*. International Development Research Centre: Ottawa.
Katyal, J. C. and Randhawa, N. S. (1983) *Micronutrients*. Fertilizer and Plant Nutrition Bulletin 7 FAO: Rome.
Kay, D. E. (1973) *Root Crops*. TPI Crop and Product Digest 2. Tropical Products Institute: London.
____ (1979) *Food Legumes*. TPI Crop and Product Digest 3. Tropical Products Institute: London.
Kee, N. S. (1972) *The Oilpalm, its Culture, Manuring and Utilisation*. International Potash Institute: Berne.
Kohnke, H. (1982) *Soil Physics*. McGraw Hill: New York/Tata McGraw-Hill: New Delhi.

Koorevar, P., Menelik, G. and Dirksen, C. (1983) *Elements of Soil Physics*. Developments in Soil Science, 13. Elsevier: Amsterdam, New York.
Kowal, J.M. and Kassam, A.H. (1978) *Agricultural Ecology of Savanna. A Study of West Africa*. Clarendon Press: Oxford.
Kramer, P.J. (1972) *Plant and Soil Water Relationships: A Modern Synthesis*. Tata McGraw-Hill, New Delhi.
Lal, R. and Greenland, D.J. (eds.) (1979) *Soil Physical Properties and Crop Production in the Tropics*. John Wiley: Chichester and New York.
Landon, J.R. (ed.) (1984) *Booker Tropical Soil Manual*. Longman/Booker Agriculture International: London.
Lathwell, D.J. (ed.) (1979) *Phosphorus Response on Oxisols and Ultisols*. Cornell International Agriculture Bulletin 33. Cornell University: Ithaca, New York.
___ (1979) *Crop Response to Liming of Ultisols and Oxisols*. Cornell International Agriculture Bulletin 35. Cornell University: Ithaca, New York.
Lee, K.E. and Wood, T.C. (1971) *Termites and Soils*. Academic Press: London.
Lock, G.W. (1969) *Sisal*. 2nd edn. Longman: London.
Longman, K.S. and Jenik, J. (1974) *Tropical Forest and its Environment*. Longman: London.
MacDonald, L.H. (ed.) (1982) *Agroforestry in the African Humid Tropics/Agroforesterie en Afrique Tropicale Humide*. United Nations University: Tokyo.
Maignien, R. (1966) *Review of Research on Laterites*. Natural Resources Research, IV. Unesco: Paris.
Mengel, K and Kirkby, E.A. (1978) *Principles of Plant Nutrition*. International Potash Institute: Berne.
Mokwunye, A.U. and Vlek, P.L.G. (eds) (1986) *Management of Nitrogen and Phosphorus Fertilisers in sub-Saharan Africa*. Martinus Nijhoff: Dordrecht.
Monteith, J.L. (1972) Solar radiation and productivity in tropical ecosystems. *Journal of Applied Ecology*, **9**, 747–66.
Moormann, F.R. and Breeman, N. van (1978) *Rice: Soil, Water, Land*. IRRI: Los Banos.
Moss, R.P. (ed.) (1968) *The Soil Resources of Tropical Africa*. Cambridge University Press: London.
Nair, P.K.R. (1984) *Soil Productivity Aspects of Agroforestry*. ICRAF: Nairobi.
National Academy of Sciences (1972) *Soils of the Humid Tropics*. Washington DC.
National Research Council (1984) *Leucaena: Promising Forage and Tree Crop for the Tropics*. 2nd edn. National Academy Press: Washington DC.
Norman, M.J.T., Pearson, C.J. and Searle, P.G.E. (1984) *The Ecology of Tropical Food Crops*. Cambridge University Press: Cambridge.
Nye, P.H. and Greenland, D.J. (1960) *The Soil under Shifting Cultivation*. Commonwealth Bureau of Soils, Technical Communication 51. Commonwealth Agricultural Bureau: Farnham Royal.
Olson, G.W. (1981) *Soils and the Environment*. Chapman and Hall: New York and London.
Olson, R.A. (ed.) (1971) *Fertiliser Technology and Use*. 2nd edn. Soil Science Society of America: Madison.
Onwueme, I.C. (1978) *Tropical Tuber Crops: Yams, Cassava, Sweet Potato, Cocoyams*. John Wiley: New York.
Orvedal, A.C. (1975, 1977, 1983) *Bibliography of Soils of the Tropics*. US Agency for International Development: Washington DC.
Overseas Development Administration (1972) *A Guide to Project Appraisal in Developing Countries*. HMSO: London.
Panabokke, C.R. (1967) *Soils of Ceylon and Fertilizer Use*. The Ceylon Association for the Advancement of Science.
Paul, C.L. and DeWalt, B.R. (eds) (1985) *El sorgo en Sistemas de Produccion en America Latina*. INTSORMIL: Londres, Mexico.
Pereira, H.C. (1973) *Land Use and Water Resources*. Cambridge University Press: London.
Polhamus, L.G. (1962) *Rubber*. Leonard Hill: London.
Ponnamperuma, F.N. (1972) The chemistry of submerged soils. *Advances in Agronomy*, **24**, 29–96.
Purseglove, J.W. (1972) *Tropical Crops: Monocotyledons. Tropical Crops: Dicotyledons*. Longman: London.

Robertson, G. P., Herrera, R. and Rosswall, T. (eds) (1982) *Nitrogen Cycling in Ecosystems of Latin America and the Caribbean*. Martinus Nijhoff: Dordrecht.

Roose, E. (1980, 1981) *Quelques Conclusions des Recherches Françaises sur la Dynamique Actuelle des Sols en Afrique Occidentale*. Cahiers ORSTOM, série Pédologie, XVIII, 3–4, 285–96.

Rosswall, T. (ed.) (1980) *Nitrogen Cycling in West African Ecosystems*. SCOPE/UNEP International Nitrogen Unit: Stockholm.

Russel, R. S. (1977) *Plant Root Systems*. McGraw-Hill: London.

Ruthenberg, H. (1980) *Farming Systems in the Tropics*. 3rd edition. Oxford University Press: London.

Salmon, S. C. and Hanson, A. A. (1964) *The Principles and Practice of Agricultural Research*. Leonard Hill: London.

Sanchez, P. A. (1976) *Properties and Management of Soils in the tropics*. John Wiley: New York, London.

Sanchez, P. A. (ed.) (1977) *A Review of Soils Research in Tropical Latin America*. Technical Bulletin 219. North Carolina Agricultural Experiment Station: Raleigh.

Sanchez, P. A. and Tergas, L. E. (1979) *Pasture Production in Acid Soils of the Tropics*. CIAT: Cali, Colombia.

Shaner, W. W., Philipp, P. F. and Schmehl, W. R. (1982) *Farming Systems Research and Development*. Westview Press: Boulder, Colorado.

Silver, W. S. and Schröder, E. C. (eds) (1984) *Practical Application of Azolla for Rice Production*. Martinus Nijhoff: Dordrecht.

Sivarajasingham, S., Alexander, L. T., Cady, J. G. and Cline, M. G. 1962 'Laterite' in *Advances in Agronomy, 14*. Academic Press Inc: New York.

Skinner, F. A. and Uomala, P. (1986) *Nitrogen Fixation with Non Legumes*. Martinus Nijhoff: Dordrecht.

Smyth, A. J. (1966) *The Selection of Soils for Cocoa*. Soils Bulletin 5. FAO: Rome.

____ (1970) *The Preparation of Soil Survey Reports*. Soils Bulletin 9. FAO: Rome.

Smyth, A. J. and Montgomery, R. F. (1962) *Soils and Land Use in Central Western Nigeria*. The Government of Western Nigeria: Ibadan.

Soil Management Support Services (1987) *Keys to Soil Taxonomy*. US Dept of Agriculture, SMSS Technical Monograph 6. Cornell University.

Soil Management Support Services/ Cornell University (1981) *Soil Resource Inventories and Development Planning*. US Dept of Agriculture: Washington DC.

Soil Science Society of America (1975) *Glossary of Soil Science Terms*. Soil Science Society of America: Madison. **Soil Science Society of America** (1985) *Potassium in Agriculture*. Soil Science Society of America: Madison.

Soil Survey Staff (1951) *Soil Survey Manual*. Agricultural Handbook 18. US Dept of Agriculture: Washington DC.

____ (1975) *Soil Taxonomy. A Basic System of Soil Classification for Making and Interpreting Soil Surveys*. Agricultural Handbook 436. US Dept of Agriculture: Washington DC.

Surre, C. and Ziller, R. (1963) *Le Palmier à Huile*. G-P Maisonneuve et Larose: Paris.

Swindale, L. D. (ed.) (1978) *Soil Resource Data for Agricultural Development*. University of Hawaii: Honolulu.

Sys, C. and Riquier, J. (1980) Ratings of FAO/ Unesco soil units for specific crop production. In *Report on the Second FAO/UNFPA Expert Consultation on Land Resources for Populations of the Future*, 55–96. FAO: Rome.

Tandon, H. L. S. (1984) *Sulphur Research and Agricultural Production in India*. Fertilizer Development and Consultation Organisation: New Delhi.

Truong, B., Pichot, J. and Bernard, P. (1978) Caractérisation et Comparaison des Phosphates Naturels Tricalciques d'Afrique de l'Ouest en vue de leur Utilisation Directe en Agriculture. *Agronomie Tropicale*, **XXXIII**, 2, 136–45.

Uexküll, H. R. von (1986) *Efficient Fertiliser Use in Acid Upland Soils of the Humid Tropics*. FAO Fertilizer and Plant Nutrition Bulletin 10. FAO: Rome.

Van den Abeele, M. and Vandenput, R. (1956) *Les Principales Cultures du Congo Belge*. Direction de l'Agriculture, des Forêts et de l'Elevage: Brussels.

Van Wambeke, A. (1974) *Management Properties of Ferralsols*. Soils Bulletin 23. FAO: Rome.

—— (1981/82) *Calculated Soil Moisture and Temperature Regimes of South America/Africa*. Soil Management Support Services, US Dept of Agriculture: Washington DC/Dept of Agronomy, Cornell University: Ithaca, New York.

Vizier, J. F. (1983) *Etude des Phénoménes d'Hydromorphie dans les Sols des Regions Tropicales à Saisons Contrastées*. ORSTOM: Paris.

Vlek, P. G. L. (ed.) (1985) *Micronutrients in Tropical Food Crop Production*. Martinus Nijhoff: Dordrecht.

Walker, B. (1983) *A Glossary of Agricultural Terms Relating Top Soil and Water*. Merlin Books: Braunton.

Webster, C. C. and Wilson, P. N. (1980) *Agriculture in the tropics*. Longman: London.

White, F. (1983) *Vegetation Map of Africa 1:5 Million, with Descriptive Memoir*. Unesco Natural Resources Research 20. Unesco: Paris.

White, R. E. (1979) *Introduction to the Principles and Practice of Soil Science*. Blackwell Scientific Publications: Oxford.

Wild, A. (ed.) (1988) *Russell's Soil Conditions and Plant Growth*. 11th edn. Longman: London.

Williams, C. N., Chew, W. Y. and Rajaratnam, J. A. (1980) *Tree and Field Crops of the Wetter Regions of the Tropics*. Longman: London.

Wills, J. B (ed) (1962) *Agriculture and Land-use in Ghana*. Oxford University Press: London.

Wrigley, G. (1982) *Tropical Agriculture: The Development of Production*. Longman: London.

Yoshino, M. M. (ed) (1984) *Climate and Agricultural Land-use in Monsoon Asia*. University of Tokyo Press: Tokyo.

Young, A. (1976) *Tropical Soils and Soil Survey*. Cambridge University Press: London.

—— (1985) *The Potential of Agroforestry as a Practical Means of Sustaining Soil Fertility*. Working Paper 34. ICRAF: Nairobi.

Zimmerman, R. (1982) *Environmental Impact of Forestry*. FAO Conservation Guide 7. FAO: Rome.

Index

Reference should also be made to the over 200 definitions in the Glossary (pages vi to xii).